Digital Economies at Glob

Digital Economies at Global Margins

edited by Mark Graham

The MIT Press
Cambridge, Massachusetts
London, England

International Development Research Centre
Ottawa • Amman • Montevideo • Nairobi • New Delhi

© 2019 Contributors

This work is licensed under a Creative Commons Attribution 4.0 (CC-BY 4.0) International License.

Published by the MIT Press. MIT Press books may be purchased at special quantity discounts for business or sales promotional use. For information, please email special_sales@mitpress.mit.edu.

A copublication with
International Development Research Centre
PO Box 8500
Ottawa, ON K1G 3H9
Canada
www.idrc.ca/ info@idrc.ca

The research presented in this publication was carried out with the financial assistance of Canada's International Development Research Centre. The views expressed herein do not necessarily represent those of IDRC or its Board of Governors.

ISBN 978-1-55250-600-4 (IDRC e-book)

This book was set in ITC Stone Sans Std and ITC Stone Serif Std by Toppan Best-set Premedia Limited. Printed and bound in the United States of America.

Library of Congress Cataloging-in-Publication Data

Names: Graham, Mark, 1980- editor.
Title: Digital economies at global margins / edited by Mark Graham.
Description: Cambridge, MA : MIT Press, [2018] | Includes bibliographical
  references and index.
Identifiers: LCCN 2018010198 | ISBN 9780262535892 (pbk. : alk. paper)
Subjects: LCSH: Small business--Technological innovations. | Electronic commerce.
  | Marginality, Social. | Social marketing.
Classification: LCC HD2341 .D54 2018 | DDC 384.309172/4--dc23 LC record
  available at https://lccn.loc.gov/2018010198

10 9 8 7 6 5 4 3 2 1

Contents

Acknowledgments ix

1 Changing Connectivity and Digital Economies at Global Margins 1
   Mark Graham

Opening Essays 19

   Marginal Benefits at the Global Margins: The Unfulfilled Potential of Digital Technologies 21
   Uwe Deichmann and Deepak Mishra

   Toward the Transformative Power of Universal Connectivity 25
   Bitange Ndemo

   A Data-Driven Approach to Closing the Internet Inclusion Gap 29
   Robert Pepper and Molly Jackman

   Digital Services and Industrial Inclusion: Growing Africa's Technological Complexity 33
   Calestous Juma

   Platforms at the Margins 39
   Jonathan Donner and Chris Locke

   Digital Economies at Global Margins: A Warning from the Dark Side 43
   Tim Unwin

   Digital Globality and Economic Margins—Unpacking Myths, Recovering Materialities 47
   Anita Gurumurthy

I Digitalization at Global Margins 53

2 Making Sense of Digital Disintermediation and Development: The Case of the Mombasa Tea Auction 55
Christopher Foster, Mark Graham, and Timothy Mwolo Waema

3 Development or Divide? Information and Communication Technologies in Commercial Small-Scale Farming in East Africa 79
Madlen Krone and Peter Dannenberg

4 Digital Inclusion, Female Entrepreneurship, and the Production of Neoliberal Subjects—Views from Chile and Tanzania 103
Hannah McCarrick and Dorothea Kleine

5 "Let the Private Sector Take Care of This": The Philanthro-Capitalism of Digital Humanitarianism 129
Ryan Burns

6 The Digitalization of Anti-poverty Programs: Aadhaar and the Reform of Social Protection in India 153
Silvia Masiero

7 The Myth of Market Price Information: Mobile Phones and the Application of Economic Knowledge in ICTD 173
Jenna Burrell and Elisa Oreglia

II Digital Production at Global Margins 191

8 Hope and Hype in Africa's Digital Economy: The Rise of Innovation Hubs 193
Nicolas Friederici

9 Hackathons and the Cultivation of Platform Dependence 223
Lilly Irani

10 Meeting Social Objectives with Offshore Service Work: Evaluating Impact Sourcing in the Philippines 249
Jorien Oprins and Niels Beerepoot

11 Digital Labor and Development: Impacts of Global Digital Labor Platforms and the Gig Economy on Worker Livelihoods 269
Mark Graham, Isis Hjorth, and Vili Lehdonvirta

12 Geographic Discrimination in the Gig Economy   295
    Hernan Galperin and Catrihel Greppi

13 Margins at the Center: Alternative Digital Economies in Shenzhen, China   319
    Jack Linchuan Qiu and Julie Yujie Chen

14 African Economies: Simply Connect? Problematizing the Discourse on Connectivity in Logistics and Communication   341
    Stefan Ouma, Julian Stenmanns, and Julia Verne

Author Affiliations   365
Index   367

Acknowledgments

This book was only brought into being with the support, guidance, and contributions of many friends and colleagues. I wish to thank, first of all, the book's authors, all of whom put together their contributions with great effort, care, and thoughtfulness, cooperatively and patiently revising their chapters.

This volume was born of conference sessions at the 2015 meeting of the American Association of Geographers and the 2015 Global Conference on Economic Geography. These sessions, and the initial ideas for this book, were put together in collaboration with my colleagues Nicolas Friederici, Heather Ford, Chris Foster, and Isis Hjorth. I am very grateful for the energy and creative guidance they each provided.

The set of concerns that guide this book also find a supportive home among two overlapping groups of scholars at the Oxford Internet Institute: the "Geonet" and "Digital Inequality" research clusters. I wish to thank Mohammed Amir Anwar, Grant Blank, Margie Cheesman, Stefano De Sabbata, Martin Dittus, Nicolas Friederici, Fabian Braesemann, Iginio Gagliardone, Khairunnisa Haji Ibrahim, Sanna Ojanperä, Joe Shaw, David Souter, Ralph Straumann, Michel Wahome, Jamie Woodcock, and Alex Wood for helping to build such an inspiring community focused on digital inequalities.

I also wish to thank David Sutcliffe for his ever-attentive role as an editor. Sanna Ojanperä and Mohammed Amir Anwar also provided critical reviews on individual chapters. The book is much improved as a result of their help.

At all stages of putting the book together, I also relied heavily on the constant encouragement, support, and guidance offered by Kat Braybrooke. Her ability to act as a sounding board, creatively think through ideas, and offer constructive guidance has been invaluable. Thank you Kat.

Finally, I am very grateful to the International Development Research Centre (IDRC) and MIT Press, who have allowed this book to be released under an open license, ensuring that it will be accessible to far more people than is usual for an academic volume. IDRC grant (107384-001), Leverhulme Prize grant PLP-2016-155, and the European Research Council under the European Union's Seventh Framework Programme for Research and Technological Development (FP/2007–2013) (grant agreement 335716) helped to fund my time on this project.

In the end, this book was put together not just to understand the digital transformations taking place at the world's economic margins, but also to help shape them. If you are reading these words, you have likely played, or will play, a role in doing just that. I therefore dedicate this book to those of you who are striving to build fairer digital futures.

Mark Graham

1 Changing Connectivity and Digital Economies at Global Margins

Mark Graham

This book emerges in a moment of changing connectivity at the world's economic margins. In Manila, Manchester, Mogadishu, the banlieues of Marseille, and everywhere in between, the world is becoming digital, digitized, and digitally mediated at an astonishing pace. Most of the world's wealthy have long been digitally connected, but the world's poor and economically marginal have not been enrolled in digital networks until relatively recently. In only five years (2012–2017), over one billion people became new Internet users (ITU 2016). In 2017, Internet users became a majority of the world's population. The networking of humanity is thus no longer confined to a few economically prosperous parts of the world. For the first time in history, we are creating a truly global and accessible communication network.

As ever more people and places join this globe-spanning digital network, this book asks what digitalization and digital production can mean for the world's economic margins. Places that were once economic peripheries can potentially transcend their spatial, organizational, social, and political constraints. An Indian weaver, a Chinese merchant, and a Kenyan transcriber all have opportunities to instantly interact with markets outside their local contexts. In other words, possibilities now exist for fundamentally transformed economic geographies.

This book brings together new scholarship that addresses what increasing digital connectivity and the digitalization of the economy means for people and places at economic margins. As you read through the book, you might find it useful to think about the roles digital connectivity plays in transforming these economically peripheral areas: whether digital tools and technologies are simply amplifying existing inequalities, barriers, and constraints, or allowing them to be transcended; who is actually benefitting from processes of digitalization and practices of digital engagement; who engages in digital production and where does it occur; whether changes in digital economies at the margins really match up to our expectations for change; and ultimately who are the winners and losers in our new digital and digitally mediated economies.

Digital Economies

Digital technologies, and digitized modes of communication, have driven hugely transformative changes in the global economy. But most of the available evidence on digital economies remains focused on high-income economies, with relatively little known about the implications of the digital for those at the global margins.[1] And yet, optimism abounds about the potentials of digital economies to transform livelihoods in low-income countries. Commentators, policymakers, development organizations, and many others are increasingly promoting and funding plans and projects that aim to support or create digital economies. But without sustained and critical inquiry into how digital economies are being envisioned and enacted, as well as into the effects of digital economies in these countries, it is difficult to move beyond hype and hope. The diverse chapters of this book interrogate these increasingly digital economies in two ways. Instead of seeing the digital only as a discrete end product, we recognize how digital information, services, and goods are always embedded in, and part of, broader sociotechnical systems. No end product is therefore purely digital. We can thus think of the digital economy as producing outcomes on a spectrum. On one end of the spectrum, digital information is used to alter constellations of value creation and capture, by enhancing, complementing, or replacing economic transactions and processes that have traditionally been analog, a process called digitalization. This digitalization of goods, productions, and services is crucial to an ever-increasing amount of economic value creation. A growing body of research in economics, economic sociology, economic geography, and economic anthropology is pointing to the potential advantages that can be gained in global production networks through producing, capturing, manipulating, and moving all sorts of digital and digitized information.

On the other end of the spectrum, a key component of the end product or service might itself be digital or digitally transmittable (e.g., software development, graphic design, writing and editing, etc.): what we refer to in this book as digital production. Places in every corner of the planet now aspire to become centers of digital production. Attempts to emulate the Silicon Valley model abound in places like the Silicon Glen, Silicon Savannah, Silicon Cape, Silicon Fjord, Silicon Roundabout, Silicon Prairie, and even the Silicon Swamp (Graham and Mann 2013). Meanwhile, alternative models of digital capitalism are emerging, from *jugaad* (innovative hacks) practices in India to *shanzhai* (copying) in China (Braybrooke and Jordan 2017). These two trends of proliferating digitalization and digital production form the twin pillars of newly emerging digital economies, raising questions for both digital enterprises and digital laborers about who controls, owns, and can access these new modes of economic production (Foster et al. 2017; Foster and Graham 2016; Weber 2017; Murphy and Carmody 2015).

Digital information, or data, is one of the fuels of the new economy (Kitchin 2014). Data is often cheap, nonrivalrous, and ubiquitous, which raises the question: In a global economy with world-spanning production networks, should the openness and transmittability of data be maximized, or should digital information be seen as a key resource in production processes that needs to be better protected and governed to avoid strengthening global cores at the expense of peripheries? (See Weber 2017 for a fuller treatment of these two questions.) Issues around whether trade in data is a positive-sum game, and who controls and benefits from new modes of digital and digitally enabled production, intersect with a need to better understand the digital connectivity mediating all this economic activity.

Changing Connectivities

At the time this book went to press, the world had over three and a half billion Internet users and five billion mobile phone users (GSMA 2017). Ninety-five percent of the world's population live in a place that is covered by a mobile-cellular network, and 84 percent of people on the planet live somewhere covered by mobile broadband networks (ITU 2016).

As figure 1.1 illustrates, a majority of the world's Internet users now live in low- or middle-income countries. The Internet is no longer a network that connects just the Global North. Furthermore, most of the world's new

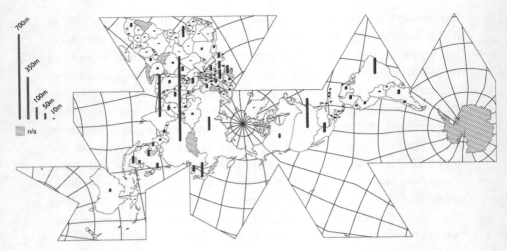

Figure 1.1
Total number of people with access to the Internet, 2015 (most recently available data at time of publication). Data sources: World Bank, Natural Earth. Visualization by Ralph Straumann, Geonet, http://geonet.oii.ox.ac.uk/.

growth in Internet users is coming from these low- and middle-income countries, in part because high-income countries have mostly reached saturation levels (see figure 1.2), and in part because access is increasingly possible through cheap mobile devices (Donner 2015).

As figure 1.3 shows, however, not all parts of the world with low Internet penetrations have high growth rates of Internet connectivity. There are still places with small Internet-using populations, low penetrations, and slow growth rates. These connectivity black spots have been the focus of several plans and programs, developed by governments, international organizations, and corporations, to connect the currently disconnected (Friederici, Ojanperä, and Graham 2017). Facebook, for instance, with its Internet.org partnership, aims to connect the planet. The company's website even explicitly clarifies that this means the whole world, not just some of us. It aims to do this through a combination of free apps (including Facebook) and unmanned aircraft that can deliver Internet access to remote areas.

Google has similar ambitions with its Project Loon, an initiative to use balloons floating through the stratosphere to provide Internet access to rural areas. Friederici and colleagues (2017) outline how such plans are not

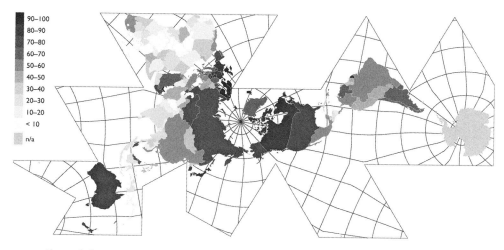

Figure 1.2
Internet penetration percentages in 2015 (most recently available data at time of publication). Internet penetration is the proportion of individuals who have used the Internet in the last twelve months. Data sources: World Bank, Natural Earth. Visualization by Ralph Straumann, Geonet, http://geonet.oii.ox.ac.uk/.

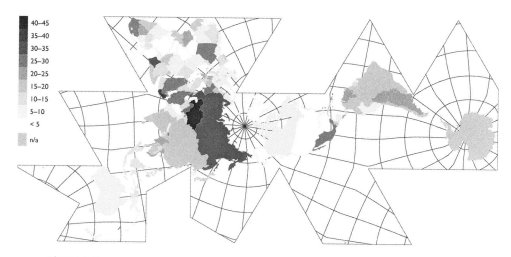

Figure 1.3
Percentage growth in Internet penetration, 2010–2015. Data sources: World Bank, Natural Earth. Visualization by Ralph Straumann, Geonet, http://geonet.oii.ox.ac.uk/.

the sole preserve of digital corporate behemoths. The African Development Bank claims that US$55 billion has been pledged for its Connect Africa initiative, and the World Bank has invested over a billion dollars in broadband infrastructure projects.

In all these plans, however, connecting people is not an end goal. It is a means to another end: fulfilling a vision of economic development sparked by increased digital access, helping the world's poorest in the process (Galperin and Fernanda Viecens 2017; Friederici, Ojanperä, and Graham 2017). But why do we think changing connectivity will make the poor less poor?

Shifting Geographies

Elsewhere (Graham, Andersen, and Mann 2015; Graham 2015), I have argued that a key reason such visions exist is that digital technologies alter what Eric Sheppard (2002, 308) refers to as "positionalities," that is, "the shifting, asymmetric, and path-dependent ways in which the futures of places depend on their interdependencies with other places."

This is not to imply that visions of technologies altering positionalities are in any way new (Kern 2003). Technologies as varied as print, photography, and rail have been mechanisms to "annihilate time and space" (Solnit 2003, 4). The Victorians, for instance, were amazed by the potentials of the newly constructed telegraph system, which would allow people on different sides of the planet to instantaneously communicate with one another. Romance, crime, and of course new economic interactions blossomed on what Standage (1998) refers to as "the Victorian Internet." A proposal in the mid-nineteenth century to connect cities on both sides of the Atlantic led one commentator to speculate that now "all of the inhabitants of the earth would be brought into one intellectual neighborhood, and be at the same time perfectly freed from those contaminations which might under other circumstances be received" (Marvin 1988, 201). In the early twentieth century, the rush by European powers to build roads and railways in their colonies was similarly framed as a move that would connect economic centers to margins and, in doing so, bring economic development and prosperity to those margins (Graham, Andersen, and Mann 2015).

The thesis that remote and inaccessible economic positionalities hinder the economic development of a region has been an ongoing narrative. Thinkers from Adam Smith (1776) to Jeffrey Sachs, Andrew Mellinger, and John Gallup (2001) all point to economic perils for places at the margins. As the Internet began to be globalized, and access reached today's economic peripheries, the hope cycle began anew. Newspaper stories appeared about African farmers or Asian weavers who were suddenly able to sell their products to the wider world (cf. Graham 2010; Kisambira 2009). National development plans such as Kenya's predicted that "by 2030, it will become impossible to refer to any region of our country as remote" (Government of Kenya 2007, 6). And presidents and prime ministers expressed visions of a newly digital world auguring new beginnings. For instance, Rwanda's president, Paul Kagame (2006, 5), famously noted:

> Just as it is clear that growth in the 19th and 20th centuries was driven by networks of railways and highways, growth and development in the 21st century is being defined and driven by digital highways and ICT-led value-added services. In Africa, we have missed both the agricultural and industrial revolutions and in Rwanda we are determined to take full advantage of the digital revolution. This revolution is summed up by the fact that it no longer is of utmost importance where you are but rather what you can do—this is of great benefit to traditionally marginalized regions and geographically isolated populations.

Margins are defined as margins, after all, because they are not in a center. But if technologies can change positionalities by allowing us all to interconnect, irrespective of where we are, they can bring the margins to the center and the center to the margins. Information and communication technologies (ICTs), by changing geographies and topologies, offer the chance to change the world's economic positionalities.[2]

Nonetheless, because people have very different types of control over the modes and methods of connectivity, ICTs do not necessarily shrink distance or bring a digitally shared space into being. Seeing the world in those limited ways might not allow us to visualize what Doreen Massey (2005) refers to as "power-geometries"—an idea that "time–space compression for some may be time–space expansion for others" (Warf 2001, 11). Or, said differently, "ascribing new connectivities with associated economic and political change allows (and allowed) those forms of change to be ascribed an air of inevitability: a teleological trick that serves to depoliticize the very processes that technology, and the changing connectivities that it is thought to enable, mediate" (Graham, Andersen, and Mann 2015, 345). Different places, in other words, all have their own emergent paths and will not necessarily follow the historical trajectories of anywhere else.

Digital enterprises, digital entrepreneurs, and digital workers from every corner of the world might all be able to connect using the same network, but this does not necessarily mean that they can all use it to alter their positionalities or level playing fields in the same ways, leading to a "thintegration" into global economies that does little to change positions of economic dependency (Carmody 2013). We therefore need to develop more nuanced, grounded, and historicized accounts of the mergings of technology and connectivity if we are to understand how they intersect with economic development.

Information and Communication Technologies for Development

Much of the research that has been done on this topic has been conducted under the banner of information and communication technologies for development (ICT4D). The ICT4D literature has often engaged with this topic through a developmental lens, looking at how to design interventions to benefit the world's poor. The term "development" itself comes with a lot of conceptual baggage and means very different things to different people. But, at its core, it always implies a focused intervention: a focus on transforming one thing into another.

Many of these interventions have traditionally had overtly economic purposes and have been exclusively concerned with economic growth (Unwin 2009). Gross domestic product (GDP), for instance, is often used as a proxy for understanding welfare or development. Yet GDP, as a measure of economic growth, does not necessarily tell us much about how well a nation is performing, developing, or increasing welfare. This is not to say that alternative orientations do not exist. Bhutan, for instance, has pioneered the measurement of gross national happiness (GNH) in place of gross domestic product (GDP) as a way of understanding welfare. Scholars from Amartya Sen (2001) to Dorothea Kleine (2013), while recognizing the usefulness of economic growth, have spoken of the need for a focus, in development, on freedoms for people to achieve their capabilities. Anita Gurumurthy (2011), relatedly, has pointed to the dangers of how ICTs have extended global spaces of power and contestation. She argues for explicitly feminist visions of the network society that can better allow the potentials and dangers of digital transformations to be realized. As Buskens and Webb (2009, 5) note, "the use of ICTs to enhance one's life presupposes a measure of control over one's space and time"; therefore, pre-existing hierarchies (like gender hierarchies) can result in digital societies paying insufficient attention to the circumstances of women as those societies develop.

Also important to note is that development often fails even on its own terms. Perhaps because ICT4D is centered on technology as a central agent of change, it tends to focus on the effects of technology at the individual or organization level, rather than on structural characteristics and changes (an argument strongly made by Murphy and Carmody 2015). This myopic approach can lead to ICT4D plans and programs falling far short of their lofty ambitions. Echoing anthropologist James Ferguson's (1994) famous question, "What do aid programs do besides fail to help poor people?," we can ask what else ICT4D projects do.[3] Unwin (2017) argues that not only have ICTs increased inequality in the pursuit of development (see also Carmody 2012), but they have turned ICT4D on its head: "Instead of 'ICTs for Development' (ICT4D) we have become increasingly and surreptitiously enmeshed in a world of 'Development for ICTs' (D4ICT) where governments, the private sector, and civil society are all tending to use the idea of 'development' to promote their own ICT interests" (Unwin 2017, 9).

Digital Economies at Global Margins

This book provides new empirical and theoretical raw material to help guide you through some of the above-mentioned debates. *Digital Economies at Global Margins* does not seek to take a singular position, but rather aims to bring together a diverse range of cutting-edge research to focus on the dynamic interplays between economic peripheries and the contemporary global and digital economy using both micro and macro levels of analysis. This framing enables us to arrive at explanations of how the local and global are mutually influencing, and are being influenced by, rapid changes in connectivity, the digitalization of economic activities, and digital production.

Legacies of economic imbalances and inequalities concerning capacity, power, and access to opportunities clearly persist, and they continue to affect who partakes in and benefits from emerging digital economies at the world's economic margins. A repeated theme in this book is that the unevenness between economic cores and peripheries is rarely leveled by ICTs. At this moment of change, when so many look to connectivity to bring about sustainable, inclusive digital economic development, we need to bring together the voices of those who have thought carefully and critically about digital economies outside global centers. This book shows how those processes are inherently political, socially embedded, path dependent, highly uneven—and contested.

The book opens with seven introductory reflections from key voices who take divergent positions about the potentials of digital economies at the world's margins. This collection of opinions from economic and development thought leaders, who are themselves based variously in the public sector, the private sector, and academia, illustrates some of the diversity of positions on the topic.

We begin with Uwe Deichmann and Deepak Mishra from the World Bank. The authors of the *World Development Report 2016: Digital Dividends* argue that using digital technologies at global margins confers massive benefits but that those benefits are highly unevenly distributed. Bitange Ndemo, a professor at the University of Nairobi, follows with a piece that draws on his time as the former permanent secretary in Kenya's Ministry of Information and Communication. He points to some of the transformational powers of digital connectivity in East Africa and highlights how new technologies can help measure (and thus achieve) the United Nations' Sustainable Development Goals (SDGs).

Robert Pepper and Molly Jackman from Facebook's Global Connectivity division similarly emphasize opportunities for businesses and individuals brought about by greater connectivity, concluding with some concrete steps to achieve more inclusive connectivity. Returning to a focus on Africa, Calestous Juma, Harvard University's professor of the practice of international development, addresses some of the current policy-level issues surrounding industrial transformation in the technology sector. Jonathan Donner and Chris Locke from Caribou Digital address the significant power now wielded by platforms in the digital age. Platform companies shape much of what happens in the digital economy, and therefore understanding (and changing) digital economies relies on grasping the reach of these companies. Tim Unwin, professor emeritus of geography at Royal Holloway,

University of London, then offers a warning that digital technologies should be used by those who most need them—rather than by those who most need to justify more digital investment. Anita Gurumurthy closes the section with a warning about what she calls the "digital-financial" assemblage and concludes with a plea to reimagine how digital tools can be used for the well-being of all.

The book is then organized into two key sections. The first, "Digitalization at Global Margins," explores how digitalization and increasing connectivity affect value creation in traditionally nondigital production networks. Economic actors at the margins in global production networks traditionally have limited direct access to inputs and markets. Ever more connectivity, however, could result in improved access to markets and new efficiencies in economic exchanges, ultimately leading to potentially greater impacts for smaller and more marginal producers.

In chapter 2, Christopher Foster, Mark Graham, and Timothy Mwolo Waema address some of these issues in the context of disintermediation in the East African tea sector. Using digital technologies to create more direct channels between buyers and sellers does not necessarily help tea growers, or even disintermediate production networks, as many predicted it would. Chapter 3, by Madlen Krone and Peter Dannenberg, in contrast, illustrates some of the risks of digital exclusion and marginalization. The chapter begins with a focus on how the ICT4D literature can highlight potential income gains to low-income communities. With a case study of commercial small-scale horticultural farmers in Kenya and Tanzania, the authors point to the risks of marginalization and long-term loss of commercial markets for those who lack access to new communication technologies. Chapter 4, by Hannah McCarrick and Dorothea Kleine, builds on the previous chapter by pointing out further risks—not of disconnectivity, but of connectivity itself. With case studies from Chile and Zambia, they show that by introducing distinctly neoliberal paradigms of development into the lives of women, ICT interventions may ultimately risk impoverishing them. Chapter 5 turns our attention to how, by focusing on emerging digital humanitarianism, digital tools are used to incorporate private sector logics into domains where they were previously less prominent. The author, Ryan Burns, shows how even humanitarianism is becoming a new site for capital accumulation. In chapter 6, Silvia Masiero also addresses the use of ICTs in active development interventions. Focusing on antipoverty policies in India, she demonstrates some of the effects of ICT infrastructures on user entitlements. Jenna Burrell and Elisa Oreglia conclude the section in chapter 7 by pointing to the additional disconnect between overstated visions of ICT potentials and their actual effects on the ground. They use ethnographic research to challenge the oft-repeated notion that farmers using mobile phones to acquire market information can raise their incomes and participate positively in more efficient markets.

The second section, "Digital Production at Global Margins," moves the book in a significantly different direction. Enhanced connectivity offers the potential for digital goods and services to be produced at the world's economic margins, with newly connected people able to engage in work like software development and online freelancing, and to produce locally adapted digital content, services, and applications. This, in turn, affords new ways of creating and capturing economic value. This part of the book examines various types of digital production at different levels of analysis and across a wide range of geographies.

Chapter 8 begins with an examination of digital entrepreneurship and the rise of so-called innovation hubs in Africa. Nicolas Friederici shows that while development organizations have attempted to position innovation hubs as key infrastructures for Africa's digital economy, many technology entrepreneurs have dismissed these hubs as ineffective interventions. Hackathons have always played an important part in digital entrepreneurship around the world, and chapter 9, by Lilly Irani, focuses specifically on what hackathons do at global economic margins. She shows that those invested in the Internet can make the hackathon format a global technology. Hackathons are a way of incorporating a world of diversity, but not necessarily without privileging certain stakeholders and outcomes over others. While the previous two chapters center on relatively high-skilled services in the digital economy, chapter 10 turns our attention to the other end of the spectrum with an analysis of what is referred to as "impact sourcing" (i.e., socially responsible outsourcing). Jorien Oprins and Niels Beerepoot show that impact sourcing brings digital jobs to new parts of the world but struggles to reach truly marginalized workers. In chapter 11, Mark Graham, Isis Hjorth, and Vili Lehdonvirta similarly focus on the lower-skilled and precarious end of the digital labor market. They illustrate that despite important and tangible benefits for a range of workers, multiple risks and costs unduly affect the livelihoods of digital workers. Hernan Galperin and Catrihel Greppi, in chapter 12, also examine digital jobs that can, in theory, be done from anywhere. Using data from one of the world's largest online labor platforms, they show that even though much digital work can be done from anywhere, workers outside global economic cores can suffer from discrimination. They are less likely than their European counterparts to win contracts, even after a range of characteristics are accounted for. Jack Linchuan Qiu and Julie Yujie Chen argue, however, that being in an economic periphery is not always a disadvantage. In chapter 13, they use case studies of Chinese ride-hailing platforms and shanzhai mobile phone manufacturing to contend that digital innovation may allow cores and margins to shift and even reverse positions. Finally, chapter 14, by Stefan Ouma, Julian Stenmanns, and Julia Verne, problematizes one of the core questions of this book: that of connectivity. To lay the groundwork for a more progressive politics of connectivity, the authors unpack current articulations of how development thinking is wrapped together with thinking about connectivity.

The chapters of this book follow no single narrative about the positive or negative effects of digitalization and increased connectivity. The authors hail from five of the world's continents and bring with them a diverse range of findings and arguments. That diversity (and divergence) in stories, in many ways, is the purpose of this book. Overly simplistic narratives about the potentials of the digital to transform economically underdeveloped peoples and places have too often been casually repeated and reproduced.

As the chapters in this volume show, looking to standardize positive or negative effects of digital technologies is to unproductively fetishize the digital. Digital technologies are a diverse range of tools that have myriad uses in myriad contexts. The purpose of making this argument, however, is not to throw our hands in the air as a response to this sort of complexity. Nor is it a position of ambivalence or neutrality about the yawning gap between rich and poor in our digital age. It is instead a plea to look to the commonalities that we know actually exist. Instead of seeing "the digital" as a force that can drive economic outcomes in one way or another, we can deconstruct what the digital actually is. When talking about the digital, are we talking about the ability to compute, the ability to connect, or something else entirely? Are we looking at mechanisms of information storage, transfer, or manipulation? Digital tools allow connections where previously none existed. They can form an important part of the underlying infrastructures needed to bring markets into being. They can augment physical processes with additional information. And they can facilitate both synchronous and asynchronous modes of communication.

The chapters in this book allow us to do just that—to look to the commonalities. Every chapter provides both a synthesis and an original contribution at the intersections of thinking about digital economies, digitalization, digital production, and global margins. By taking nuanced positions about the roles played by connectivity and digital technologies, the chapters point to some ways we can ask questions about who benefits and who doesn't from changes to the ownership of data in commodity chains, algorithmic governance in labor markets, or the availability of market price information. In other words, a focus on these underlying digital and digitally mediated mechanisms enables us to look for risks, costs, benefits, and opportunities in the context of changing positionalities.

This book does not offer definitive answers to all these questions. Instead, this volume shows that there are no straightforward answers. To echo Melvin Kranzberg's "laws of technology," ICTs are neither good nor bad; nor are they neutral. Digital tools are just that—tools—which can amplify intent. They can make the powerful more powerful and increase the reach of that power. But, in some cases, ICTs facilitate a story of convergence rather than divergence by offering opportunities to those at economic margins. We need to learn from those relatively untypical cases, asking not just who benefits and who doesn't, but also *why* they do or don't. Said differently, we need to move away from overprivileging technology and connectivity as primary agents of change, and instead focus on the forces using those tools as a medium. And we need that evidence to inform, challenge, and build new kinds of contemporary theories that can make sense of the networking of not just global cores, but also of the poorest people and places on our planet.

At some point in 2017, somebody (most likely from a low- or middle-income country) accessed the Internet and became the world's first person to connect to a global network that includes more than half of humanity. This signal that we are entering a world in which connectivity is the norm rather than the exception should lead us to redouble our efforts to understand what digital connectivity and the digitalization of the economy mean for the economic geographies at the world's economic margins.

This should not be just an academic exercise. As this book shows, there is a significant disconnect between expectations about the potentials of digital economies in global margins and their actual effects. Many people are left out, marginalized, and even harmed by the shift to a more digital world. By deploying a diverse collection of empirical cases and building a broad range of theories, this book's authors help reset our expectations and develop more realistic visions of what is possible and probable. Together, the chapters in this volume lay the foundations for a more just and equitable digital world. And they provide an integral starting point for those of us who are keen not just to understand change at global margins, but also to participate constructively in it.

Notes

1. Economic inequalities within countries can render traditional definitions of "developing" and "developed" economies an unproductive heuristic for understanding the uneven and complex global distribution of wealth and power. Instead of using "developing countries" as the book's object of focus, we focus instead on global margins—the people, places, and processes that have not been able to occupy central positions in transnational networks of production and value creation. This focus on "margins" allows us to question whether globally changing connectivities are reconfiguring or reinforcing existing balances of economic power.

2. Visions of altered positionalities have tended to manifest in two primary ways: first, as the idea that ICTs shrink relative distances, for instance, in the sense that the cost or time distances between Lagos and London have been shrunk; second, that the distance itself has ceased to matter, and the world has become a flat "global village" (Graham 2015).

3. See Easterly (2014) and Moyo (2009) for related arguments from economists.

References

Braybrooke, K., and T. Jordan. 2017. Genealogy, Culture and Technomyth: Decolonizing Western Information Technologies, from Open Source to the Maker Movement. *Digital Culture and Society* 3 (1): 25–46. doi:10.14361/dcs-2017-0103.

Buskens, I., and A. Webb. 2009. *African Women and ICTs: Investigating Technology, Gender, and Empowerment.* London: Zed Books; Ottawa: IDRC.

Carmody, Pádraig. 2012. The Informationalization of Poverty in Africa? Mobile Phones and Economic Structure. *Information Technologies and International Development* 8 (3): 1–17.

Carmody, Pádraig. 2013. A Knowledge Economy or an Information Society in Africa? Thintegration and the Mobile Phone Revolution. *Information Technology for Development* 19 (1): 24–39.

Donner, Jonathan. 2015. *After Access: Inclusion, Development, and a More Mobile Internet*. Cambridge, MA: MIT Press.

Easterly, W. 2014. *The Tyranny of Experts*. New York: Basic Books.

Ferguson, James. 1994. The Anti-Politics Machine: Development, Depoliticization, and Bureacratic Power in Lesotho. *Ecologist* 24 (5): 176–181.

Foster, Christopher, and Mark Graham. 2016. Reconsidering the Role of the Digital in Global Production Networks. *Global Networks* 17 (1): 68–88. doi:10.1111/glob.12142.

Foster, C., M. Graham, L. Mann, T. Waema, and N. Friederici. 2017. Digital Control in Value Chains: Challenges of Connectivity for East African Firms. *Economic Geography* 94 (1): 68–86.

Friederici, Nicolas, Sanna Ojanperä, and Mark Graham. 2017. The Impact of Connectivity in Africa: Grand Visions and the Mirage of Inclusive Digital Development. *Electronic Journal of Information Systems in Developing Countries* 79 (2): 1–20.

Galperin, Hernan, and M. Fernanda Viecens. 2017. Connected for Development? Theory and Evidence about the Impact of Internet Technologies on Poverty Alleviation. *Development Policy Review* 35 (3): 315–336. doi:10.1111/dpr.12210.

Government of Kenya. 2007. *Kenya Vision 2030: A Popular Version*. Nairobi: Government of Kenya.

Graham, Mark. 2010. Justifying Virtual Presence in the Thai Silk Industry: Links Between Data and Discourse. *Information Technologies and International Development* 6 (4): 57–70. http://itidjournal.org/itid/article/view/642/277.

Graham, Mark. 2015. Contradictory Connectivity: Spatial Imaginaries and Technomediated Positionalities in Kenya's Outsourcing Sector. *Environment and Planning A* 47 (4): 867–883. doi:10.1068/a140275p.

Graham, Mark, Casper Andersen, and Laura Mann. 2015. Geographical Imagination and Technological Connectivity in East Africa. *Transactions of the Institute of British Geographers* 40 (3): 334–349. doi:10.1111/tran.12076.

Graham, Mark, and Laura Mann. 2013. Imagining a Silicon Savannah? Technological and Conceptual Connectivity in Kenya's BPO and Software Development Sectors. *Electronic Journal of Information Systems in Developing Countries* 56 (2): 1–19.

GSMA. 2017. *The Mobile Economy 2017*. London: GSMA. http://www.gsma.com/mobileeconomy/.

Gurumurthy, Anita. 2011. Feminist Visions of the Network Society. *Development* 54 (4): 464–469. doi:10.1057/dev.2011.82.

ITU. 2016. *ICT Facts and Figures 2016*. Geneva: ITU. http://www.itu.int/en/ITU-D/Statistics/Documents/facts/ICTFactsFigures2016.pdf.

Kagame, Paul. 2006. *The NICI-2010 Plan: An Integrated ICT-Led Socio-Economic Development Plan for Rwanda 2006–2010*. Kigali: Government of Rwanda.

Kern, Stephen. 2003. *The Culture of Time and Space*. Cambridge, MA: Harvard University Press.

Kisambira, E. 2009. East Africa: Seacom Fibre Optic Goes Regional. *East African Business Week*, July 25. http://allafrica.com/stories/200907271215.html.

Kitchin, Rob. 2014. *The Data Revolution: Big Data, Open Data, Data Infrastructures and Their Consequences*. London: Sage.

Kleine, Dorothea. 2013. *Technologies of Choice?: ICTs, Development, and the Capabilities Approach*. Cambridge, MA: MIT Press.

Marvin, C. 1988. *When Old Technologies Were New: Thinking About Electric Communication in the Late Nineteenth Century*. New York: Oxford University Press.

Massey, Doreen. 2005. *For Space*. London: Sage.

Moyo, D. 2009. *Dead Aid: Why Aid Is Not Working and How There Is a Better Way for Africa*. New York: Farrar, Straus and Giroux.

Murphy, James T., and Pádraig Carmody. 2015. *Africa's Information Revolution: Technical Regimes and Production Networks in South Africa and Tanzania*. Oxford: John Wiley and Sons.

Sachs, J. D., Andrew D. Mellinger, and John Luke Gallup. 2001. The Geography of Poverty and Wealth. *Scientific American* 284 (3): 70–75.

Sen, Amartya. 2001. *Development as Freedom*. Oxford: Oxford University Press.

Sheppard, Eric. 2002. The Spaces and Times of Globalization: Place, Scale, Networks, and Positionality. *Economic Geography* 78 (3): 307–330.

Smith, Adam. 1776. *An Inquiry into the Nature and Causes of the Wealth of Nations*. London: W. Strahan and T. Cadell.

Solnit, R. 2003. *Motion Studies: Time, Space, and Eadweard Muybridge*. London: Bloomsbury.

Standage, Tom. 1998. *The Victorian Internet: The Remarkable Story of the Telegraph and the Nineteenth Century's Online Pioneers*. London: Weidenfield and Nicolson.

Unwin, Tim. 2009. ICT4D: *Information and Communication Technology for Development*. Cambridge: Cambridge University Press.

Unwin, Tim. 2017. *Reclaiming ICT4D*. Oxford: Oxford University Press.

Warf, Barney. 2001. Segueways into Cyberspace: Multiple Geographies of the Digital Divide. *Environment and Planning B: Planning and Design* 28:3–19.

Weber, Steven. 2017. Data, Development, and Growth. *Business and Politics* 19 (3): 397–423. Previously published online April 17, 2017, 1–27. doi:10.1017/bap.2017.3.

Opening Essays

Marginal Benefits at the Global Margins: The Unfulfilled Potential of Digital Technologies

Uwe Deichmann and Deepak Mishra

With digital technologies seemingly affecting all aspects of life in advanced economies today, an important question is whether they also benefit those at the global margins. Technology enthusiasts have long pronounced that connectivity via mobile phones and Internet-connected devices will bring massive development gains—that it will be a "game changer in the field of education," "the best thing anyone can do to improve quality of life around the world," and "the most transformative technology of economic development in our time," and that it will "help lift people out of poverty and give them a freedom from want."[1] There are indeed many examples of poor farmers and small-scale entrepreneurs using technology to raise their incomes. But, as argued in the World Bank's *World Development Report 2016: Digital Dividends* (which we led the preparation of), those benefits have so far fallen short and been unevenly distributed—modest for those at the global margins while often massive for those best prepared to take advantage of these new opportunities.

At the global margins are those living in extreme poverty. What Paul Collier in 2008 called the "bottom billion" are today the 767 million people living below the absolute poverty line of $1.90 per day (in 2013 PPP dollars; World Bank 2016a). Of those, 80 percent live in rural areas, more than 60 percent are engaged in agriculture, 40 percent are under the age of fifteen, and a similar share has no education. These are large numbers, but they are declining. Since 1990, about 1.1 billion people have escaped extreme poverty as economies have become more urban and market oriented.

We do not know how much new technologies have contributed to poverty reduction. Most of the discussion about their impact relies on case studies or anecdotal evidence. Research in West Africa showed how simple information technologies improved learning, reduced producer price uncertainty, and generated time savings for poor farmers. Mobile money, accessed through basic feature phones, has spread even to remote communities in countries from Bangladesh to Kenya. This has made the transfer of remittances cheaper and more reliable, with direct welfare consequences for the poor. And research in Peru showed that mobile phone access boosted household consumption by 11 percentage points between 2004 and 2009 as well as reducing poverty by 8 percentage points.[2]

Even those who do not own a mobile phone or computer can benefit from technological advances. One example is the rollout of digital identification systems. In India and elsewhere, biometric IDs have increased the government's efficiency—removing sixty-two thousand ghost workers from public payrolls saves Nigeria US$1 billion per year—and its capacity to deliver services to remote and often disadvantaged communities. Perhaps even more important, digital identity can empower individuals, for instance, by making it easier to access basic financial services and participate in democratic processes.

Yet, there is reason to believe that these benefits have not had the massive impact on poverty reduction that many expected. Even though more people in developing countries have access to a mobile phone than to electricity or safe water and sanitation in their homes, most of the poor still do not own a mobile phone or computer. Exact figures are hard to come by, but even among the bottom 40 percent of the income distribution in sub-Saharan Africa, for instance, only a little more than half owned a mobile phone in 2014, and only 5 percent had access to the Internet. Even considering the rapid cost reductions and technological advances, achieving universal access will be hard for mobile phones and even harder for Internet access. This should not be surprising. Consider that 200 years after the conception of universal schooling, about 40 percent of adults in low-income countries remain illiterate; 150 years after electricity was discovered, more than 1.3 billion people—almost 20 percent of the world's population—lack reliable access to the grid; and 100 years after the first automobile rolled off an assembly line, more than two-third of households in the world don't own a car (Poushter 2015).

But there is another reason to be skeptical of ICT poverty reduction claims. Like many previous technological innovations, ICTs tend to be productivity biased, skill biased, and voice biased. Those who are already successful, talented, or better connected tend to benefit most. The problem then is not just access but also capability. In countries like Niger and Afghanistan, 70 percent of adults are illiterate. In Mali and Uganda, 75 percent of third graders can't read. Even low-income people who can afford a mobile phone (among some groups of African phone owners, the median expenditure on phone service is 13 percent of their income) will benefit far less than those already better off.[3] Rather than being a great equalizer, digital technologies risk amplifying existing inequalities—echoing Isaac Asimov's prediction when, in 1964, he looked forward to the world in 2014: "Not all the world's population will enjoy the gadgety world of the future to the full. A larger portion than today will be deprived and although they may be better off, materially, than today, they will be further behind when compared with the advanced portions of the world. They will have moved backward, relatively" (Asimov 1964, n.p.).

Add to this the likely implications of widespread automation. The traditional path to development has been through labor-intensive manufacturing—arguably the greatest contributor to poverty reduction in China and elsewhere. Reduced employment opportunities in these sectors will put large pressure on wages in remaining low-skill occupations. And if future jobs are knowledge intensive, major shortcomings in early childhood development in many developing countries, in education but also in nutrition, become even more damaging.

What are the implications for development policy? Clearly the goal of universal affordable Internet access is still important. Even just the private benefits of easier communication with friends and family and access to useful information justify public policies that ease ICT infrastructure investments. Distortions in the telecom markets rather than a lack of capital more often hold back such investments or keep prices high, including in remote and sparsely populated areas that are often also the poorest.

But policymakers also need to realize that the Internet is not a shortcut to high-income status, even if it can be an enabler and perhaps an accelerator of development. Technology by itself can become a placebo, making us feel better in the short term, while delaying the deeper changes required to solve the real underlying problems. *Digital Dividends* focuses on three areas where complementary improvements are necessary: (1) strengthening the business environment, especially competition policies to curb excessive concentration of market power in a handful of digital platforms but also in other ICT-enabled sectors; (2) improving skills development—not just ICT skills but, equally important, the "soft skills" that will not be easily replaced by computers; and (3) improving accountability in the public sector, so technology is deployed to empower the poor, not to strengthen control. These are the foundations of economic development—the business climate, human capital, and governance—and though the Internet and mobile phones can help improve these foundations in many ways, new technologies are not a substitute.

Notes

1. Quotations from former US education secretary Arne Duncan, Google chairman Eric Schmidt, economist Jeffrey Sachs, and former US secretary of state Hillary Clinton, respectively, as cited in Toyama (2015).

2. See World Bank (2016b) for a discussion of these and the subsequent examples and specific references.

3. World Development Report 2016 team calculations based on Research ICT Africa surveys in a sample of countries (various years) (World Bank 2016b).

References

Asimov, Isaac. 1964. Visit to the World's Fair of 2014. *New York Times*, August 16, 1964. http://www.nytimes.com/books/97/03/23/lifetimes/asi-v-fair.html.

Collier, Paul. 2008. *The Bottom Billion: Why the Poorest Countries Are Failing and What Can Be Done About It*. Oxford: Oxford University Press.

Poushter, Jacob. 2015. Car, Bike or Motorcycle? Depends on Where You Live. *Fact Tank*, Pew Research Center. http://www.pewresearch.org/fact-tank/2015/04/16/car-bike-or-motorcycle-depends-on-where-you-live/.

Toyama, Kentaro. 2015. *Geek Heresy*. New York: Public Affairs.

World Bank. 2016a. *Poverty and Shared Prosperity 2016: Taking on Inequality*. Washington, DC: World Bank.

World Bank. 2016b. *World Development Report 2016: Digital Dividends*. Washington, DC: World Bank.

Toward the Transformative Power of Universal Connectivity

Bitange Ndemo

Barely nine years ago, Africa was struggling with broadband access and affordability. At the time, there was only one undersea fiber optic cable connecting the western part of Africa to South Africa and Asia. Virtually all countries used satellite for connectivity, which was expensive. A regional initiative by twenty-two East African countries to link an undersea fiber optic cable to the rest of the world was taking too long to realize because of bickering and competing priorities. It thus appeared as though the region would have to wait much longer to witness the transformative power of modern information and communication technologies (ICTs).

I was the permanent secretary of ICT in Kenya at that time. Mutahi Kagwe, the minister for ICT, and I convinced President Mwai Kibaki that Kenya needed to go it alone and land its own undersea cable to fast track the benefits of these new technologies to the region. The president accepted. Within eighteen months, the East African Marine Systems (TEAMS) cable landed on the shores of the Indian Ocean.

It was a risky proposition. We were committing meager national resources to a venture whose returns could only be guessed. In hindsight, however, the decision changed the fortunes of the region. A few East African countries have connected to the Kenyan cable, while others are barely managing to stay abreast of the latest developments. Those that have connected have extracted huge advantages, and everyone agrees that the challenge now lies in improving the access and affordability of broadband.

In Kenya, we devised deliberate policy measures that hugely paid off. These included subsidizing fiber optic connectivity to universities and access to devices; removing taxes on both broadband and devices; enabling open data to fuel innovative applications; softening the regulatory framework to allow greater risk taking leading up to transformative mobile money; and liberalizing the industry to allow for greater competition.

The difference between transformational change and a lack of such change is the level of connectivity. At a minimum, the world today needs universal 3G connectivity while improving terrestrial fiber optic networks. The time has come for each country to work toward greater access and affordability of the key driver of ICTs—broadband Internet—and enable it to be a tool for rapid economic growth, poverty reduction, and better monitoring of progress toward the Sustainable Development Goals (SDGs).

It is no longer a question of whether ICTs will transform public services in Africa—because they are doing so every day. In East Africa, for example, the advent of money applications for mobile phones has demonstrated the ability of ICTs to increase the financial inclusivity of millions.

Kenya's largest slum, Kibra, once a dark spot on Nairobi's city maps, is now more connected than ever.[1] For years, government as well as aid agencies could not see the potential of this two-square-mile slum of more than three hundred thousand people.

But a group of young mappers alongside Kibra residents have begun to turn things around using ICT tools. Using an open mapping platform, cheap tools, connectivity, and the Global Positioning System (GPS), they created a digital map, giving every asset in the slum an address.

Prior to this initiative, nearly everybody thought that more than 1 million people lived in the slum. Now, with accurate population measures, policymakers are able to make more informed decisions and design much more effective policies for the residents of Kibra. For example, a policy that allowed only public primary school students into elite public high schools was reversed when the mapping clearly showed that most schools in Kibra were privately run by church groups or other donor agencies.

In Uganda, the United Nations' Global Pulse is producing data of all kinds from satellites to create predictive weather solutions for farmers and track disease patterns. Soon some of the satellite pictures will be used to estimate country poverty levels—something that has been impossible until now.

A small startup in Nairobi, Gro Intelligence, aggregates both structured and unstructured data to help farmers better understand what factors affect their productivity. Never before in any country in Africa has specific economic information been analyzed in this manner and disseminated to rural farmers to help them increase their yields. If governments were to cooperate and support the expansion of such ICTs in agricultural practices, the perennial challenge of food insecurity on the continent could be significantly reduced.

In virtually all other sectors, ICTs are transforming communities. In Kenya, for example, ICTs have eliminated intermediaries in the agricultural sector, thus enabling farmers to realize value for their produce; enabled the poor to access healthcare in remote places; brought education closer to the very poor, who could not afford to buy textbooks; and enabled pastoralist communities to sell their livestock through the mobile phone application WhatsApp, reducing how often they need to travel long distances to marketplaces.

More importantly, the adoption of ICTs continuously proves wrong the critics who argue that access and affordability are not sufficient if capacity building is not present. In most transformative applications that solve critical problems, adoption rates have been higher without any training. No wonder, then, that Albert Einstein said, "It has become appallingly obvious that our technology has exceeded our humanity." He was perhaps alluding to the phenomenon of internalization, which suggests that technology has enslaved us to the extent that when we need it, we learn it without undergoing any special training.

A recent study, *Game of Phones: Deloitte's Mobile Consumer Survey* in Africa, shows that the uptake of digital transactions is on an upward trend led by Kenya.[2] The report says that more than 33 percent (a significant number of whom are either illiterate or semi-illiterate) of the population in Kenya today uses a digital money application. What is puzzling is how this substantial part of the population came to learn this fairly complex use of a mobile platform. Although encouraging, further studies are needed to establish the evolution of this trend. Findings from such studies would help other countries reap the economic benefits of ICTs.

As Africa celebrates the transformative nature of ICTs, all parties should continuously revisit the question of its inclusiveness. Emerging data analytics show that virtually all big data research focuses erroneously on people with an existing "digital footprint" (people who have experience with the Internet and other ICTs). This flawed approach could lead to new forms of discrimination. Yet, if universal connectivity were to become a reality, this discrimination could be avoided.

What's more, data and ICTs are crucial if the world is to be effective in measuring the SDGs that were adopted in 2015. This is the message from development institutions and other nongovernmental organizations (NGOs) currently grappling with the question of how to measure progress and ensure success.

To transform communities and achieve the SDGs globally, the world must strive toward universal connectivity that enables access to ICTs. These technologies now help us measure what we could not measure previously. With better measurement, we can better implement development.

Notes

1. Kibra is the correct spelling of what used to be (and is commonly) called Kibera.

2. Deloitte, *Game of Phones: Deloitte's Mobile Consumer Survey—The Africa Cut 2015/2016* (Johannesburg: Deloitte, 2016).

A Data-Driven Approach to Closing the Internet Inclusion Gap

Robert Pepper and Molly Jackman

Connectivity creates opportunities for businesses and individuals to participate directly in globalization (Friedman 2005) and is associated with job creation, productivity gains, and GDP growth (Deloitte 2014). Connectivity is not just a byproduct of progress—it is a crucial enabler. The majority of the world's people are unconnected, and many are using the Internet less than they would if it were cheaper and faster. Connectivity must be improved for the Internet to be globally inclusive and beneficial.

For individuals, the Internet provides a pathway out of poverty. It enables people to share information, access education, transfer funds, and identify savings in a globally competitive marketplace. According to one study focused on the poorest population in East Africa, people with ICT access gained approximately twenty-one dollars more per month than those without access and narrowed their income gap with others in higher income brackets (May, Dutton, and Manyakazi 2014). A similar study based in Peru found that individuals who gained Internet access between 2007 and 2009 obtained household incomes 19 percent higher than those who failed to gain access (Pepper and Garrity 2015).

The economic benefits of the Internet do not exist just at the individual level but apply to businesses as well. Twenty years ago, only large multinationals and some governments could buy and sell products at an international scale. The Internet has democratized globalization (Manyika and Lund 2016). Never has it cost less to connect with and sell to customers around the world, so small businesses can scale. Advances in advertising technologies enable small businesses and entrepreneurs to reach out to niche as well as mass audiences who will find their products useful and relevant.

As a consequence, entrepreneurs and small businesses that may have struggled to grow at a local level can now participate meaningfully in a truly global economy. Moreover, groups that have traditionally been excluded can engage with broader markets. Take the example of Kalpana Rajesh, who started a business in India selling wedding headdresses (*poola jada*) after the birth of her son. She created a Facebook page and started sharing photos and boosting posts to reach people around the world. Soon, her business, Pelli Poola Jada, grew to include forty-five branches and 250 employees—all women, who, because of their work and earnings, gained greater respect from their families and communities.

As Kalpana's example shows, connectivity can provide businesses with the tools they need to grow and, in so doing, enables the creation of new jobs. Yet, particularly in underconnected markets, small businesses may have a hard time filling job openings because qualified candidates have no way to learn about openings or to match their skills to positions. In Kenya, for instance, small and medium enterprises (SMEs) report difficulty finding qualified candidates despite a population of ten million underemployed youth. A significant challenge is connecting labor supply with business demand—a problem that could be solved through technology. Duma Works, a startup based in Nairobi, uses an algorithm to match job seekers to employers (particularly under-resourced SMEs) online or offline using SMS. The company has already helped nearly three thousand Kenyans find jobs at over 250 companies.

At a more macro level, the Internet enables economies of scale, creating more efficient markets for consumers and producers. Traditionally, most Colombian food vendors were required to wake up at 3 a.m. to buy the day's inventory at expensive local markets. Agruppa allows these vendors to place their orders via SMS or app, aggregating them into one large order to buy wholesale, reducing prices by up to 30 percent. Agruppa receives the goods and distributes them to the vendors. As a result, consumers enjoy healthy, fresh food at affordable prices, while small vendors enjoy bigger profits and better working hours.

These are just a few of the ways in which connectivity is being used to solve some economic challenges in developing markets. There are countless more examples just like them.

When we consider them in the aggregate, a picture forms of how connectivity can transform the world and move us closer to achieving sustained economic growth in the least developed countries. Research that unpacks the relationship between connectivity and development is needed to inform policy decisions and to guide the allocation of resources. The chapters in this volume provide a tremendous contribution to our understanding of the complex and endogenous relationship between connectivity and development. They add to the growing body of evidence demonstrating that connectivity is a catalyst of economic growth, not the coincident.

But, perhaps ironically, the people and communities that can benefit most from connectivity are currently least connected. Today, 4.1 billion people do not have access to the Internet, and 90 percent of the unconnected population lives in the developing world. The largely market-driven growth of connectivity has brought us to this point. If we continue on this trajectory, connectivity may never reach those whose lives it can transform the most.

Ensuring the equal flow of connectivity throughout the world is therefore extremely important, lest we run the risk of exacerbating existing inequalities (Pepper and Garity 2015). But united, the private sector, governments, communities, and civil society have an opportunity to close the inclusion gap and break down global inequalities in a way that is historically unprecedented.

First, we must ensure access to the unconnected. This will require solving difficult technical challenges to bring infrastructure to low-density and remote corners of the world—for example, building last-mile links where nonexistent, and providing high-capacity backhaul and middle-mile infrastructure through new technologies and satellites. Second, we need to *improve* connections for the underconnected. This means expanding and extending the next generation of Wi-Fi so that people can upgrade from 2G to 3G and 4G networks. Third, increasing relevant content in local languages requires incentives for local content production and translation of existing content. Finally, we must make it a priority to close the connectivity gender gap. Many reasons explain why women are not online—including affordability, relevance, and readiness. To be sure, narrowing the gender gap will require long-term, multi-pronged, and context-specific approaches, but we can start by improving educational opportunities available to women, creating greater awareness of the Internet and the benefits it can bring, and making it more affordable and acceptable for women to obtain connected devices.

We share a responsibility and global imperative to ensure that the benefits of connecting to and using the Internet do not flow exclusively to those who have historically benefited from technological advances. We need to come together as a global community to bring greater access to the unconnected and, just as important, better access for the underconnected.

References

Deloitte. 2014. *Value of Connectivity: Economic and Social Benefits of Expanding Internet Access*. London: Deloitte. https://www2.deloitte.com/content/dam/Deloitte/ie/Documents/TechnologyMediaCommunications/2014_uk_tmt_value_of_connectivity_deloitte_ireland.pdf.

Friedman, Thomas. 2005. *The World Is Flat*. New York: Farrar, Straus and Giroux.

Manyika, J., and S. Lund. 2016. Globalization for the Little Guy. McKinsey Global Institute, McKinsey & Company. January 2016. http://www.mckinsey.com/business-functions/strategy-and-corporate-finance/our-insights/globalization-for-the-little-guy.

May, J., V. Dutton, and L. Manyakazi. 2014. Information and Communication Technologies as a Pathway from Poverty: Evidence from East Africa. In *ICT Pathways to Reduction: Empirical Evidence from East and Southern Africa*, edited by Edith Ofwona Adera, Timothy M. Waema, Julian May, Ophelia Mascarenhas, and Kathleen Diga, 34–52. Rugby: Practical Action Publishing; Ottawa: IDRC.

Pepper, Robert, and John Garrity. 2015. ICTs, Income Inequality, and Ensuring Inclusive Growth. In *The Global Information Technology Report 2015: ICTs for Inclusive Growth*. World Economic Forum.

Digital Services and Industrial Inclusion: Growing Africa's Technological Complexity

Calestous Juma

The expansion of mobile services in Africa is heralded in popular narratives as an example of how the continent can leapfrog technologically—participating in the global economy through local dynamic and innovative enterprises. The disruption of landlines by mobile phones is at times described by researchers as technological leapfrogging, but without the associated industrial transformation (Aker and Mbiti 2010). In this short reflection, I argue that this focus limits the discussion on digital inclusion to the use of services. The challenge instead should be to expand economic inclusion through local industrial development.

The international expansion of Kenya's mobile money transfer system, M-Pesa, has been heralded as an example of what is possible, with the technological platform also being applied to other areas, such as energy, water, and sanitation. Some of the most transformative applications are in agriculture (Juma 2015). The promise, however, is misplaced because the mobile revolution has hardly been a stimulus for economic inclusion via industrial expansion. Africa still lags behind other regions of the world in manufacturing, and it has not made major steps to move to the production of mobile-related technologies. Doing so would help to broaden the base for economic inclusion beyond the core provision of telecommunications services.

As the M-Pesa example shows, the wide adoption of mobile phones in Africa has created remarkable enthusiasm for technology on the continent. It symbolizes the great potential that lies in technological catch-up, that is, by leapfrogging along the path blazed by South Korea to break out of the middle-income trap (Lee 2012). While inspirational, however, the mobile revolution appears to have had little effect on the industrial policies of

African countries, partly because of misinterpretation of what the revolution actually entailed.

Popular narratives of the mobile revolution focus on access devices and the services that the sector provides. The mobile handset, especially in the hands of ordinary Africans, has become the symbol of the revolution. There is a good basis for this imagery. The business model that made it possible for Africa to rapidly adopt mobile telephony involved the availability of low-cost handsets. What often goes unstated, however, is that the mobile revolution was fundamentally about telecommunications infrastructure (Batuo 2015). The spread of mobile phone towers across Africa is the outer manifestation of a complex engineering system that enables mobile communication.

Creating such a system involved reforming laws across Africa to create the entrepreneurial space for the new infrastructure (Ndemo 2016). The policy champions of this disruptive technology confronted many issues, including opposition from the existing landline industry (Juma 2016). But the public and private entrepreneurs introduced new business models—included prepayments and low-cost handsets—that enabled the poor to be included in the revolution.

The infrastructure revolution has also undergone dramatic changes. Early mobile phone systems were connected to the rest of the world via satellite links. Until 2009, only a small number of West African cities had access to undersea fiber optic cables. Today all of continental Africa and the Indian Ocean states have access to marine fiber optic cables with significantly higher bandwidth. Terrestrial connectivity is the latest investment frontier.

The challenge for many countries is how to leverage broadband infrastructure for economic transformation. In some countries, telecoms operators have yet to migrate from a reliance on satellite links to using fiber optics cables. As a result, the promise of low communication costs has yet to be realized. Even where the migration to fiber has occurred, access charges remain prohibitive. As a result, the infrastructure is not being fully utilized to foster innovation and development. This is not just a telecoms issue but indicates the lack of a complementary evolution in innovation policy.

There is great optimism over the emergence of information technology (IT) hubs in major urban areas across Africa (Adesina, Karuri-Sebina, and Resende-Santos 2016). These hubs have become a symbol of youth entrepreneurship in Africa. Indeed, many of them are producing new technologies designed to solve Africa's problems. But their appearance away from centers of research and learning also signals the need to foster more integrated innovation ecosystems that bring business, academia, and government together. The hubs have also exposed the need to improve the overall funding and policy environment for technology-based ventures.

The definition of mobile inclusion in Africa needs to broaden to cover industrial development. This includes the potential to manufacture devices, equipment, and infrastructure components. It also entails strengthening human capacity in the related engineering fields. This industrial expansion can be supported by linking industrial development directly to current efforts to grow African markets through regional integration (Mangeni and Juma, forthcoming).

Take the case of Taiwan. In the early 1960s, the country was a world leader in mushroom exports, a high-volume, low-value perishable commodity. Taiwan then took advantage of the emerging semiconductor industry to redefine itself as an industrial player. Taiwan's Industrial Technology Research Institute, which spawned many of its leading semiconductor firms, was created by consolidating four dilapidated research centers left behind by Japanese occupiers (Shih 2005). The case of Taiwan illustrates how a country can move from the initial use of existing technologies to generating increasingly diverse products through industrial policy. Many African countries have greater research capacity than Taiwan did when it entered the semiconductor field. The difference is that Taiwan viewed it as an opportunity for industrial growth, not just the provision of services.

An effective industrial policy will entail continuous interactions among government, industry, and academia. Many of the critical elements needed for this process to work are emerging. For example, several African countries have created higher education institutions to train new professionals for the digital sector, including new telecoms universities in Egypt, Kenya, and Ghana. The impetus for creating these institutions came from telecoms rather than education ministries. Countries such as Ethiopia have started local assembly of mobile handsets. This could help widen the base for industrial development. The rise of IT hubs in African cities such as Nairobi and Lagos is another node in a potential industrial ecosystem based on mobile technology.

One of the key policy challenges is that shifting from mobile services to industrial development entails higher-level coordination. This involves facilitating interactions between ministries, higher research and technical institutions, and civil society actors in a wide range of sectors, including finance, telecommunications, industry, services, education, and marketing. This coordination will need to be pursued with the support of systematic science and innovation advice, which is often missing in African countries. At the very least, heads of state and government offices will need to be guided by the best available technical advice to incrementally forge industrial policy that is suited to the task.

On the whole, the key lessons from the mobile revolution are yet to be fully learned and implemented as a basis for inclusive industrial transformation. Until they are learned, the popular call for technological leapfrogging and the associated industrial development will remain a mirage. This situation is due to a failure to appreciate the scale and scope of the reforms needed to shift Africa from its current focus on users of services to a focus on contributors to the transformation of local industry.

References

Adesina, O., Geci Karuri-Sebina, and João Resende-Santos, eds. 2016. *Innovation Africa: Emerging Hubs of Excellence*. Bingley, UK: Emerald.

Aker, J. C., and I. M. Mbiti. 2010. Mobile Phones and Economic Development in Africa. *Journal of Economic Perspectives* 24 (3): 207–232.

Batuo, M. E. 2015. The Role of Telecommunications Infrastructure in the Regional Economic Growth of Africa. *Journal of Developing Areas* 49 (1): 313–330.

Juma, C. 2015. *The New Harvest: Agricultural Innovation in Africa*. New York: Oxford University Press.

Juma, C. 2016. *Innovation and Its Enemies: Why People Resist New Technologies*. New York: Oxford University Press.

Lee, K. 2012. *Schumpeterian Analysis of Economic Catch-up: Knowledge, Path-creation and the Middle Income Trap*. Cambridge: Cambridge University Press.

Mangeni, F., and Juma, C. Forthcoming. *Emergent Africa: The Evolution of Regional Economic Integration*. Cambridge: Cambridge University Press.

Ndemo, B. 2016. Inside a Policymaker's Mind: An Entrepreneurial Approach to Policy Development and Implementation. In *Digital Kenya: An Entrepreneurial Revolution in the Making*, edited by B. Ndemo and T. Weiss, 239–267. London: Palgrave Macmillan.

Shih, C., ed. 2005. *Industrial Technology and the Industrial Technology Research Institute: Visible Brain*. Taipei, Taiwan: Industrial Technology Research Institute.

Platforms at the Margins

Jonathan Donner and Chris Locke

The globe has become bathed in mobile signals and handsets are a nearly ubiquitous necessity, no longer just a convenience for the prosperous. The resulting benefits of widespread connectivity are many, and the associated enthusiasm about it is deserved. But these gains, and these narratives, also risk obscuring a shift in the political economic action. As we have written elsewhere (Donner 2015), the power of digital technologies to structure social and economic lives increasingly resides not in the handsets, nor even in the towers that connect them, but with the platform companies at the heart of the Internet. The rise of platforms and the pervasiveness of mobile are inextricably linked. As the shift to mobile operating systems, networks, and apps has occurred—surpassing an Internet of PCs, unmetered connections, and open browsers—platform companies have developed and leveraged new points of control. Google runs Android as the window into the Google services within it. Facebook is a "mobile first" company, soaking the attention of nearly two billion monthly users, the majority via mobile interfaces. Apple's mobile iOS (and the app store platform) touches more people than its PCs ever did. Newer behemoths like the ride-sharing platform Uber are "mobile only" affairs, owing their existence to the particular ways in which smartphones scramble space and time. A more mobile Internet may touch more people. But it seems to do so in a way that depends on (or at least affords) centralization, scale, and standardization (Caribou Digital 2016).

The ways in which these platforms alter the Internet could fill volumes. Our goals in these remarks are twofold. First is simply to signal that dynamics are at play between this age of a platform-driven, more mobile Internet and the prospects for digital inclusion at a global scale. The second is to raise a concern that the decades-old and optimistic frame of ICTs for development (ICT4D) is increasingly naïve in accounting for the impact of these platforms—while smartphones and apps may have massively democratized the means of production, they have correspondingly intensely focused ownership of the means of distribution.

The 2016 World Development Report was a clarion call here, sensitizing a broad community to how these new digital dynamics altered and complicated what we thought we knew about social and economic development (World Bank 2016). The same unnerving, problematic dynamics debated in the Global North are at play at the margins. Algorithmic culture, "fake" news, the rise of surveillance, and the reorganization of digital labor are worldwide phenomena, partly because those platforms are worldwide.

But this is just the beginning—we suspect that even in the Global South, relationships between individual, state, and market will change more in the next ten years than they have over the past fifty, thanks to a more ubiquitous, more mobile Internet. The changes will be rife with unanticipated and complex consequences. For example, in "digital identity," the biometric Aadhaar system, and the remarkable India Stack of interrelated services will bring millions of Indians into formal financial and state systems. But this is playing out against the shocks of rapid demonetization and considerable concerns over the erosion of user privacies. These tradeoffs and disruptions, in finance, media, supply chains, even culture, are part of an endless Schumpeterian march of innovation and disruption (Scherer 1986), but at this moment, the disruptions are particularly rapid, and particularly global, with especially challenging prospects for the trajectories of social and economic development.

Consider Facebook. Once it offered social networking to sell advertisements. It still does. But as we write this note in 2017, Facebook has become one of the world's leading news hosts and aggregators. It is also in the access industry: selling airtime in some countries and discounting "zero-rating" airtime in dozens of others. It offers financial services by facilitating peer-to-peer money transfers via Messenger. It mediates identity on its own terms by promoting a single log-on and a single digital ID, as well as increasingly drawing on artificial intelligence to offer services via chatbots and voice interfaces.

Our challenge to readers, and our enthusiasm for this volume, centers on how traditional ICT4D frames are largely unequipped to address the considerable breadth and depth of the involvement of Facebook, or Uber, or Mechanical Turk, or any other digital platform in the structure of the economic and social spheres of developing economies. The platforms are not saviors, nor inherently evil; however, they are logics unto themselves. The algorithms birthed in Menlo Park and Mountain View, California, are experienced in Mombasa and Mumbai in ways that may not be optimal for individual users, for regional economies, or even for national sovereignties. Shifts to digital advertising may stifle local news media; shifts in digital opinion may strain domestic politics; shifts in digital identify may alter the relationships between states and citizens; and shifts in digital labor may remove jobs as quickly as it can make them.

Thus, despite all the hope over the last decade for the birth of a Silicon Savannah, it's not enough to put down new offices in the hope of incubating "the next Facebook" in the Global South—we have to understand and ultimately seek to influence how Silicon Valley's platforms are structuring the information age at a global level. There is an urgent need for new circuits of ethical engagement beyond and outside regulation, which is often too late to the party and does little other than retrospectively charge fines after the damage is done. We need to work harder to bring more perspectives of power and exclusion into the broader discourse on technology and development. Critical, engaged scholarship is part of the puzzle, and we see receptivity in industry, policy, and research, but that job is just beginning. Without the kinds of critiques of the global digital platforms found within this volume, we risk standing aside as the digital extractive industries exert power in this century much as the physical extractive industries did in the last one.

References

Digital, Caribou. 2016. *Winners and Losers in the Global App Economy*. Farnham, UK: Caribou Digital. http://cariboudigital.net/wp-content/uploads/2016/02/Caribou-Digital-Winners-and-Losers-in-the-Global-App-Economy-2016.pdf.

Donner, Jonathan. 2015. *After Access: Inclusion, Development, and a More Mobile Internet*. Cambridge, MA: MIT Press.

Scherer, F. M. 1986. *Innovation and Growth: Schumpeterian Perspectives*. Vol. 1. Cambridge, MA: MIT Press. https://ideas.repec.org/b/mtp/titles/0262691027.html.

World Bank. 2016. *World Development Report 2016: Digital Dividends*. Washington, DC: World Bank. http://elibrary.worldbank.org/doi/book/10.1596/978-1-4648-0671-1.

Digital Economies at Global Margins: A Warning from the Dark Side

Tim Unwin

Global digital connectivity is widely seen as essential for economic growth (World Bank 2016) and as having significant potential to help attain the Sustainable Development Goals (SDGs; ITU 2016). Many examples of successful projects seem to support such arguments for digital development at global margins, from the use of mobiles for financial transactions to healthcare interventions, the provision of timely information for farmers, and the use of tablets connected to the Internet in schools (GSMA 2016). There is thus good evidence that some poor people do indeed benefit economically and socially from greater connectivity. Yet, all too often, such initiatives do not go to scale, or are unsustainable, and therefore larger numbers of poor people more generally do not benefit appropriately from such digital interventions.

Information and communication technologies (ICTs) have transformed most aspects of human life over the last twenty-five years for those who have access to them, can afford such access, and know how to use them. Yet, the enormous potential that such technologies can enable also means that those who do not have access to them are left relatively more disadvantaged than they were previously. Even those who only have 2G phone connectivity, for example, are now being left far behind by those with smartphones and 4G access. If poverty is defined in a relative sense, digital technologies can thus be seen as increasing relative poverty. They are a powerful accelerator of difference and inequality. This is not to suggest abandoning attempts to use ICTs to contribute to development but it is to argue that at least as much attention needs to be paid to issues of inequality (SDG 10) as to the use of ICTs for economic growth.

Margins are not just geographic. Although connecting remote rural areas to the Internet is more difficult, and thus more costly, many people living in well-connected areas cannot afford such connectivity or are prevented from using the Internet for their empowerment. In patriarchal societies, women are often marginalized in their usage of, and benefit from, ICTs; children living on the streets of major cities fail to benefit from the digital revolution taking place in schools; those with disabilities are widely forgotten. These dimensions of social, political, economic, and cultural marginalization are at least as important as geographic marginalization, and they imply that digital solutions to poverty reduction must be much more subtle and sophisticated than just ensuring that everywhere has connectivity at a reasonably affordable price. Without any connectivity, no one can benefit from the full potential that ICTs can offer, but more needs to be done to support the poorest and most marginalized in their use of ICTs once connectivity is in place.

Recent research by the Organisation for Economic Co-operation and Development (OECD 2016) has shown how those with higher socioeconomic status tend to use the Internet for activities that will enhance their status and careers, whereas those from disadvantaged backgrounds use it mainly for chatting or playing games, thereby perpetuating a digital divide based on socioeconomic status. If the poor and marginalized, wherever they are found, are to benefit from connectivity, much more needs to be in place to support and empower them. This goes way beyond the standard arguments surrounding affordability of access, local content, digital literacy, and the provision of infrastructure (World Economic Forum 2016). Above all, it requires all those involved in delivering such interventions to focus primarily on the needs and interests of the poorest and most marginalized, rather than on ensuring that everyone is connected. Rather than advocating connecting the next billion, we should focus first on connecting the "bottom billion," those I prefer to call the "first billion."

There is little evidence of sufficient global will to enable this agenda to be realized, largely because the private sector, governments, and even civil society tend to be focused mainly on using the idea and practice of development primarily to serve their own ICT interests (i.e., Development for ICT, D4ICT), rather than on using ICTs for development (ICT4D). This was typified, for example, in the rush of applications by civil society organizations and others to develop Internet-based solutions for Ebola during the outbreak in Sierra Leone between 2014 and 2016, when less than 5 percent of the country had Internet access. Likewise, the interests of many private sector companies are primarily in generating profits from expanding Internet usage, rather than in enabling poor people to use the Internet effectively to enhance their lives and livelihoods.

The poorest and most marginalized are also more likely to suffer disproportionally from some of the darker aspects of Internet connectivity. As yet, scant research explores the effects of digital crime and abuse on different sections of society, but theft of small amounts of money in online financial transactions will clearly affect someone with little money more dramatically than someone who is richer. Likewise, poor children may be more likely to be targets of abuse through online pornography than are richer children. Women in patriarchal societies are subject to online sexual harassment more than men. Marginalized ethnic groups are particularly vulnerable to ethnic cleansing by governments that increasingly have good digital records about their citizens.

None of this is to suggest that efforts to connect the unconnected should not continue. Without such connectivity, people do not even have the chance to benefit from the potential of ICTs. Nonetheless, it is in everyone's interests to ensure that the poorest and most marginalized are indeed able to be empowered through such technologies. An increasingly digitally marginalized and disenfranchised population is not only morally wrong, it is also a danger to the sustained economic growth that dominates global rhetoric on development. This situation certainly requires appropriate policies by governments at all scales, particularly through their regulatory mechanisms, but above all, the private sector and civil society need to focus more on ICT4D than they do on D4ICT.

References

GSMA. 2016. *Unlocking Rural Coverage: Enablers for Commercially Sustainable Mobile Network Expansion. Connected Society*. London: GSMA.

ITU. 2016. ICTs for a Sustainable World #ICT4SDG. Accessed October 29, 2018. http://www.itu.int/en/sustainable-world.

OECD. 2016. Are There Differences in How Advantaged and Disadvantaged Students Use the Internet? *PISA in Focus* 64 (July). http://www.keepeek.com/Digital-Asset-Management/oecd/education/are-there-differences-in-how-advantaged-and-disadvantaged-students-use-the-internet_5jlv8zq6hw43-en#.V7GjFLWM-d1#page1.

World Bank. 2016. *World Development Report 2016: Digital Dividends*. Washington, DC: World Bank.

World Economic Forum. 2016. *Internet for All: A Framework for Accelerating Internet Access and Adoption*. Cologne: World Economic Forum, in collaboration with the Boston Consulting Group.

Digital Globality and Economic Margins—Unpacking Myths, Recovering Materialities

Anita Gurumurthy

Digital Economy: Mission Unbounded

Built on algorithmic mathematics, and revealing a sophistication that hides its brutality (Sassen 2017), the digital-financial nexus is emerging as the preferred modus operandi of information age capitalism.

The global financial crisis of 2007–2008 was propelled by the logic of finance and had nothing to do with traditional banking and credit. The sole aim of the information age capitalism that fueled the crisis was to extract a contract (the mortgage agreement) from low-income households (a majority of whom could have qualified for regular mortgages; Newman 2009) to increase capital flows to investors, and not for the purpose of mortgage repayment itself. Algorithms were able to slice, dice, and redistribute credit over a chain of balance sheets to a point where nobody really *knew* where regulated liabilities ended and unregulated liabilities began.[1]

An important feature of what can be called the digital-financial assemblage—constitutive of contemporary capitalism—is its systemic nature. It is a self-propelling juggernaut exploiting and generating precarity and debt. It feeds off a division of labor in which specialists in law, statistics, and business intelligence at the higher end provide over-valorized expert labor, and invisible and alienated individuals at the lower end contribute the grist to keep the extraction mill going (Chen 2014).

For the digital-financial assemblage, venture capital is the chosen instrument for control over market share. The value of digital behemoths is largely dependent on speculations of potential profits the company can make out of the huge pool of user data. Although the viability of their business model is yet to be proven, the promise of financial market capitalization propels these corporations on their path.

Digital Globality: Material and Metaphoric

Current metrics and meanings of the digital economy are far too narrow to adequately capture and delineate its material and symbolic moorings. In digital globality, the economy is not a fixed idea. It is fluid. In conventional terms, economic activity is organized into sectors or industries bound by institutional norms, rules, and practices concerning production and distribution of goods and services for that sector. The emergence of data and digital intelligence as the means of production destabilizes the very logic of such organization. Amazon is held up as an exemplar for its "courage to cannibalise itself" (World Economic Forum 2016) and reinvent its product and service portfolio. As Amazon redefines its territory from digital content to delivery drones, and now, to brick-and-mortar food retail, market share battles are becoming intensified in newer sectors. When Amazon announced price cuts in its food business, "shares of rival grocers tumbl[ed]" (Shepardson and Baertlein 2017). The material consequences of the digital-financial apparatus are without doubt likely to reverberate through the retail supply chain, suppressing payments of farm produce and annexing new sites of primitive accumulation (Roberts 2017).

The acquisitions and mergers in the digital-financial landscape suggest an extractive raison d'être that will take over lives and livelihoods, through a postmodern version of colonial sub-infeudation—the system by which land-holding tenants carved out new and distinct tenures by further subletting parts of their lands. Technology giants like Google, Apple, and Facebook are poised to enter the financial services business beyond mere experiments in mobile wallets (McCormack 2017), suggesting an economic future of hyperconsolidation. There are no traditional industry silos here. In this new architecture of economic organization, intelligence is organized *across* economic territories *for* runaway finance.[2] The fintech-philanthropy-development complex is the latest in the digital solutionism championed by global policy institutions (Soederberg 2013).[3] Thanks to fintech, global finance can profile poor households into generators of financial assets, while states can sharpen their surveillance gaze (Gabor and Brooks 2017).

Discursive formations of the digital economy play a pivotal role in its material manifestations. Big platform players are emerging as transnational sovereigns who speak with independent voices in emerging global policies—from trade negotiations to proceedings conducted by Internet standard-setting bodies. Their tactics to avoid liability include invocations of privileges and immunities, as well as aggressive construction of public debate through lawyers and lobbyists (Cohen 2017).

As digital intelligence blends with finance, the myth of rationality assumes the exalted status of an economic model, applicable anywhere, everywhere. The absurdities of political prediction markets in the "war against terror" based on intuition and hunches about security (Aitken 2011), for instance, reveal the fragile and questionable foundations of socioeconomic modeling.

The grand narrative of the connected digital worker flattens the idea of work, reducing it to a poststructural narrative of enterprise and creativity, erasing the difference between flexibility and precarity. It occludes the extraordinary risk taking and resilience of the toiling poor in the Global South, including the gig worker with little choice in the economy (Hussenot 2017).

Disassembling the Present, Assembling the Future

As the global economic and political elite peg their hopes on the fourth industrial revolution, a hard look at unqualified optimism about the digital economy is urgently warranted. The World Economic Forum recommends we embrace the disruptive unknown of artificial intelligence, acknowledging on the one hand the real risk of "artificial stupidity," but on the other, investing extraordinary faith in industry self-regulation and the transparency of digital companies (World Economic Forum 2016).

Equality in future society is contingent on normative frameworks for the public regulation of finance and the network-data complex. The governance response to information flows can no longer take its cue from the self-serving pseudo-wisdom of the digital-financial hegemons. Neither should it be guided by the state's knee-jerk repression of civic freedom. New legal-institutional measures for a global social contract that can bring about deep systemic change are necessary. In a politically polarized world, where human destinies are tied to the global, policies at the national and local levels that can support a transformative social contract depend on an idea of justice that is multiscalar. Such an idea necessitates an alternative economic model, in which the digital is reimagined for the well-being of people and the planet.

Notes

1. The End of Banking, last updated February 7, 2018, http://www.endofbanking.org/.

2. Scholars like Appadurai (2001) point to the "runaway quality"—the elusiveness—of global finance in the neoliberal paradigm. Finance, in the current context, is noted as being untouched by traditional constraints of information transfer, national regulation, industrial productivity, or "real" wealth in any particular society, country, or region.

3. Fintech refers to financial technology innovations that use digital methods for the delivery of financial services, for example, banking or credit through smartphones.

References

Aitken, Rob. 2011. Financializing Security—Political Prediction Markets and the Commodification of Uncertainty. *Security Dialogue* 42 (2): 123–141. https://doi.org/10.1177/0967010611399617.

Appadurai, Arjun. 2001. *Globalization*. Durham, NC: Duke University Press.

Chen, Yujie. 2014. Production Cultures and Differentiations of Digital Labour. *Triple C* 12 (2). http://www.triple-c.at/index.php/tripleC/article/view/547/626.

Cohen, Julie E. 2017. Law for the Platform Economy. *UC Davis Law Review* 51:133–204.

Gabor, Daniela, and Sally Brooks. 2017. The Digital Revolution in Financial Inclusion: International Development in the Fintech Era. *New Political Economy* 22 (4): 423–436.

Hussenot, Anthony. 2017. Freelancing May Be the Future of Employment—Though It's Not Always as Glamorous as It Sounds. *Scroll.in*, August 21, 2017. https://scroll.in/article/847531/freelancing-may-be-the-future-of-employment-though-its-not-always-as-glamorous-as-it-sounds.

McCormack, E. 2017. The Marriage of High Tech and High Finance. *Economist*, June 27, 2017.

Newman, Kathe. 2009. Post-Industrial Widgets: Capital Flows and the Production of the Urban. *International Journal of Urban and Regional Research* 33:314–331. doi: 10.1111/j.1468-2427.2009.00863.x.

Roberts, Wayne. 2017. Amazon's Move on Whole Foods Is Primal, More Than "Disruptive." *Medium*, June 19, 2017. https://medium.com/@wayneroberts/why-amazon-should-not-be-allowed-to-take-over-the-whole-food-industry-2524e792671.

Sassen, Saskia. 2017. Predatory Formations Dressed in Wall Street Suits and Algorithmic Math. *Science, Technology & Society* 22 (1): 6–20. https://doi.org/10.1177/0971721816682783.

Shepardson, David, and Lisa Baertlein. 2017. Amazon's Announcement of Whole Foods Price Cuts Sends Shares of Rival Grocers Tumbling. *Christian Science Monitor*, August 25, 2017. https://www.csmonitor.com/Business/2017/0825/Amazon-s-announcement-of-Whole-Foods-price-cuts-sends-shares-of-rival-grocers-tumbling.

Soederberg, Susanne. 2013. Universalising Financial Inclusion and the Securitisation of Development. *Third World Quarterly* 34 (4): 593–612. doi:10.1080/01436597.2013.786285.

World Economic Forum. 2016. Artificial Intelligence: Improving Man with Machine. *Digital Transformation of Industries: Digital Enterprise*, January 2016. http://reports.weforum.org/digital-transformation/artificial-intelligence-improving-man-with-machine/.

I   Digitalization at Global Margins

2 Making Sense of Digital Disintermediation and Development: The Case of the Mombasa Tea Auction

Christopher Foster, Mark Graham, and Timothy Mwolo Waema

Introduction

Tea is a key part of the economy in East Africa and a major export earner for countries such as Kenya and Rwanda. Twice a week, buyers and sellers come together in the coastal city of Mombasa in Kenya to trade East Africa's tea. They do so as part of the Mombasa tea auction, the primary link between local tea from East Africa and international tea firms who sell their products throughout the world (figure 2.1).

The tea auction emerged during the colonial era, and with its antiquated traditions, slow speed, and frequent accusations of corruption, there has been strong demand for it to move online. The auction itself is relatively predictable, and with falling costs of online access in the region, digitization of the auction seemed almost inevitable. An online auction offers the potential to speed up the processes of tea trading and bypass various brokers, warehouses, and traders. Reduced costs and improved efficiency would ensure that East Africa remains competitive with its rapidly advancing competitors in Asia.

Yet, processes of digitization and disintermediation have not taken place as expected in the East African tea sector. A proposed "e-auction" was abandoned. Only selected aspects of tea trading have been digitized, and only larger multinational tea firms appear to be moving toward disintermediation. The effects on more marginal tea firms and producers have been limited, and they continue to trade in the Mombasa auction.

The case study in this chapter points toward a more complex picture of digital development than is usually presented, compelling us to reconsider how we tackle disintermediation in theory and practice. In the clamor to promote improvement of firms, digitization is becoming an end in itself,

Figure 2.1
The Mombasa tea auction. *Source*: Wikimedia commons.

with issues of equality and impact falling off the agenda. Yet, the issues are vital in making sense of digitally enabled disintermediation and development. This case prompts us to refocus on crucial considerations around the development consequences of digitization projects. Does disintermediation lead to economic gains? Which firms are able to disintermediate digitally, and who is excluded?

Disintermediation and Transaction Costs

The Hopes of Disintermediation

Digitally enabled disintermediation is the process by which digital or online systems allow the removal of intermediaries involved in transactions (Bambury 1998; Chircu and Kauffman 1999). The term is often associated with digital networks, whereby product creators or service providers can link more directly to consumers or buyers in many sectors (Gellman 1996). Disintermediation has often been articulated as one of the key results of Internet connectivity in low-income countries, centered on the idea that connectivity would disintermediate the old restraining monopolies, intermediaries, and incumbents to enable dynamic market activity (Graham 2011; UNCTAD 2001; World Bank 2016).

Earlier empirical work exploring disintermediation in low-income countries has suggested fairly limited effects (e.g., Molla and Heeks 2007; Moodley 2003; Surborg 2009). Firms and producers often came up against barriers associated with the digital divide, such as the high costs of ICT ownership, lack of Internet access, limited digital skills, and a dearth of appropriate online services to support activity. With cheaper Internet access in recent years (ITU 2017), alongside the emergence of connectivity-enabled applications for low-income users (e.g., mobile money, apps, SMS tools, online platforms; UNCTAD 2015), the perceived barriers associated with the digital divide are decreasing (Foster et al. 2018). Thus, we have seen a second generation of research on disintermediation exploring the richer use of ICTs and connectivity (e.g., Aker 2010; Muto and Yamano 2009; Paunov and Rollo 2015; Zanello, Srinivasan, and Shankar 2014).

Yet, for all the wealth of literature, detailed accounts of digital disintermediation tend to be unclear in discussions of the potentially uneven consequences of disintermediation. Qualitative research rarely digs into the details of disintermediation, while quantitative research tends to build models that do not conceptualize uneven outcomes (Foster et al. 2018). Thus, it is appropriate to re-examine the concept of digital disintermediation in low-income countries to build a clearer knowledge of these processes.

Introducing Transaction Cost Models

To analyze digital disintermediation, we draw on transaction cost models, a large field of economic study that explores the costs involved when firms transact. Transaction cost models underlie how digital disintermediation has been conceptualized (Humphrey et al. 2003; Molla and Heeks 2007). While often seen as a fairly homogeneous concept from outside the field, transaction cost models have been the subject of two differing perspectives, which Allen (1999) calls the "property rights" and the "neoclassical" approaches. Exploring these two perspectives provides a clear understanding of how transaction costs are used, as well as highlighting potential gaps in analysis of digital disintermediation.

From the "property rights" perspective, analysis tends to explore the legal and institutional underpinnings of transactions. Property rights approaches thus focus on a wider set of "rights," both formal (legal rules, organizational structures, contracts, partnerships) and informal (norms, trust), which shape how transactions are undertaken. For instance, firms are more likely to undertake market transactions where enhanced protection (e.g., laws, regulation) reduces risks. Conversely, where these safeguards are not in place, firms may face high transaction costs to protect themselves. Indeed, they may prefer to transact in other ways, for example, through contracts or internal firm exchange (Williamson and Winter 1993).

In contrast to the property rights perspective, the neoclassical approach is more focused, homing in on analyzing the costs specifically related to the actual market transactions. This perspective tends to align with neoclassical economics. Key concerns of this literature are the drivers and characteristics of market transactions, for example, exploring links between the volumes of market trade and transaction costs such as transportation, market discovery, contracting, and so on (Benham and Benham 2000). As such, the scope is often limited to a set of factors and how these influence market exchange (Allen 1999).

As shown in figure 2.2, both perspectives on transaction costs look to explore factors that influence transactions. Yet they signal quite different ways to consider transactions, one exploring the direct drivers of exchange, the other more holistically studying conditions that orient transactions. Explorations of digital disintermediation in low-income countries tend to lean toward neoclassical approaches, which can lead to incomplete analysis of the reality of the process.

Transaction Costs, Information, and Digital Technologies

Many aspects of transactions are information rich (e.g., communication between firms, searching, contracting, monitoring). Thus, an important component of transaction costs is the outlay for finding, gathering, and using information, referred to as *information costs*.

Digital technologies can reduce information costs and thus affect transaction costs. Digital information flows enable rapid discovery of buyers and sellers as well as communication about transactions, even at a distance (Allen 1999). Beyond digitally enabled information flows, digital platforms can play an important role in transaction costs by aggregating buyers and sellers and facilitating transactions online, which reduces coordination expense (Sarkar, Butler, and Steinfield 1995; Wigand 1997).[1]

Making Sense of Digital Disintermediation and Development

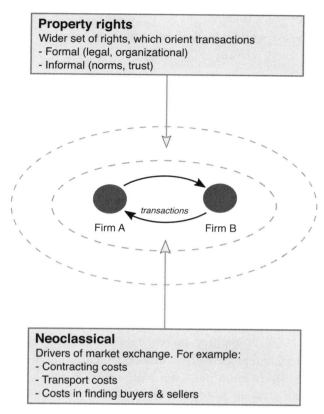

Figure 2.2
Different perspectives on transaction costs. *Source*: Authors.

Enhanced digital information also affects the role of intermediaries, which are often conceptualized to exist because of high information costs in transactions. They emerge when knowledge and information are scarce and provide services to reduce information costs (Sarkar, Butler, and Steinfield 1995). For example, a broker with knowledge of an industry can save firms the expense of searching for suppliers (Malone, Yates, and Benjamin 1987). With the growth of digital information and particularly digital platforms, the reduction of information costs can lead to a state where intermediaries and their knowledge are less crucial.

This simple idea of digital information flows and platforms reducing information costs underlies disintermediation. This is not the complete picture, however, and the transaction cost literature, particularly when taking a property rights perspective, has explored a set of broader considerations around transactions. Below we highlight three aspects discussed in the literature that are particularly relevant to this study.

The Nature of the Transaction   It is important to consider the conditions of transactions that influence whether firms decide to trade internally or buy externally (Benjamin and Wigand 1995; Malone, Yates, and Benjamin 1987). Key aspects relate to properties of the transaction and its complexity, often discussed in the transaction cost literature using the terminology *asset specificity* (referring to the interdependency of assets in production, such as resources, time limitations, skills) and *complexity of product specificity* (referring to the costs of ascertaining product information, such as in requirements and monitoring).

Thus, even where transaction costs decrease with improved information flows, the underlying nature of the goods or services being transacted may affect how a transaction is undertaken (Malone, Yates, and Benjamin 1987). Firms may continue with nonmarket transactions, or may use only digital forms when in a trusted relationship (Clemons, Reddi, and Row 1993; Dedrick, Xu, and Zhu 2008). Humphrey and colleagues (2003) highlight an example of how transaction properties can influence digitization in the garments sector in low-income countries. Even with several digital initiatives, this sector is quite resistant to being integrated into market-based platforms. One reason is that clothing quality is often determined by the "feel" of goods, and this complexity in assessing goods limits digital platforms and disintermediation. Buyer firms will thus only transact at a distance with trusted suppliers, continuing to use intermediaries when this is not possible.

Institutional Frameworks of Transactions   As mentioned, property rights approaches pay closer attention to the institutional contexts underlying transactions. In low-income countries, a widespread lack of institutional frameworks often shapes the types of transactions that occur. For example, the risk that one side may break a transaction without any consequences can reduce trust. A dearth of such institutions may limit the potential of market-based exchange on digital platforms, or lead to firms needing to make additional investments in monitoring or contracting (Clemons, Reddi, and Row 1993).

Strong or well-established institutions can also orient transactions, potentially making them resistant to change even when information costs are falling. Such cases have been documented among low-income groups introduced to mobile-based platforms. For example, mobile platforms among farmers often do not lead to disintermediation where strong informal institutions are present. Intermediaries still hold power in key institutional bodies or possess high social capital, including rich relationships with farmers over long periods (e.g., by training them or providing loans; Kumar 2014; Srinivasan and Burrell 2013).

Externalities of Digitizing Transactions New challenges can emerge when digitally enabled transactions are implemented. Although they may reduce some information costs (such as for searches), their introduction can also lead to what Cordella (2006) describes as "externalities," when new transaction costs emerge. For example, platforms reduce direct search costs by providing a way to interact among a wider array of transactors. A resulting externality, however, is not knowing all these transactors, with increasing risks in transacting and in evaluating buyers and sellers as their numbers grow. The emergence of digital platforms often leads to a wider geographic spread of firms undertaking transactions. This can result in additional externalities that emerge around exporting and logistics (Cordella 2006).

Thus, even when digital resources facilitate transactions and reduce costs, it is important to explore the externalities, that is, the spillover effects that come from digitization. Most notable for our interest in disintermediation, the literature suggests that *reintermediation* rather than disintermediation is common when externalities lead to evolving roles for intermediaries as a consequence of digitization (Agrawal, Agrawal, and Singh 2006; Sarkar, Butler, and Steinfield 1995).

Summary

How do these theoretical perspectives relate to discussions of digital disintermediation and transaction costs in the Global South? (e.g., Aker 2010; Jensen 2007; Muto and Yamano 2009; Singh 2008; UNCTAD 2001; World Bank 2016). In general, accounts of disintermediation tend to follow the neoclassical model. Some accounts do acknowledge property rights perspectives within discussions of the balance between market and nonmarket exchanges and the configuration of sectoral institutions. However, they rarely dig into detail about the wider conditions shaping market transactions, such as those we presented in the previous section. This is problematic. Integrating property rights approaches is liable to be particularly important for exploring digitally facilitated exchange among marginal actors, where cultural norms, trust, and power have been well documented as key aspects orienting activity (Harriss, Hunter, and Lewis 2003).

Aspects of transaction costs discussed in this chapter are grouped in table 2.1, which provides a framework for a more systematic analysis of transactions. Certainly, the nature of the actual transaction and information

Table 2.1
Transaction cost perspectives and their use in exploring digital disintermediation

Underlying approach	Key concepts	Perspectives on digital information, transactions, and disintermediation
Neoclassical	Drivers and constraints of market exchange	Explore how transaction costs of market exchange change as a result of improved information flows • Information costs affected by digital information flows • Ability of ICT/digital connectivity to disintermediate • Disintermediation through digital platforms • Constraints in digitally enabled exchange (competition, rules)
Property rights	Nature of transactions	Examine underlying properties of transactions and how they affect digitally enabled transactions • Shared resources, product requirements • Complexity of transactions and ability to ascertain and monitor quality
	Institutional frameworks	Explore digital information within a constellation of rules, rights, and norms that orient transactions • Underlying rights and norms that characterize exchange, which affects digitally enabled transactions • Nature and makeup of institutional bodies • Potential use of strategy and power in orienting digital transactions
	Externalities	Explore the spillover effects of digital information flows and platforms • Impact on other elements of transaction costs • Digital disintermediation and reintermediation

costs in market exchanges are important in exploring disintermediation and digital technologies. The literature from the property rights approach highlights additional directions for deeper scrutiny of digital technologies in terms of the nature of transactions, underlying institutions, and digitally driven externalities, which are crucial in characterizing how disintermediation plays out in practice (Foster and Graham 2017).

Methodology

To highlight aspects of the framework in table 2.1 and explore disintermediation in more depth, we return to the case study of the tea auction in Mombasa, Kenya. Our interest in the Mombasa tea auction stems from a research study exploring the effects of Internet connectivity on three economic sectors in East Africa—a material export-oriented chain (tea), a material service-oriented chain (tourism), and a more immaterial chain (business process outsourcing). During our research in the tea sector, multiple respondents spoke at length about the attempted introduction of a Mombasa "e-auction," which provides substantial *explanatory insight* into disintermediation (following Miles and Huberman 1994).

Our research on the tea sector occurred between September 2012 and March 2014 and involved seventy-five semistructured interviews in the tea sector as well as four focus groups, analyzing how different actors (farmers, intermediaries, and large firms) were using digital networks to improve production. We have drawn particularly on fifteen interviews with actors who discussed the history of the auction at length (e.g., tea auction brokers, tea auction buyers, tea warehousers in Mombasa, large firms involved in East African tea, and policy-making actors in both Kenya and Rwanda). In the next section, we focus on some of the empirical findings around the tea auction before using these findings for a more conceptual analysis.

The Evolution of the Mombasa Tea Auction

Over the last decade, the tea sector has undergone a process of change in East Africa. Historically, the key link between regional tea processors (sellers) and international buyers has been the Mombasa tea auction, where processors mainly sell unpackaged but processed loose black tea produced

Figure 2.3
Tea produced in the highlands of Rwanda (left) is transported to be processed (middle). Tea in the region is often blended, but some high-quality tea is sold as value-added tea (right). *Source*: Christopher Foster (left), Laura Mann (middle), and Birchall website screenshot (right).

in highland areas of East Africa, to buyers who mostly work for firms based outside the region (as illustrated in figure 2.3).

Tensions in the Auction
From the time of colonialism, the tea auction system has been the core institution for buying and selling East African tea, but the auction is increasingly struggling under the demands being placed on it. Goods are transported from the factories of tea processors in the region (mainly located in highland regions of Kenya, Uganda, and Rwanda) to be stored in warehouses in Mombasa while they wait to be auctioned. The auction occurs twice weekly, and only nominated sellers (brokers) are allowed to sell. Once the tea is sold, buyers make payments to the auction, at which point they are able to collect the tea to export. This whole process takes a minimum of a few days, but tea often remains in warehouses longer, accruing costs.[2] Intermediaries between the tea buyers and sellers provide useful services, particularly by supplying tea samples to buyers for tasting prior to the auction (to determine quality) and by ensuring that full payment is made after sales (some intermediary roles are shown in figure 2.4).

Increasingly, however, international tea buyers consider elements of the auction to be "backwards" or "quaint" in the modern market, as outlined by one large exporter in Mombasa: "For me they [brokers] are a complication. ... What is happening is we cannot buy directly from the producer at the auction; we have to buy from a broker. The brokers are the only people

Making Sense of Digital Disintermediation and Development

Figure 2.4
Roles of intermediaries in the tea value chain. The Tea Trade Center in Mombasa, home to the tea auction (left); tea tasting undertaken by tea brokers (middle); and storage of tea lots awaiting auction (right). *Source*: All images courtesy of Laura Mann.

who sell tea at the auction, so they actually control the auction. That is why there is the perception of a cartel."

Such mistrust also occurs in the relationship between tea processors and brokers. For instance, tea processors wished to understand the reason behind recent declines in the price of their goods, but, as outlined by one manager of a tea processor in Rwanda, they mistrusted the information they got back from brokers: "It's nice to know what's happening in the market. … The last two years have been bad. Last year's been very poor, and it's going further down this year. So, I wonder why. We do keep getting information from people, but sometimes I think it's rubbish that comes. What everybody does is justifies his position."

Concerns about the suitability of the auction have been raised with particular reference to the growing demands for data. Tea buyers want to better track auction prices, specifically the availability of particular tea grades (i.e., quality), for internal planning, including integrating with information systems, so that they purchase the right amount of each grade of tea at the best price. Further, with the increasing importance of tea origin and ethical production marks to value-added tea, firms would like to receive complete information that they can digitize to aid their planning.

Thus, there is growing pressure to reform the auction, and in a competitive global tea market with an excess supply of tea (Bird 2007), there has also been pressure to replace the auction with an agile electronic system.[3] Online tea auctions already take place in competitor countries in Asia (for example, in India and Sri Lanka). As outlined by an East African consultant, the e-auction would allow more integrated and agile engagement for international buyers and would be crucial for the region's competitiveness in tea: "The online auction breaks the boundaries so people in the US will be able to access the information … so they don't have to come all the way to Mombasa to buy tea, they can access our tea from our systems, trading can be done online, and [we] will ship the tea, and they will wait for the tea on the other side."

The Emergence of the e-Auction
In 2012, the East Africa Tea Trade Association (EATTA) attempted to introduce an electronic auction system, called the auction management information system (AMIS), including running a full trial of the system with key tea actors. The AMIS online tea auction was intended to replace the face-to-face activity. Many intermediary firms vehemently opposed the e-auction and continue to do so today. For instance, brokers have argued for the importance of the auction's face-to-face exchange, as outlined by one manager in a brokerage firm who was directly involved in auction activity: "My business is much better when I can see physically; I can know whether you are giving two or three dollars, but if I can sit [in the office] here I can't know your body language—to know if by looking at you I can get that one more dollar. I think that this is something we're going to lose if we go that way."

Further criticism of the e-auction has sprung from a fear that a lack of presence might imply collusion among tea buyers behind closed doors. One manager of a brokerage firm described this common concern: "The resistance [to the e-auction] was based on the fear that the buyers may collude; you know they are seated behind a machine like this … in an office somewhere. Being traders they may want to buy teas at the lowest prices possible, and it is easy for five of them to come together and say, 'Hey, you buy for us we are not going to push you,' and then tomorrow somebody else does the same, and the next week somebody else does the same."

International buyers however, remain less convinced by the brokers' arguments. This was best put by a manager for one of the largest exporters in Mombasa, who jokingly suggested that these risks were overstated for commodity trading: "They were saying the human factor, negotiating a price cannot be replaced … like you would find if you are selling a piece of art and everyone is raising their bids, or people are looking excited just from the facial expression. You would think that this piece of art is very expensive, and they would give probably a very high bid on it. So they [brokers] are using the same kind of logic."

The EATTA consulted extensively with stakeholders throughout the value chain to explain the form of the new e-auction and to run and evaluate the trial. After the trial, however, a majority of EATTA members voted against permanently implementing the e-auction (EATTA 2012). While there isn't a transparent record of the voting choices, our research suggests that opposition was highest among intermediaries—tea agents, brokers, and warehousers—all based in Mombasa. Not only were these groups most vocal in resisting the e-auction reform, but they were most active in campaigning against the e-auction. This campaigning seemed to influence particularly those on the supply side in East Africa (i.e., tea producers and tea processors), whose main linkage into the value chain is through their relationships with brokers. Most of these actors also voted against the e-auction. Opposition was especially strong from smaller associations, representing growers in the eastern regions of Kenya, as well as from Rwanda and Uganda. In discussions, the concerns of these groups often mirrored those of the brokers—fear of collusion in the e-auction, apprehension about the viability of online systems, and expressions of the value of face-to-face trade. Thus, the face-to-face auction in Mombasa survived.

Although we do not discount the genuine concerns about the e-auction, seeing how similar views became widespread was noteworthy. Our research suggests that the influential role of intermediaries was a key factor in the decision not to implement the e-auction. Intermediaries, particularly the brokerage organizations, also play a key role in the governance of EATTA, which is itself the governing institution for tea in East Africa.

Commercial pressures for a more agile Mombasa auction remain. Over time, the face-to-face auction has been supplemented by incremental additions of digital technologies, all of which were initially opposed by intermediaries. Two examples are online auction catalogs, which had previously been available only as paper copies, and an electronic payment system for quick payment of auction costs, simplifying management and logistics. The direct benefit of these innovations is not negligible; for instance, several tea processors told us that the improved efficiency of electronic payments had led to direct savings because of lower warehousing costs in Mombasa. The innovations did not disrupt the long-standing form and institutions of the auction; however, some intermediaries, such as brokers, have shifted roles, even if they have not been fully disintermediated. With the electronic auction payment system, brokers have less work to do related to payments, and they have begun to play an important role in other areas, such as collecting and sharing price and auction information with tea processors.

New Channels of Disintermediation

The slow digitization of the auction has led to the growth of alternative channels of tea trading. In the past, virtually all the tea trade in East Africa would go through the Mombasa auction, but in recent years, private sales have grown between tea processors and international firms' buyers to sidestep the limitations of the auction. This is referred to in the trade as direct sales. The channel of direct sales is shown in figure 2.5.

Tea statistics are extremely difficult to interpret; nevertheless, several indicators suggest a growth of direct sales. For instance, Rwandan data suggest that direct sales have grown to around 23–24 percent of tea being sold outside the auction in 2012 and 2013 (NAEB 2013). Amalgamating Kenyan Tea Board data with Mombasa data suggests that direct selling has grown in Kenya, fluctuating between 33 and 47 percent between 2010 and 2014 (Africa Tea Brokers 2015; TBK 2015).

For many tea processors in East Africa, direct sales are advantageous, and thus processors are keen to increase direct sales. The price paid for direct sales is likely to be higher than the auction price. Equally as important is that processors involved in direct sales will receive quicker payment in comparison to the sluggish turnaround time of the auction. Direct sales are also desirable in that they reduce the costs of brokerage and warehousing fees associated with the auction.

When examining who was selling through these more disintermediated channels, we found that most direct sales were made by tea processors who were subsidiaries of international tea firms, or who operated in close partnership with them. As the tea sector has become increasingly led by the private sector in East Africa, multinational firms (such as Unilever, McLeod Russel, and Jayshree) have pushed into the region, taking control of certain local tea processors.[4] Subsidiary tea processors tend to be more integrated with their parent firms, and there is a move for more integrated digital processing and tracking in many of these subsidiaries. Direct sales also fulfill

Making Sense of Digital Disintermediation and Development

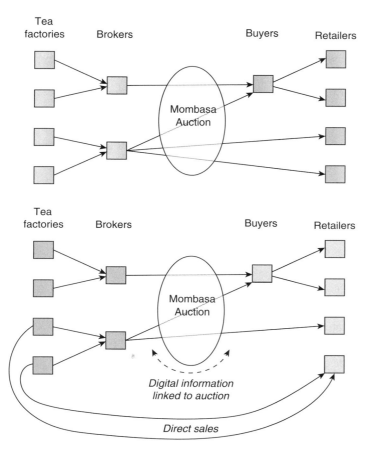

Figure 2.5
Evolution of tea trading channels. The auction remains, but with some aspects of digitalization, new direct sales routes have become important, disintermediating those involved in the auction. *Source*: Authors' fieldwork.

the needs of international buyers. They can quickly gain information about what types of tea are being processed, and, in the future, they will be able to dynamically plan and manage direct sales, aided by integrated digital systems. As outlined by one regional tea manager of a multinational firm, some companies are even considering going further. One is thinking of building an internal auction system that mirrors the idea of the e-auction: "We've been debating on the idea of selling all our product online. We would have an [internal] auction as well. … We would have it on a portal where, you know, this is the type of tea we have. And we send samples to so many buyers worldwide anyway."

In sum, as shown in figure 2.5, tea sales channels have become divided between direct sales and the Mombasa auction. Direct selling revolves around disintermediated trade and tends to be for goods at a premium, such as ethical teas. A driver of direct sales is the growth of information systems and digital data, and there is a clear trajectory toward automation and potentially new auction platforms. The Mombasa auction continues to take place every week, with tea lots sold through open face-to-face bidding. It is still an access point for tea processors to access global buyers, but the auction can be unpredictable, particularly in recent years, as the global oversupply of tea has increased (Bird 2007). Thus, we predict in the long term that (in contrast to direct sales) the auction may become a channel for lower-grade bulk "commodity" tea sold at lower prices.

Inevitably, direct selling is not for all tea producers. Direct selling is emerging out of privatization policies in East Africa, where global tea producers have bought stakes in lucrative tea processors. In these relationships, digitally facilitated disintermediation emerges—in larger tea processors who have built trust, who have invested in meeting the requirements of tea quality or certification, and who are integrating digitally with their parent companies.

Epilogue

Ironically, given the brokers' insistence that the e-auction would lead to collusion, a recent controversy in the tea sector centers on collusion in the face-to-face auction:

I want the Chairperson to state whether he is aware that the Kenya Tea Development Agency (KTDA) as reported in the Tea Industry Status Report of May, 2014, is accused of the following acts: (1) Colluding with cartels to manipulate tea prices. (2) Conducting direct sales with big markets outside auction venues. (3) Buying tea directly from factories at lower prices and then importing cheap tea. (4) Colluding with various players to create the impression that there is excess tea in the market in order to maintain low prices.[5]

The above accusations, made in the Kenyan Parliament in 2014, were picked up by the regional press and led to fierce criticism and recrimination among Kenyan politicians. The aforementioned Tea Industry Status Report (which is unpublished and not publicly available) alleged that several bodies, including the influential KTDA, a large umbrella organization that supposedly represents Kenyan smallholder tea farmers, had been engaging in practices that reduced the prices for those same smallholders. One regional governor even launched a KSh87 billion (US$1 billion) lawsuit on behalf of his smallholder tea-producing constituents following these accusations (Tanui 2015).

While these accusations remain unproven, suspicion of the Mombasa tea auction has been reignited, exacerbated by the perceived lack of reform and activities that remain shrouded in mystery to the outsider. The ensuing fallout of these accusations has revived political pressure for the tea auction to become digitized, which for politicians would remove corrupt elements through transparency. Members of the EATTA and the Kenyan Tea Board were summoned to appear before Kenyan ministers. To date, they have resisted any move toward an e-auction but have begun digitizing additional aspects of the auction to improve transparency. For instance, the EATTA has recently made live webcam coverage of the Mombasa auction available online (Xinhua 2014).

Discussion

Constraints in Transactions

Using the transaction cost framework, we can highlight the key drivers and constraints to disintermediation in the Mombasa tea auction case. From an *economic perspective of market transitions*, the e-auction seemed, on the surface, to be viable. Allowing more efficient buyer/seller discovery, transactions, and exchange in tea markets appear to be key steps for regional competitiveness. Introducing a digital platform would reduce transaction costs by disintermediating broker intermediaries, thus saving warehousing fees and other costs associated with the auction in Mombasa. Overall the nature of market exchanges also appears coherent in terms of formal rules and contracts in the transaction, which are guided by the regional tea body EATTA, with strong interest from national export boards. Economic theory would suggest these regular and fairly controlled transactions would be suitable for a digital platform.

In terms of the underlying *nature of the transaction*, tea can be regarded as a relatively standardized commodity good. Compared to more complex goods or innovations, tea quality is relatively simple to ascertain.

Nonetheless, with the recent growth in value-added tea and interest in the quality of tea, successful platforms are likely to need to digitize and integrate a wider range of parameters and properties.

A deeper analysis of *the nature of underlying institutions* highlights institutional path dependency and well-established rules that conflict with the goals of disintermediation. In the tea sector, the close alignment between intermediaries and sectoral governance is a particular concern, and these have been important factors in resistance to the e-auction.

Multiple *externalities from digital technologies* have emerged in this case. Even as some intermediaries were being marginalized by the growth of digital payments in the auction, they were able to use their social capital to reintermediate themselves into aspects of transactions that were becoming more information intensive. Brokers were able to become key providers of information and intelligence that supported the increasing market focus of tea growers and processors in East Africa.

The categories from the framework also highlight policy approaches that could support future initiatives to reduce information costs. The Mombasa auction still faces a set of constraints on digitalization related to the nature of transactions and the growing complexity of exchange. Stakeholders doubted it was possible for online systems to codify face-to-face activities, for instance, the quality and grading of tea. Such concerns need to be considered as part of design decisions in a digital platform (e.g., by making quality and ethical-mark data visible in the exchanges). Offline activities such as quality testing by a trusted party also might support disintermediation.

Constraints around institutions suggest that the e-auction may emerge only in hand with wider institutional reform, and with a strong political push by national tea boards for reform. Many actors' concerns were related to fears around maintenance of quality and trust. Establishing clearer rules for auction-related contracting or embedding collusion-detection algorithms within the software could be further steps to supporting full disintermediation.

Strategic Actions and Transaction Costs

In the Mombasa auction, the empirical outcome of digitization has not been desirable for all. We found evidence of disintermediation in direct sales, facilitated, in part, by online access, with the improved ability for buyers and sellers to interlink. Transactions in direct trade were marked by greater trust, clearer expectations on quality, and potential ongoing contracts between specific buyers and sellers. Thus, digitally supported disintermediation did occur but only for certain actors under certain conditions. Some processors thus trade through disintermediated private channels while others continue to trade in auction markets—indeed some firms simultaneously use both. The properties of these different channels—and who is disintermediated—are constantly in flux as channels are refined and improved.

Direct sales are liable to benefit only the best-linked and highest-quality tea producers. In the longer term, without an agile and efficient digital platform for trade (that is also open to all), marginal producers may miss out on the potential to make international linkages and reduce their ability to trade in value-added tea. These differential outcomes should prompt us to further reflect on the variation in trust, skills, and power of different actors in these transactions, and how this has led to differential outcomes of digitization in the tea sector. Key firms and those powerful intermediaries being disintermediated were not passive—they formed strategic coalitions that were strong enough to resist the introduction of the e-auction. Indeed, we can argue that resistance came from the strategic use of institutional and transaction weaknesses (e.g., notions of collusion, role in EATTA) to defend against disintermediation.

This perspective aligns with a direction taken by the "new institutional" literature, which looks to move institutional analysis and transaction costs toward exploring power, politics, and strategic activity as key to shaping institutions and driving institutional resistance (Khan 2010; North, Wallis, and Weingast 2009; Oliver 1991). Nonetheless, it is rare to see disintermediation linked to the complexities of strategic and institutional analysis in the literature on digital connectivity and disintermediation. Such work would help to introduce some clear concepts of power back into understandings of transactions and digital connectivity.

Conclusion

As we have shown in the East African tea sector, in-depth analysis of transaction costs highlights complexities that are rarely detailed in the literature on digital technology and disintermediation. Theoretically, mainstream approaches to transaction costs highlight key drivers and restraints of digitally enabled exchange in markets, and thus highlight recommendations in transactions (e.g., rules, contracts) that can support more agile market exchange. Models of transaction costs from a property rights perspective supplement this analysis by supporting a clearer understanding of the nature of transactions, their institutional basis, and the ways in which digitally enabled disintermediation may become an additional constraint for market participation. We have also highlighted constraints to efficient e-auction platforms by certain aspects of trust and quality that contributed to suspicions and rejection of an e-auction.

Our empirical findings have highlighted the privileged digitally enhanced direct sales implemented for smaller groups of buyers and sellers; reintermediation as intermediaries evolve; and institutional resistance. As shown in the East African tea sector, the eventual benefits of disintermediation may be liable to come to those already in privileged relations. Thus, we suggest that a greater awareness of how actors exert power and strategically use institutional resources is important in understanding the wider developmental impacts of digital disintermediation. In Mombasa, digitally enabled transactions are not at present transforming the tea sector, but, through the strategic activities of more powerful actors, these transactions are solidifying the relationships of those who are already well linked and able to capture resources.

Acknowledgments

We would like to acknowledge Laura Mann and Charles Katua, who played critical roles in the fieldwork connected with this chapter. This study was based on research funded by the UK Economic and Social Research Council (ESRC) and the Department for International Development (DFID). Grant reference (RES-167-25-0701) and ESRC reference (ES/I033777/1).

Notes

1. A range of terms is used in the literature to describe digitally enabled platforms for transactions: electronic marketplaces, cybermediaries, infomediaries, information exchanges, e-business systems, and platforms. In this chapter, we use the term digital platforms.

2. A minimum of 3 percent of the tea cost is paid in brokerage and warehousing fees, but once indirect costs are taken into account, this is likely to be higher, probably in the region of 5–10 percent of the auction price.

3. Our findings show that many brokers and processors in the region now use the Internet regularly in their activities, so access limitations are likely to be less problematic.

4. East African tea was traditionally organized so that farmers received a fixed "farm gate price" for tea as specified by the government. Processing factories were run by the government, with state-owned marketing boards responsible for international sales. Privatization has led to factories or shareholdings being sold to the private sector, where farm gate and processed tea prices are determined by the market. Government now takes a back seat, with a supporting role through development boards.

5. Kennedy Mong'are Okong'o, MP for Nyamira, statement in the Kenyan Parliament, August 2014.

References

Africa Tea Brokers. 2015. Tea Charts. Africa Tea Brokers. March 25, 2015. http://www.atbltd.com/Docs/graph.

Agrawal, Durgesh K., Dev P. Agrawal, and Deepali Singh. 2006. Internet Based Distribution Systems: A Framework for Adoption. *Decision* 33 (1): 21–46.

Aker, Jenny C. 2010. Information from Markets Near and Far: Mobile Phones and Agricultural Markets in Niger. *American Economic Journal. Applied Economics* 2 (3): 46–59.

Allen, Douglas W. 1999. Transaction Costs. In *Encyclopedia of Law and Economics*, edited by B. Bouckaert and G. De Geest, 893–926. Cheltenham, UK: Edward Elgar. http://ecsocman.hse.ru/data/008/450/1217/0740book.pdf.

Bambury, Paul. 1998. A Taxonomy of Internet Commerce. *First Monday* 3 (10). doi:10.5210/fm.v3i10.624.

Benham, Alexandra, and Lee Benham. 2000. The Costs of Exchange. In *Institutions, Contracts and Organizations: Perspectives from New Institutional Economics*, edited by Claude Ménard, 367–375. Cheltenham, UK: Edward Elgar.

Chircu, Alina M., and Robert J. Kauffman. 1999. Strategies for Internet Middlemen in the Intermediation/Disintermediation/Reintermediation Cycle. *Electronic Markets* 9 (1–2): 109–117.

Clemons, Eric K., Sashidhar P. Reddi, and Michael C. Row. 1993. The Impact of Information Technology on the Organization of Economic Activity: The "Move to the Middle" Hypothesis. *Journal of Management Information Systems* 10 (2): 9–35.

Cordella, Antonio. 2006. Transaction Costs and Information Systems: Does IT Add Up? *Journal of Information Technology* 21 (3): 195–202. doi:10.1057/palgrave.jit.2000066.

Dedrick, Jason, Xin Xu Sean, and Kevin Xiaoguo Zhu. 2008. How Does Information Technology Shape Supply-Chain Structure? Evidence on the Number of Suppliers. *Journal of Management Information Systems* 25 (2): 41–72. doi:10.2753/MIS0742-1222250203.

EATTA. 2012. *Final Validation Report on the Proposed Auction Management Information System Project.* Nairobi: East African Tea Trade Association.

Foster, Christopher, and Mark Graham. 2017. Reconsidering the Role of the Digital in Global Production Networks. *Global Networks* 17 (1): 66–88. doi:10.1111/glob.12142.

Foster, Christopher, Mark Graham, Laura Mann, Timothy Waema, and Nicolas Friederici. 2018. Digital Control in Value Chains: Challenges of Connectivity for East African Firms. *Economic Geography* 94 (1): 68–86.

Gellman, Robert. 1996. Disintermediation and the Internet. *Government Information Quarterly* 13 (1): 1–8. doi:10.1016/S0740-624X(96)90002-7.

Graham, Mark. 2011. Disintermediation, Altered Chains and Altered Geographies: The Internet in the Thai Silk Industry. *Electronic Journal of Information Systems in Developing Countries* 45 (1): 1–25. https://doi.org/10.1002/j.1681-4835.2011.tb00321.x.

Harriss, John, Janet Hunter, and Colin Lewis. 2003. *The New Institutional Economics and Third World Development.* London: Routledge.

Humphrey, John, Robin Mansell, Daniel Paré, and Hubert Schmitz. 2003. *Reality of E-Commerce with Developing Countries.* London: LSE. http://eprints.lse.ac.uk/3710/.

ITU. 2017. *World Telecommunication/ICT Indicators Database.* Geneva: International Telecommunications Union.

Jensen, Robert. 2007. The Digital Provide: Information (Technology), Market Performance, and Welfare in the South Indian Fisheries Sector. *Quarterly Journal of Economics* 122 (3): 879–924.

Khan, Mushtaq. 2010. *Political Settlements and the Governance of Growth-Enhancing Institutions.* London: SOAS. http://core.ac.uk/download/pdf/2792198.pdf.

Kumar, Richa. 2014. Elusive Empowerment: Price Information and Disintermediation in Soybean Markets in Malwa, India. *Development and Change* 45 (6): 1332–1360. doi:10.1111/dech.12131.

Malone, Thomas W., Joanne Yates, and Robert I. Benjamin. 1987. Electronic Markets and Electronic Hierarchies. *Communications of the ACM* 30 (6): 484–497. doi:10.1145/214762.214766.

Miles, M. B., and A. M. Huberman. 1994. *Qualitative Data Analysis: An Expanded Sourcebook*. London: Sage.

Molla, Alemayehu, and Richard Heeks. 2007. Exploring E-Commerce Benefits for Businesses in a Developing Country. *Information Society* 23 (2): 95–108. doi: 10.1080/01972240701224028.

Moodley, Sagren. 2003. The Promise of E-Business for Less Developed Countries. *International Journal of Electronic Business* 1 (1): 53–68.

Muto, Megumi, and Takashi Yamano. 2009. The Impact of Mobile Phone Coverage Expansion on Market Participation: Panel Data Evidence from Uganda. *World Development* 37 (12): 1887–1896.

NAEB. 2013. *Statistics 2013 on Tea, Coffee and Horticulture*. Kigali, Rwanda: National Agricultural Export Development Board.

North, Douglass C., John Joseph Wallis, and Barry R. Weingast. 2009. *Violence and Social Orders: A Conceptual Framework for Interpreting Recorded Human History*. Cambridge: Cambridge University Press.

Oliver, Christine. 1991. Strategic Responses to Institutional Processes. *Academy of Management Review* 16 (1): 145–179. doi:10.2307/258610.

Paunov, Caroline, and Valentina Rollo. 2015. Overcoming Obstacles: The Internet's Contribution to Firm Development. *World Bank Economic Review* 29 (Supplement): S192–S204. doi:10.1093/wber/lhv010.

Sarkar, Mitra Barun, Brian Butler, and Charles Steinfield. 1995. Intermediaries and Cybermediaries. *Journal of Computer-Mediated Communication* 1 (3): 1–14.

Singh, Nirvikar. 2008. Transaction Costs, Information Technology and Development. *Indian Growth and Development Review* 1 (2): 212–236. doi:10.1108/17538250810903792.

Srinivasan, Janaki, and Jenna Burrell. 2013. Revisiting the Fishers of Kerala, India. In *Proceedings of the Sixth International Conference on Information and Communication Technologies and Development: Full Papers*, vol. 1, 56–66. New York: ACM. http://dl.acm.org/citation.cfm?id=2516618.

Surborg, Björn. 2009. Is It the "Development of Underdevelopment" All over Again? Internet Development in Vietnam. *Globalizations* 6 (2): 225–247. doi:10.1080/14747730902854182.

Tanui, Nikko. 2015. Great Storm Brewing in Tea Industry as County's Landmark Sh87b Suit Kicks Off. *Kenya Standard*, March 23, 2015. https://www.standardmedia.co.ke/business/article/2000155691/great-storm-brewing-in-tea-industry-as-county-s-landmark-sh87b-suit-kicks-off.

TBK. 2015. Kenya Tea Export. Tea Board of Kenya. March 25, 2015. http://www.teaboard.or.ke/statistics/exports.html.

UNCTAD. 2001. *E-Commerce and Development Report 2001*. Geneva: UNCTAD.

UNCTAD. 2015. *Information Economy Report 2015—Unlocking the Potential of E-Commerce for Developing Countries*. Geneva: UNCTAD.

Wigand, Rolf T. 1997. Electronic Commerce: Definition, Theory, and Context. *Information Society* 13 (1): 1–16. doi:10.1080/019722497129241.

Williamson, Oliver E., and Sidney G. Winter. 1993. *The Nature of the Firm: Origins, Evolution, and Development*. New York: Oxford University Press.

World Bank. 2016. *World Development Report 2016: Digital Dividends*. Washington, DC: World Bank.

Xinhua. 2014. Tea Auction Starts Live Data Feeds to Increase Transparency and Efficiency. *Kenya Standard*, November 20, 2014. https://www.standardmedia.co.ke/business/article/2000141934/tea-auction-starts-live-data-feeds.

Zanello, Giacomo, Chittur S. Srinivasan, and Bhavani Shankar. 2014. Transaction Costs, Information Technologies, and the Choice of Marketplace among Farmers in Northern Ghana. *Journal of Development Studies* 50 (9): 1226–1239. doi:10.1080/00220388.2014.903244.

3 Development or Divide? Information and Communication Technologies in Commercial Small-Scale Farming in East Africa

Madlen Krone and Peter Dannenberg

Introduction

Over the last decade, there has been much debate about the potentials of information and communication technology for development (ICT4D). This debate has been fueled by the rapid distribution of mobile phones and access to the Internet in most low-income countries, including in remote areas and among low-income groups (e.g., Okello et al. 2013). Various studies have come to the conclusion that ICTs are improving economic and social development (Kirk et al. 2011; Qiang et al. 2011; UNDP 2012). According to these studies, ICTs create new modes of connectivity and enable the integration of thus-far marginalized businesses and regions into commercial value chains in the globalized world. As a result, large programs and activities have been brought into being: by private firms who want to develop new markets for mobile phone companies, by donors who support small businesses in low-income countries, and by national governments who want to use ICTs to better integrate their export-oriented production into global markets (Kirk et al. 2011; Okello et al. 2013; UNDP 2012; Qiang et al. 2011).

Examining the scholarly literature of recent years, however, makes it obvious that this optimistic perspective is partly driven by wishful thinking and lacks a fundamental theoretical and empirical foundation (see also the critique by Donner and Escobari 2010; Heeks 2010). Without a clear scientific foundation, and given the fact that the mid- and long-term effects of these developments are unpredictable, some of the optimistic studies mentioned above might be misleading. In fact, some recent studies on mobile phone use indicate negative economic effects for segments of the population (e.g., a digital divide; Carmody 2012; Murphy, Carmody, and Surborg 2014).

There is, therefore, a danger that practitioners and policymakers are rushing to embrace and translate conceptual thinking into their agendas before the theory is sufficiently robust and, as a result, are failing to achieve their stated objectives (see the debate on "theory led by policy"; Yeung 2015; Lovering 1999).

In this chapter, we aim to contribute a more robust scholarly foundation to this question by analyzing examples from small-scale fresh fruit and vegetable (FFV) farmers in Tanzania and Kenya and their use of ICT to integrate themselves into commercial value chains. Based on a quantitative and qualitative study of FFV farmers in the Mt. Kenya region (Kenya) and Mwanza region (Tanzania), we discuss to what extent and under which conditions ICTs lead to positive or negative effects. In this way, we contribute to a more differentiated and specific conceptual and empirically based framework for the assessment of ICT4D, in contrast to previous relatively one-sided perspectives (e.g., Kirk et al. 2011; Qiang et al. 2011). We first identify distinct ICT usage types of farmers and discuss this usage in relation to their characteristics and capabilities. We then identify and analyze the effects that these distinct ICT usages can have on farmers.

ICT for Development and Its Relevance for Small-Scale Farming

Positive Aspects in General and for African Small-Scale Farming

The organization and coordination of business activities worldwide is largely supported by ICT-based solutions like web-enabled management tools in complex logistics systems (Lasserre 2004). In low-income countries, the rise of ICTs—albeit still often based on simple hardware and software—has dramatically increased access to, as well as volume and richness of, business-relevant knowledge (Unwin 2009) and has opened up various ways of overcoming remoteness and spatial barriers for social and economic interactions (Pfaff 2010). This has also led to positive expectations about the impacts of ICTs on economic, social, and political developments in low-income countries, including agricultural production systems and their integration into professional and international markets (Unwin 2009; Loh 2013).

These expectations have resulted in the realization of public and private programs and projects funded by international banks and donor organizations, such as USAID, the Food and Agriculture Organization (FAO), and the World Bank, under the term ICT4D. Studies of such initiatives often highlight the positive effects of ICTs, including economic recovery through leapfrogging (by skipping the stage of landline telephones) and improving business as well as social connectivity (Graham 2011b; Friederici, Ojanperä, and Graham 2017). Aker and Mbiti (2010) identify five potential mechanisms through which ICT can provide economic benefits: (1) improving access to and use of information; (2) improving productive efficiency, allowing businesses to better manage their supply chains; (3) creating new jobs to address the demand for ICT-related services; (4) facilitating communication within social networks in response to shocks; and (5) facilitating the delivery of financial, business, health, and educational services (see also Carmody 2012).

Access to commercial markets is an especially great challenge for small-scale farming in Africa (see Bbun and Thornton 2013) and for small-scale resource-based businesses in low-income countries in general. Qiang and colleagues (2011) argue that access to text-messaging services or websites bears the potential to fundamentally increase small-scale farmers' access to market links, distribution channels, financial services, and extension services that have previously been unavailable to them. Dannenberg and Lakes (2013) show that mobile phone use can support farmers in linking up with local organizations and extension officers to access knowledge and to fulfill the process requirements of their respective buyers. The authors further indicate that simple information (e.g., simple facts on weather or prices) can be exchanged easily by farmers via ICTs, although this is much more difficult or limited in the case of complex knowledge exchange (e.g., production techniques).[1] Generally, the opportunity to exchange knowledge and interact depends on type of ICT usage (e.g., the Internet can be used to exchange large volumes of codified complex knowledge like manuals, which is not possible in telephone calls or text messages).

Humphrey (2002) argues that ICT use can help small agricultural producers reduce information asymmetries with their buyers (e.g., regarding export market prices), which are common features in global value chains, and that, therefore, the use of ICTs can strengthen farmers' bargaining positions. Mukhebi and coauthors (2007) see a transforming potential as a result of Internet- and mobile phone–related innovations, which may result in new forms of organization and access to markets in agriculture.

Critical Perspectives on ICT4D and Their Bearing on African Small-Scale Farming

While the positive achievements and potentials of ICTs have dominated the applied debate, critical visions of ICTs for farming in low-income countries indicate that the effects of ICTs might be overestimated or lead to negative developments (Donner and Escobari 2010; Murphy, Carmody, and Surborg 2014; Murphy and Carmody 2015). Within this debate, scholars criticize the promises of disintermediation and the transformative potentials of ICTs. While Donner and Escobari (2010) predict that progressively increasing use of ICTs in low-income countries will affect production and distribution systems, they argue that these developments will not necessarily change the underlying mechanisms and structures within value chains (in particular, the unfavorable power relations of farmers with their buyers). They see as more likely a consolidation of intermediaries, who themselves use ICTs, especially mobile phones. Similarly, Murphy, Carmody, and Surborg (2014, 264) see a "thintegration," in which only a few benefits arise for small businesses in low-income countries, while power relations and structures that sustain extraversion and underdevelopment persist (Murphy and Carmody 2015; Foster and Graham 2016).[2]

Humphrey (2002) argues not only that the potential positive effects of ICTs might be limited, but that ICTs may even bring about negative outcomes. One negative effect is an increasing digital divide. ICT use depends on the openness of the necessary physical infrastructure and software, which can create entry barriers not only to new ICT-based knowledge flows and transactions, but also to non-ICT knowledge flows and additional transactions that are shifting to an ICT-based exchange. In this context, as Bbun and Thornton (2013) outline, commercial markets are "not a level playing field" but a competition between competitors with different backgrounds and capabilities to adapt to market dynamics or changes in the sociotechnological regime for small-scale enterprises. Given that whether farmers use ICTs depends on each person's different capabilities and characteristics (e.g., financial capability, age, education), the proliferation of ICTs is likely to increase inequalities (e.g., between wealthier enterprises and poorer enterprises that are not able to afford a mobile phone; Heeks 2014).

Carmody (2012) also outlines unequal competition on a global scale as a further negative outcome for southern producers. He argues that if ICTs help connect to international markets, they also lead to increasing worldwide competition, in which enterprises in low-income countries often struggle to compete against strong international competitors. Another negative development from the spread of ICTs in low-income countries—particularly for producers in international value chains—is the increase of dependencies. The assumption that ICTs can improve the coordination and control of international value chains also means that these technologies are likely to increase the power of companies that already possess coordination and control functions (e.g., the so called "lead firms," which in the case of FFV are usually supermarkets; Dannenberg 2012). Given the asymmetric bargaining position in the chains, the broad dissemination of ICT generally may even lead to a new sociotechnological regime (see, e.g., Wiskerke 2003) in which farmers and traders are forced to adjust or be excluded. This was partly the case after the private standard GlobalGAP was introduced in Kenyan horticultural value chains, when large numbers of farmers and exporters had to reorganize themselves to meet the new requirements (Dannenberg 2011). ICTs could have transformative, but negative, effects (e.g., entry barriers and increased dependencies) on many African small-scale farmers, while other, more powerful, better-skilled actors in the chains gain from the same effects.

Based on these observations, we argue that ICTs have an important influence on farming (especially on knowledge access). They provide the possibility of using different distribution channels and altering the bargaining position of the farmer. We also argue, however, that the influence of ICTs on farming is dependent on different variables, including, at a minimum, the type of ICT usage and the different capabilities of the farmers (e.g., education). To test these assumptions, we posit two research questions:

1. What different ICT usage options do farmers have for business purposes in relation to the characteristics and capabilities of the farmers?
2. How does the different usage of ICT influence farmers' knowledge access, distribution channels, and bargaining positions?

Methodology

Case Study Selection

The Kenyan FFV sector is a forerunner and one of the few success stories of integrating small-scale farmers and traders into the European Union's international value chains (Dannenberg and Nduru 2013).[3] In Tanzania, such a large-scale export orientation in the FFV sector has not yet taken place, but Tanzanian farmers are increasingly supplying professional commercial retailers for domestic and African export markets (König et al. 2011). Thus, the Kenyan and the Tanzanian FFV sectors are two cases that show distinct stages of integration into different distribution systems. Despite these differences in value chain integration, FFV farmers in both countries have experienced the rapid spread of ICT use in their sector over the last few years. To connect with commercial markets and value chains, an increasing number of small-scale businesses in Kenya and Tanzania (as well as in many other low-income countries) are starting to use ICTs, including mobile phones and the Internet (Donner and Escobari 2010).

The data for this chapter are drawn from two rounds of fieldwork in the Mt. Kenya and Mwanza regions in 2013 and 2015 (qualitative interviews and quantitative surveys in autumn 2013; completing qualitative interviews in spring 2015). We selected these two regions based on the large numbers of small-scale FFV farmers producing for commercial value chains as well as the relatively good and affordable mobile phone and Internet networks in both regions (see also Molony 2008a). Instead of focusing on a comparison between the two regions, we have used data from both to provide a broad statistical population, which gives a more differentiated picture of the varying types of commercial small-scale farms and their integration into different distribution systems.

Data Collection

We applied a mixed method approach, including 61 semi-structured qualitative interviews and 368 quantitative surveys with farmers. We selected commercial small-scale farmers involved in horticulture (fresh fruit and vegetables) for the surveys and the qualitative interviews. In Kenya and Tanzania, the horticultural sector has been one of the most dynamic agricultural subsectors over the last ten years, with comparatively high incomes (Krone, Dannenberg, and Nduru 2016). This sector is especially interesting for studying ICT use since it is characterized by the high perishability of products, which require fast trading transactions. Thus, rapid delivery processes and prompt communication are needed (Molony 2008b). To measure the influence of ICT use on agriculture, we interviewed both ICT users and nonusers, contacting the first interviewees through informal leaders (gatekeepers). After we established contact with selected farmers, snowball sampling followed (Flick 2009).

We conducted further semistructured qualitative interviews with intermediaries, traders, exporters, and external experts (scientists and agricultural extension officers), although the primary focus of the research was on farmers. Interviews were conducted in English or Swahili (by local research assistants accompanied and supervised by the principal researchers).

Data Analysis

The survey questionnaire contained precategorized and partially categorized questions oriented toward particular aspects of the core research questions, including questions concerning ICTs, small business characteristics, and businesses capabilities (table 3.1). Additionally, we asked for the farmers' subjective assessment of access to knowledge, power relations between the individual actors and the buyers (e.g., bargaining position), and the distribution system.

Following the primary survey, we analyzed the collected data with a software package using descriptive analyses and applied statistical testing procedures, including χ^2 tests, to prove the significance of the results. We differentiated the results according to ICT usage types or distribution

Table 3.1
Overview of quantitative interviews

Characteristics of interviewees ($N = 368$)	%
Residence in Mt. Kenya	52
Residence in Mwanza region	48
ICT user	91
Non-ICT user	9
<30 years	20
30–50 years	63
>50 years	17
Primary education	69
Higher than primary	31
Female	31
Male	69

Source: Authors' findings (Krone, Dannenberg, and Nduru 2016).

channels and tested the significant differences regarding farming or ICT characteristics with χ^2 tests (see Bahrenberg, Giese, and Nipper 2013). Additionally, we used logistic binary regression analysis to identify associations between indicators for the expected dimensions of ICT-driven effects and indicators for the different characteristics of the outlined variables. An advantage of the binary (logit/probit) method over linear regression is that the distribution of binary variables can be correctly modeled. We used the logit method, because the regression coefficients can be more easily interpreted in terms of odds ratios. The analysis of the semistructured qualitative interviews followed the principles of qualitative content analysis (Mayring 2004). The qualitative data were mainly used to interpret the quantitative results.

Empirical Results and Discussion

Different ICT Usage Types in Relation to the Farmers' Characteristics and Capabilities

Table 3.1 gives an overview of the main characteristics and capabilities of the surveyed farmers (in total 368 smallholder farmers). Unexpectedly, the majority of respondents were ICT users (91 percent), all of whom at the very least used mobile phones for farming business.[4] This was a remarkably high share, given official data showing that, in 2014, only 71 percent of all Kenyans and 56 percent of all Tanzanians used mobile phones (see ITU 2016).

We identified three types of ICT usage among the farmers: (1) voice-only user; (2) voice and text user; and (3) voice, text, and Internet user (see table 3.2).[5] The least complex ICT usage type comprised farmers using their phones for calls only (13 percent). These farmers usually would call to immediately access information on a particular topic, such as market prices. Among the different ICT usage types, the combination of text and voice was dominant (67 percent). These farmers often used text messages to confirm business deals that had been negotiated previously. Additionally, 64 percent of all respondents used mobile payment systems like M-Pesa via text messaging (see figure 3.1).[6] This payment system is commonly used to store and transfer value in a mobile account (with simple handsets and without being connected to the Internet). Texting for crucial knowledge

Table 3.2
Overview of ICT usage types ($N = 361$)*

ICT usage types	% (no.) of interviewees
No ICT usage	9 (32)
Voice only	13 (45)
Voice and text	67 (243)
Voice, text, and Internet	11 (41)

Source: Authors' findings (Krone, Dannenberg, and Nduru 2016).

*The total numbers of N vary between the different tables as not all farmers answered each question; see also tables 3.3 and 3.4.

Figure 3.1
M-Pesa shops for mobile payment in Mt. Kenya. *Source*: Authors.

transfer, however, was often described as unreliable as an immediate means of communication.

Only 11 percent of all farmers used the Internet. This usage was mainly basic, including email, web searching (e.g., for price information or production techniques), and business-relevant Facebook groups. The limited usage was partly related to such factors as lack of awareness of the Internet, limited knowledge about using it, and prohibitive connectivity and hardware costs (see also Dannenberg and Lakes 2013). Nevertheless, some respondents who used the Internet replied that they do use it to access knowledge—such as on specific pesticides and their usage, or on prices outside the region—and generally saw it as a valuable source of information.

We next ran a binary logistic regression analysis to determine how significantly ICT usage types correlated with farmers' characteristics and capabilities, such as age, gender, monthly income, and level of education (table 3.3). The results from the model were only partly significant (at $p < .1$ and $p < .05$).

Predictably, using the Internet in East Africa requires not only being able to read and write but also having a command of the English language, while

Table 3.3
Effects of characteristics and capabilities on ICT usage types

Characteristics and capabilities (odds ratios)	Nonuser of ICT	Voice-only user	Voice and text user	Voice, text, and Internet user
Characteristics				
Age: 30–50 years (1 = yes)	−1.088*	−0.665	1.014**	−0.104
Age: >50 years (1 = yes)	−0.323	0.519	−0.200	−0.446
Gender (1 = male)	−0.082	−0.233	0.202	0.086
Capabilities				
Monthly income: >KSh20,000 (1 = yes)	−1.311	0.463	0.003	0.446
Educational level: >primary school (1 = yes)	−1.286*	0.073	0.452	1.130**
Pseudo $R^{2\dagger}$	0.050	0.035	0.060	0.041
Prob χ^2	0.103	0.063	0.088	0.077

Source: Authors' findings (Krone, Dannenberg, and Nduru 2016).
†We calculated Cox & Snell R-Quadrat.
*$p < .1$; **$p < .05$

further education is an advantage in understanding more complex written texts and applications. Thus, having more than a primary school education is especially important for using the Internet. This was also reflected in our interviews and statistical results.

Regarding age, we compared three categories: under thirty years, between thirty and fifty, and over fifty. The results showed that age was a significant factor for ICT use in general and especially for calling and texting. Farmers between thirty and fifty years old used calling and texting significantly more than both other groups. Our qualitative interviews helped explain these differences: the generation currently in middle age grew up at least in part with these technologies and are experienced enough farmers to be able to make effective use of ICT as (for example) a marketing tool (see also World Bank 2016).

Interestingly, gender was not clearly associated with farmers' use of ICTs. Neither our quantitative nor our qualitative findings supported the argument that cultural attitudes and women's multiple roles would exclude them from ICT access (as outlined by, e.g., World Bank 2016).[7]

Furthermore, we found no significant link between financial resources and access to ICTs, in contrast to findings in, for example, Dannenberg and Lakes (2013). But their study draws on data from 2008 to 2009, and since then, prices for electronic devices have gone down; farmers today can buy mobile phones for less than US$20 (figure 3.2). Furthermore, the costs for Internet access have decreased thanks to the arrival of fiber optic broadband communications cables installed in East Africa between 2009 and 2012 (Graham and Mann 2013).

Overall, our results showed that a broad variety of farmers with diverse capabilities and characteristics use ICTs. The main capability that influenced ICT usage in our study was education, but age was also an influencing characteristic. Use of the (more complex) Internet tended to be higher among well-educated and middle-aged farmers, but most farmers still did not use the Internet. Nevertheless, nearly everybody used simple ICT functions (voice), including women and low-income small-scale operators. This suggests that a strong digital divide is not taking place on the level of simple ICT usage but could increase in the future if the Internet or other more complex usage forms (e.g., tracking systems) become more important.

Figure 3.2
Front of a mobile phone shop in the Mt. Kenya region, with mobile phones going for less than KSh2,000 (US$20). *Source*: Authors.

Dimensions of Knowledge Access

Existing studies have already demonstrated that mobile phones can provide quick access to simple but relevant information, leading to improved agricultural productivity (e.g., Aker and Mbiti 2010). Farmers in our case study especially valued increased access to timely price information, as explained by one farmer: "Previous[ly] I took the product to the market without knowing the supply, the prices, and traders. Now I just call somebody at the market to get the information. With that knowledge, I am able to prepare my farm to harvest and sell the products" (Farmer 5, 2013).

Nevertheless, 57 percent of the farmers who did not use ICTs stated that they had good or very good access to simple information, revealing that ICT usage is not necessarily a precondition accessible relevant knowledge. Several farmers stated that they could acquire simple information, such as updated market prices, via face-to-face contact with other farmers and villagers at the local scale.

The advantage of using ICTs was greater when farmers needed access to complex knowledge not generally available locally, such as information on how to implement standard requirements. Such knowledge often requires access to external experts. Our data show a significant statistical correlation between access to complex knowledge and a broader combination of ICT usage types (figure 3.3). The Internet especially tended to facilitate good access to complex, codified knowledge, transferring it in written form (e.g., documents about standards). Because of farmers' limited use of the Internet, however, the impact of the Internet on their access to complex knowledge is low.

To some extent, codified complex knowledge (e.g., on the application of fertilizer) was exchanged through calls. But phones are of limited use if the knowledge is too complex or tacit to convey, like the application of process

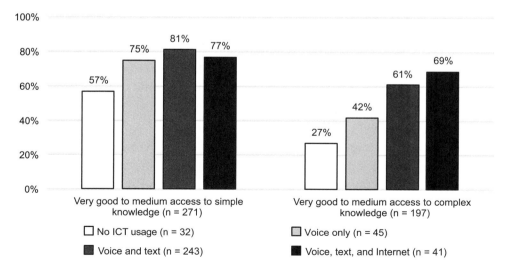

Figure 3.3
Dimensions of knowledge access according to different ICT usage types. *Source*: Authors' findings (Krone, Dannenberg, and Nduru 2016).

standards, which requires training on the job (also see Krone, Schumacher, and Dannenberg 2014). Our interviews revealed, however, that the main advantage of phones is to provide farmers with the opportunity to start and maintain the personal communication process with external experts outside their villages (e.g., with extension officers, exporters, and donors). Especially in export markets, a high degree of complex knowledge is required, which can usually be provided only by external experts. Such knowledge includes how to identify pests or the correct use of chemical inputs. It also focuses especially on the implementation and use of standards (e.g., GlobalGAP) that are crucial for market production (see also Ouma 2010).

To get and stay in contact with external experts who possess such valuable complex knowledge is much more difficult for farmers who have no access to ICTs because, without access to such technologies, they have no way to personally contact these experts. As a result, only a few farmers (27 percent) who did not use ICTs had very good to medium access to complex knowledge. The farmers we interviewed from this group were either integrated in a knowledge network (e.g., a self-help group) or had the opportunity to borrow a phone from someone else.

Distribution Channels

The farmers interviewed for this project accessed a range of different distribution channels. These distribution channels could be distinguished by different types of buyers, various levels of complexity in the buyers' requirements, the formalization needed to enter the channels, and the geographic distance between the farmer and the buyers. In total, four direct distribution channels could be identified:

1. Selling to other farmers (12 percent of all interviewed farmers), who then resell the products along with their own, was the easiest channel for many because of the deal's informality and the proximity of buyer farmers. Selling to another farmer did not require using ICTs to contact the buyer, which could usually be done directly and face-to-face within the same village.
2. Overall, farmers most often sold to local intermediaries (48 percent), who bought products without formal contracts and in small volumes.[8] For farmers aiming to sell to intermediaries, phones were important, enabling them to compare the various buyers' prices.

Development or Divide? 93

3. In Tanzania, farmers also sold in larger volumes to more formalized professional regional traders (13 percent), who either came to the farm or could be met at a wholesale market (see also Eskola 2005). In this case, farmers used their phones mainly to coordinate business activities.
4. The most lucrative, but also the most complicated and formal, channel was selling directly to exporters (27 percent).[9] While exporters usually pay the highest prices for products, they also demand the most challenging requirements. Exporters usually operate from larger cities, often prefer long-term contracted farmers, and require high standards (e.g., regarding the use of pesticides; see also Dannenberg and Nduru 2013; Graham 2011a; Dannenberg, Kunze, and Nduru 2011). The geographic distance and sophistication of this channel made ICT use, in most cases, essential for farmers.

As demonstrated in figure 3.4, our results show that farmers who use the Internet sold to exporters (42 percent) more often than those who only used phones (33 percent voice users; 26 percent voice and text users) or those who did not use ICTs (12 percent). When interacting with exporters, farmers' ICT use was important for coordination and control. Here, the exporters usually coordinated the activities. Interviewees stated that accessing exporters requires an intense communication process via ICTs

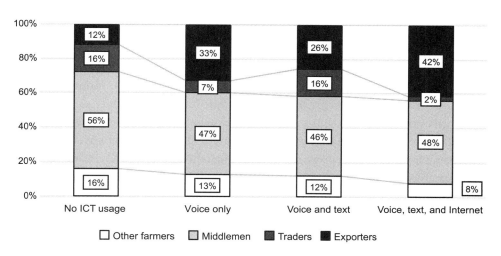

Figure 3.4
The usage of distribution channels by nonusers of ICT and different ICT usage types.
Source: Authors' findings (Krone, Dannenberg, and Nduru 2016).

because exporters do not regularly visit farms. Calling and texting have thus become necessary as coordinating activities need to be carried out over greater spatial distances (see also Boadi et al. 2007). The Internet, in particular, provides access to specific knowledge on how to produce (e.g., using chemicals) and, therefore, how to fulfill complex exporter standards. Additionally, via the Internet, exporters often provide farmers with complex knowledge about production methods for high-value production. Although direct sales to exporters usually have higher margins, our results indicate that such a sophisticated distribution channel often requires using ITCs to become integrated into the coordination and communication process and to access the necessary knowledge.

The results underline the opportunities offered by ICT usage while also revealing the problematic side, when access to ICTs becomes increasingly essential to doing business with exporters as well as with intermediaries and traders. Therefore, the risk of exclusion rises for those farmers who do not use certain ICTs. They are in danger of becoming marginalized and losing their commercial markets in the long term (see also Carmody 2012). According to our interviews, this problem had thus far been minimized, as most farmers who did not use ICTs had been able to sell via another farmer or a farmers' group/cooperative. As different farmers outlined, however, this solution came at the cost of reduced margins (as the intermediate farmers charged for their services). Furthermore, this approach can lead to new dependencies on these intermediaries.

ICTs have the potential to restructure value chains and distribution channels, for example, by introducing new actors or disintermediating intermediaries (Donner and Escobari 2010). Interviews with farmers confirmed our survey data demonstrating that ICT use improved farmers' ability to access increasingly complex and sophisticated distribution channels. For example, one Tanzanian farmer stated: "We have a larger variety of buyers and even places to sell our products. We exchange the contacts of buyers among ourselves" (Farmer 6, 2013). Farmers experienced advantages in accessing market information from different sources. In addition, ICTs made it possible to get in contact with, and compare, a large number of buyers, which led to a better selection of partners and improved the chances of higher margins.

Nonetheless, our study could not identify any restructuring or transformation of distribution channels. So far, no genuinely new actors could be identified, and middlemen continued to be common buyers. Furthermore, our interviews with farmers and exporters indicated that even though ICTs could bridge the spatial barriers to enable contact with exporters, the exporters themselves were often not interested in such direct contact, preferring intermediaries who collected larger volumes for them. Farmers were also usually not able to provide a continuous supply of bulk produce, which also hindered them from doing direct business with exporters.

Bargaining Positions

In the literature (see, e.g., Baumüller 2012), scholars often argue that improved access to information, especially market prices, can improve the bargaining position of a farmer with business partners by reducing information asymmetries. As shown in table 3.4, our statistical analyses (χ^2 test) could not clearly support this observation.[10] Regardless of the ICT usage type, using ICTs seems to have no effect on the bargaining position of a farmer, even though farmers used ICTs to compare prices. This contradictory finding can be explained by looking at the buyer side of the chain. As outlined above, exporters also use ICTs to coordinate and to control farmers (e.g., in how far they meet demanded product and process requirements). Furthermore, the introduction of ICTs has led to profound changes in the opportunities available to traders and intermediaries, which has partly reversed the improvements that enabled farmers to select a buyer. As intermediaries and traders also use phones more widely, they have been able to make cartel agreements or even increase their bargaining position.

Table 3.4
The effects of ICT usage types on bargaining position with buyers

	No use of ICTs % ($n = 32$)†	Voice-only user % ($n = 44$)	Voice and text user % ($n = 240$)	Voice, text, and Internet user % ($n = 41$)
Superior bargaining position	6 (2)*	18 (8)	18 (42)	12 (5)
Equal or inferior bargaining position	94 (30)*	82 (36)	82 (198)	88 (36)

Source: Authors' findings (Krone, Dannenberg, and Nduru 2016).
†Not all voice-only users and voice and text users responded to this question, so N differs from table 3.2.
*$p < .05$

They also had an early mover advantage, having started to use phones earlier, which gave them more time to use these technologies to organize themselves.

Nonetheless, our interviews did surface improvements in the bargaining situation of farmers that were notable even if not quantitatively relevant. One improvement is related to the seasonality of produce. In the low-supply season, real-time information on market prices from different markets via mobile phones can strengthen the bargaining position of farmers. In such cases, when the buyers urgently needed to buy, cross-checking market prices enhanced the bargaining power of farmers. Conversely, in the high-supply season, buyers decided the (generally low) price based on the high number of products available.

For example, one Tanzanian farmer stated: "Negotiating the price is depending on the period of supply. If there is high supply our tomato will not sold to buyers at the farm gate. So I must call them and to take the products to the market. They don't agree to come in the farm during high supply. But in low supply season the buyers are calling and asking for the products. And if I just say I have the products I can even get like 30 buyer in a very short time selling to good price" (Farmer 17, 2015).

In addition, a Tanzanian buyer confirmed: "According to our experience we are equal in bargaining the price. It happens that when there is low supply of watermelon, that's where you can find that the farmer increases price and we have low power in making decision" (Middleman 6, 2015).

Further, our interviews underlined that farmers who organize themselves into groups or cooperatives could achieve a better bargaining position (cf. Dannenberg and Nduru 2013). This organization could be improved through the use of phones (voice and text). Farmers could organize themselves better internally (integrating more and remote farmers, coordinating meetings and agreements more effectively) and externally (communicating with different buyers and suppliers), therefore improving their bargaining position. The cooperative usage for external business contacts also indicates that the positive effects of ICTs in farming can even spread beyond the direct users, as not every farmer in a cooperative needs his or her own phone or Internet access but can participate with the others. Such indirect participation of nonusers can help in preventing a digital divide. Nevertheless, even though our qualitative results identified some areas of improvement in the buyer relationships of farmers through the use of ICTs, the overall results could not support the argument that ICTs lead to such improvements in general.

Conclusion and Outlook

Generally, ICTs positively influence the ability to access knowledge and overcome spatial barriers to building and maintaining linkages between farmers as well as to a large variety of buyers, improving access to commercial markets. In this way, ICTs contribute to economic inclusion. The effects of ICT on farming businesses, however, are dependent on the actual types of ICT usage. While mobile phones are widespread, Internet use is still limited. Low education and advanced age are characteristics that tend to preclude using the Internet, although neither gender- nor income-related exclusion was identified for the analyzed ICT usage types. In this way, ICT usage is "not a level playing field" (Bbun and Thornton 2013), as the chances of gaining from ICT usage depend on favorable preconditions.

Regarding the different areas of potential benefits, the effects of ICT usage types differed. Even though all ICT usage forms support the exchange of simple information, Internet use significantly increased the exchange of complex knowledge and the possibility to enter more sophisticated and lucrative distribution channels and markets. Nonetheless, the use of the Internet and more complex ICTs in distribution channels can lead to a digital divide based on the existing differences in farmers' characteristics and capabilities. ICT use becoming compulsory for communication with exporters, for example, might lead to a digitally driven exclusion of less capable farmers.

Although we identified smaller structural changes in distribution systems, we did not observe transformational changes at the farm gate level, as have been mentioned in the broader ICT4D literature. Moreover, we could not demonstrate an improvement in information-related bargaining positions with buyers and suppliers (this supports the argument of a "thintegration" made by Murphy, Carmody, and Surborg 2014). Although increased access to knowledge and improved information exchange with other farmers can positively affect the bargaining position of the farmers, ICTs are also used by powerful actors in the chain to control farmers and to sustain asymmetric power relations.

In sum, the results revealed that the analysis of different ICT usages (and its associated complexity) and the different characteristics and capabilities of the farmers help explain the variegated effects of ICTs on small-scale farming. ICT4D strategies that do not take into account the different capabilities, characteristics, and buyer relations of targeted businesses (e.g., through training and educational approaches) are likely to increase existing disparities to the disadvantage of those who already have lower capabilities and who are therefore especially at risk of (further) exclusion and marginalization.

Notes

1. As explained by Lundvall and Johnson (1994) and Polanyi (1967), simple information can be defined in terms of facts ("know-what," e.g., market prices), while complex information includes different types of knowledge that go beyond, such as "know-why" and "know-how" (e.g., implementation of standards).

2. The ICT usage of firms is limited to exchanges of discrete bits of information, for example, prices or delivery dates. This information is essential for the everyday operations or success of the enterprise but usually does not result in a firm upgrading, creating value, or innovating (Murphy, Carmody, and Surborg 2014).

3. For this study, small-scale farmers are defined as farmers with less than two hectares of land.

4. We asked only about ICT use for farming business, not ICT use in general.

5. All the farmers surveyed who used the Internet also used phones for texting and voice.

6. For export production, however, farmers mainly used bank checks because of banks' formalized contract systems and higher security.

7. These findings could also point to a bias in our sample, since women who don't use ICTs were less likely to take part in our survey.

8. Farmers often trust intermediaries because of their shared cultural background and long relationships.

9. Exporters operating in Mt. Kenya mainly sell to EU markets.

10. As power relations are difficult to measure, we asked the farmers for a subjective assessment of their bargaining position related to their buyer. The answer options were given as a set of items (Likert-type scale) including inferior, slightly inferior, equal, slightly superior, and superior.

References

Aker, J. C., and I. M. Mbiti. 2010. Mobile Phones and Economic Development in Africa. *Journal of Economic Perspectives* 24:207–232.

Bahrenberg, G., E. Giese, and J. Nipper. 2013. *Statistische Methoden in der Geographie: Univariate und bivariate Statistik*. Berlin: Springer-Verlag.

Baumüller, H. 2012. Facilitating Agricultural Technology Adoption among the Poor: The Role of Service Delivery through Mobile Phones. ZEF Working Paper Series 93, 1–32. Center for Development Research, University of Bonn.

Bbun, T. M., and A. Thornton. 2013. A Level Playing Field? Improving Market Availability and Access for Small Scale Producers in Johannesburg, South Africa. *Applied Geography* 36:40–48.

Boadi, R. A., R. Boateng, R. Hinson, and R. A. Opoku. 2007. Preliminary Insights into M-Commerce Adoption in Ghana. *Information Development* 23:253–265.

Carmody, P. 2012. The Informationalization of Poverty in Africa? Mobile Phones and Economic Structure. *Information Technologies and International Development* 8:1–17.

Dannenberg, P. 2011. Wege aus der Ausgrenzung—Informeller Umgang mit dem Standard GlobalGAP im kenianischen Gartenbau. *Geographische Zeitschrift* 99:237–255.

Dannenberg, P. 2012. *Wirkung und Umsetzung von Standards in internationalen Wertschöpfungsketten*. Münster: Lit Verlag.

Dannenberg, P., M. Kunze, and G. M. Nduru. 2011. Isochronal Map of Fresh Fruits and Vegetable Transportation from the Mt. Kenya Region to Nairobi. *Journal of Maps* 2011:273–279.

Dannenberg, P., and T. Lakes. 2013. The Use of Mobile Phones by Kenyan Export-Orientated Small-Scale Farmers: Insights from Fruit and Vegetable Farming in the Mt. Kenya Region. *Economia Agro-Alimentare* 3:55–76.

Dannenberg, P., and G. M. Nduru. 2013. Practices in International Value Chains: The Case of the Kenyan Fruit and Vegetable Chain Beyond the Exclusion Debate. *Tijdschrift voor Economische en Sociale Geografie* 104:41–56.

Donner, J., and M. X. Escobari. 2010. A Review of Evidence on Mobile Use by Micro and Small Enterprises in Developing Countries. *Journal of International Development* 22:641–658.

Eskola, E. 2005. *Agricultural Marketing and Supply Chain Management in Tanzania: A Case Study*. Dar es Salaam, Tanzania: Economic and Social Research Foundation.

Foster, Chris, and Mark Graham. 2016. Reconsidering the Role of the Digital in Global Production Networks. *Global Networks* 17 (1): 68–88.

Graham, Mark. 2011b. Time Machines and Virtual Portals: The Spatialities of the Digital Divide. *Progress in Development Studies* 11:211–227.

Heeks, Richard. 2010. Do Information and Communication Technologies (ICTs) Contribute to Development? *Journal of International Development* 22:625–640.

Heeks, Richard. 2014. From the MDGs to the Post-2015 Agenda: Analysing Changing Development Priorities. Development Informatics Working Paper Series 57. Institute for Development Policy and Management, University of Manchester.

Humphrey, J. 2002. *Business-to-Business E-Commerce and Access to Global Markets: Exclusive or Inclusive Outcomes?* Brighton: Institute of Development Studies.

ITU. 2016. *Time Series by Country, Mobile-Cellular Telephone Subscriptions per 100 Inhabitants*. Geneva: International Telecommunications Union.

Kirk, M., J. Steele, C. Delbé, and L. Crow. 2011. *Connected Agriculture—The Role of Mobile in Driving Efficiency and Sustainability in the Food and Agriculture Value Chain*. London: Accenture, Vodafone, Oxfam.

Krone, M., P. Dannenberg, and G. Nduru. 2016. The Use of Modern Information and Communication Technologies in Smallholder Agriculture: Examples from Kenya and Tanzania. *Information Development* 32:1503–1512.

Lasserre, F. 2004. Logistics and the Internet: Transportation and Location Issues Are Crucial in the Logistics Chain. *Journal of Transport Geography* 12:73–84.

Lovering, J. 1999. Theory Led by Policy: The Inadequacies of the "New Regionalism" (Illustrated from the Case of Wales). *International Journal of Urban and Regional Research* 23:379–395.

Lundvall, B.-Ä., and B. Johnson. 1994. The Learning Economy. *Journal of Industry Studies* 1:23–42.

Mayring, P. 2004. Qualitative Content Analysis. In *A Companion to Qualitative Research*, edited by Uwe Flick, Ernst von Kardorff, and Ines Steinke, 266–269. London: Sage.

Molony, T. 2008a. The Role of Mobile Phones in Tanzania's Informal Construction Sector: The Case of Dar es Salaam. *Urban Forum* 19 (2): 175–186.

Molony, T. 2008b. Running out of Credit: The Limitations of Mobile Telephony in a Tanzanian Agricultural Marketing System. *Journal of Modern African Studies* 46:637–658.

Mukhebi, A., J. Kundu, A. Okolla, M. Wambua, W. Ochieng, and G. Fwamba. 2007. Linking Farmers to Markets through Modern Information and Communication

Technologies in Kenya. Paper presented at the 2nd International AAAE Conference, Accra, Ghana, August 20–22, 2007.

Murphy, J. T., and P. Carmody. 2015. *Africa's Information Revolution: Technical Regimes and Production Networks in South Africa and Tanzania.* Chichester, UK: Wiley.

Murphy, J. T., P. Carmody, and B. Surborg. 2014. Industrial Transformation or Business as Usual? Information and Communication Technologies and Africa's Place in the Global Information Economy. *Review of African Political Economy* 41:264–283.

Okello, J., E. Ofwona-Adera, L. Mbatia, and Ruth Okello. 2013. Using ICT to Integrate Smallholder Farmers into Agricultural Value Chain: The Case of DrumNet Project in Kenya. In *Technology, Sustainability, and Rural Development in Africa*, edited by Blessing Maumbe and Julius Okello, 44–58. Hershey, PA: IGI Global.

Ouma, S. 2010. Global Standards, Local Realities: Private Agrifood Governance and the Restructuring of the Kenyan Horticulture Industry. *Economic Geography* 86:197–222.

Pfaff, J. 2010. A Mobile Phone: Mobility, Materiality and Everyday Swahili Trading Practices. *Cultural Geographies* 17:341–357.

Polanyi, M. 1967. *The Tacit Dimension.* London: Routledge and Kegan Paul.

Qiang, C., S. Kuek, A. Dymond, and S. Esselaar. 2011. *Mobile Applications for Agriculture and Rural Development.* Washington, DC: ICT Sector Unit World Bank.

UNDP. 2012. *Promoting ICT-Based Agricultural Knowledge Management.* Addis Ababa: UNDP Ethiopia.

Unwin, T. 2009. *ICT4D: Information and Communication Technology for Development.* Cambridge: Cambridge University Press.

Wiskerke, J. 2003. On Promising Niches and constraining Sociotechnical Regimes: The Case of Dutch Wheat and Bread. *Environment & Planning A* 35:429–448.

World Bank. 2016. *World Development Report 2016: Digital Dividends.* Washington, DC: World Bank.

Yeung, H. W.-c. 2015. Regional Development in the Global Economy: A Dynamic Perspective of Strategic Coupling in Global Production Networks. *Regional Science Policy and Practice* 7:1–23.

4 Digital Inclusion, Female Entrepreneurship, and the Production of Neoliberal Subjects—Views from Chile and Tanzania

Hannah McCarrick and Dorothea Kleine

Introduction

Information and communication technologies (ICTs) can, in principle, support women's empowerment, defined as an expansion of their individual and collective choices (Sen 1999), in various ways. Such ways might include connecting and communicating across physical and social boundaries, receiving information from diverse sources, crafting new identities for themselves, starting to question and challenge the environments they live in, as well as improving access to health information, online learning, remittances and financial services, government services and information, and business opportunities (Buskens and Webb 2009; GSMA 2013).

Gender issues in the field of information and communication for development (ICT4D) are increasingly framed as one of the remaining frontiers of digital inclusion and exclusion. The most common approach of mainstreaming gender into ICT4D has been to start "counting women," by focusing on the number of female and male participants involved in a particular intervention. This is a step in the right direction but a crude response, grouping all women together in one homogenous category and simplifying the notion of inclusion. Thus, instead of engaging with the complex and multidimensional gendered power relations between heterogeneous groups distinguished by different gender, age, class, disability markers, in the field of ICT4D, gender is often limited to worrying about how women as a whole group can be "integrated," to use the policy language, into a "network/information/knowledge society," which is seen as a gender-neutral given.

Meanwhile, women's entrepreneurship development (WED) approaches are increasingly popular within mainstream development policy and practice, which in turn is frequently dominated by neoliberal economic approaches. Women in the Global South are reframed, moving them from a "passive victim" or female "beneficiaries" category to be recast as potential entrepreneurs. Gender-related ICT4D initiatives are increasingly focused on appropriating technology to enable, support, and advance female entrepreneurship. While this shift toward recognizing women's agency is welcome, it comes at a price. To be included, women are invited to reimagine themselves as neoliberal subjects (England and Ward 2007), who are expected to be constantly autonomous (Turken et al. 2015), flexible (Walkerdine 2006), responsibilized (England and Ward 2007), self-caring (Lemke 2001), rational, and competitive (Sennett 2006). Such framing asks the individual woman (or man) to negotiate structural economic conditions as an individual entrepreneur required to constantly rework and improve the self.

This chapter considers WED as embedded within wider "development" discourses.[1] These discourses have been criticized from several angles, including for frequently being econocentric and equating development with economic growth. This chapter, as well as the two larger studies on which it draws, is based on the alternative capabilities approach to development, which instead sees development as a process of expanding the substantive freedom of people to live the lives they themselves have reason to value (Sen 1999).[2]

Women are important participants in shaping digital economies globally, including at the margins. The figure of the female tech entrepreneur running a start-up in an African capital is an image that has been lionized and much reproduced by donor agencies. In this chapter, we move to rural Tanzania and rural Chile to examine the situation of women who have been celebrated locally by donor agencies as female entrepreneurs who benefited from women's economic empowerment (WEE) and ICT inclusion programs. In two very different contexts, we find striking similarities. We show, first, that even some of the most successful women have become entrepreneurs out of necessity rather than choice. Second, in exchange for state or donor support, they have adopted a neoliberal framing of their activities and, indeed, of their selves, which includes a far-reaching, but incomplete, embrace of tenets of neoliberal selfhood.

Based on the findings from the two case studies, we start to query the ICT4D and WED literature on two main points. First, what assumptions are being made about the ability of all women to be included and to benefit from such programs? Second, to be included and benefit, what logics and framings must women subject themselves to or be subjected to? Our analysis pays particular attention to the diversity of women's situations, perspectives, and trajectories, which shape their needs and interests. We ask whether, under a hegemonic neoliberal economic order, there are any trade-offs between what is described as women's economic empowerment and other forms of empowerment.

We argue that using ICTs to support female entrepreneurship often fits the logic that casts women as neoliberal subjects with a high level of flexibility, self-motivation, risk taking, confidence, embrace of change, and tolerance of precarity.

In our analysis, we first draw on Buskens's (2015) classification of ICT4D and gender interventions as conformist, reformist, and transformist/transformative in approach, which builds on Molyneux's (1985) interest paradigm. Second, we use the concept of neoliberal governmentality, which Ren (2005, n.p.) defines as aiming "at transforming recipients of welfare and social insurance into entrepreneurial subjects, who may be motivated to become responsible for themselves."

We draw on empirical data from two different case studies, both sites of government-funded ICT initiatives aimed at ICT literacy and entrepreneurship, where the authors were undertaking ethnographic fieldwork: First, a community center in rural Zanzibar, Tanzania, run by a local NGO that provides training in ICT, English, and entrepreneurship. One of the authors interviewed women who participated in a twelve-month women education project run at the center, as well as their teachers.

Our second case study is the town of Algun (name changed, to protect the identity of the study participants) in southern Chile, where a public telecenter offered free Internet access as well as free IT courses for local adult learners. In parallel, an EU-funded NGO was organizing a business competition, and state agencies were offering training for local "micro-entrepreneurs." All state agencies sought to include female participants in their services. One of the authors interviewed women who were using both public IT services and who were identified or self-identified as micro-entrepreneurs.

After a brief literature review, overview of methods, and introduction to the two case studies, we present and analyze findings from each before offering broader conclusions.

Gender Interests in Development Discourse

Within the literature on the role of gender within ICT4D, Molyneux's (1985) so-called interest paradigm has received significant attention. Molyneux distinguishes between women's practical and strategic gender interests in development interventions. Practical gender interests focus on making women's lives easier while leaving existing gendered power relations unchallenged. Strategic gender interests (in later work, often renamed transformative gender interests) indirectly or directly challenge power relations that disadvantage women.

While recognized as a useful distinction, this interest paradigm has also been criticized both epistemologically and politically, with scholars arguing that the two aspects cannot be easily separated (e.g., Benton 1981; Callinicos 1987; Hindess 1982; Jackson and Pearson 2005). Some critics see an implied hierarchy, suggesting that transformative gender interests are always to be preferred over practical gender interests. Molyneux (2000) has clarified that the distinction is not clear cut and that a hierarchy is not intended. Supporters have argued that while approaches supporting women's practical interests can spur transformative change processes, moving from a focus on practical interests to demands that challenge structural gender inequality can be a struggle (Molyneux 2000). Practical gender interests can be more vulnerable to co-optation through "outside agencies imposing their versions of objective interests on subject people" (Molyneux 2000, 79). This is a key point in international development, which is rife with examples imposing "development" on local people. Interpretations of the interest paradigm have in some cases conflated the concepts of interests and needs (for a critical evaluation, see Jonasdottir 1988). Needs are often more straightforwardly categorized, less politicized, more directly applicable in policy, and more often defined by others. Interests are often more politicized, intentional, and connected to women's own agency and choice (Molyneux 2000). In other words, needs are "usually deemed to exist, while interests are willed" (Molyneux 2000, 79).

An early strand of the gender literature in the 1970s, commonly summarized as the women in development (WID) discourse, pointed out that the modernization process, then often associated with development, had not liberated women from subordinated positions (Boserup 1970). WID did not criticize modernization itself but "the fact that women had not benefited from it" (Kabeer 1994, 20), and it aimed to integrate women in the mainstream modernization process through their inclusion in the market system (Jackson and Pearson 2005). WID tended to reduce larger issues of gender inequality to questions of "counting women" in mainstream development interventions. The gender and development (GAD) discourse shifted the focus from looking at women's situation to analyzing structurally manifested gendered relations between and within different groups of women and men in both development and social processes.

Further, by understanding gendered relations as intersecting with other social characteristics (Benería 2003), such as class, race, sexual orientation, disability, and so forth, GAD lends a nuanced perspective on how ICTs can facilitate and assist practical and transformative change by and for women.

Gender in the ICT4D Field

Gender in the ICT4D field has moved from being a marginal interest of a few vocal female policymakers, practitioners, and academics to being increasingly mainstreamed. This is driven by various actors and interests, including advocacy from feminist activists and an interest by international bodies to align ICT4D with a global development agenda (with notable links to the Millennium Development Goals and now the Sustainable Development Goals, where, at least on paper, gender features prominently).[3] Private sector companies have also increasingly "discovered" women in the Global South as an emerging and untapped market, including for mobile phones and services (see, e.g., the gender gap study of GSMA 2015).

Writing from a feminist perspective, expanding on Molyneux's work and applying it to ICT4D, Buskens (2015) distinguishes between conformist, reformist, and transformist approaches in gender and ICT4D. First, there are conformist ICT4D and women-focused projects that assist, support, and convince women to adapt to current realities. Second, reformist approaches strive for policies addressing gender inequalities in terms of access, use, and control over ICTs. Third, transformist/transformative approaches directly challenge underlying reasons for inequality and support women's emancipation and liberation. Buskens's taxonomy can be criticized for overlap, just as in Molyneux's distinction, and depending on context, some activities will fall into multiple categories. For instance, teaching women how to set up websites might help them work from home as web designers, thus fitting around the current gendered division of labor (conformist). That same skill, however, might be used, possibly by the same woman, to set up a website for the local women's NGO or women's equality party (transformative). Further, a rigid application of the framework risks assuming that intent of technological design will accurately predict and determine eventual usage. In this chapter, we apply Buskens's categories as a useful analytical tool while accepting that the categories might overlap and that design intent does not equate to actual outcome.

Mainstream development discourse has partly moved on from WID to GAD, and from counting women to targeting men, for instance, in the United Nations' #heforshe campaign. Unfortunately, much ICT4D discourse, including influential industry reports (e.g., Intel 2013; GSMA 2015), still largely take more of a WID approach of integrating and "counting women." Often overlooked are broader aspects of participation, impact, and the relationships between and within different groups of women and men (Buskens 2015). Some of the literature has narrowed down women's inclusion to a focus on the potential of ICTs to boost women's income and the increased female contribution to overall GDP growth (e.g., Intel 2013; UNCTAD 2011). Universal access to ICTs, this literature promises, has the ability to improve literacy, increase income, improve education, and unveil business opportunities for women, which in turn is presumed to lead to associated benefits for their communities and the wider society. Industry reports arguing for women to benefit from ICTs primarily through their "economic empowerment" and subsequent benefits resulting in overall GDP growth would most likely, through the lens of Buskens's taxonomy, be classified as conformist and reformist approaches seeking to integrate women within the hegemonic neoliberal paradigm.

Murphy and Carmody (2015) recently criticized the field of ICT4D overall as a "neoliberalized (meta-) discourse," which collectively needs to be unmasked as a neocolonial project and that is not deserving of the level of support it receives. We would not go this far; instead we recognize the potential for ICT4D work to support women's empowerment both within and beyond economic empowerment. Yet, too little attention has been paid to how ICTs can be appropriated to achieve greater gender equality in its own right. ICTs remain powerful levers to support current change and to help shape the future. The ICT4D field remains as diverse as the visions for the future held by the people participating in it. We see the role of progressive actors in the ICT4D field as amplifying the role of the less powerful in this struggle and keeping space open for the choices people themselves make about the lives they have reason to value.

Dominant neoliberal recipes lead to an unholy and outdated logical chain prevalent within parts of ICT4D that combines WID with development as economic growth and suggests that women need to be included in individualized neoliberal strategies to achieve personal betterment and national-level economic growth. At an individual level, the passport to inclusion in development then becomes the cultivation and acquisition of a neoliberal, responsibilized, and enterprising selfhood. When poverty and inequality are explained through a logic of individualization, the focus is moved from structural injustices to the individual. The individual woman shoulders the responsibility for her own success, inclusion, and development (Gonick 2006), which she is invited and expected to acquire through refashioning herself "as a successful subject: the subject of neo-liberal choice" (Walkerdine 2003, 241). This self-fashioning as a neoliberal subject then assumes a universal acquisition of characteristics such as high levels of flexibility, self-motivation, confidence, tolerance of precarity, and embrace of change. This framing frequently underlies notions of female entrepreneurship within ICT4D projects.

Women's access to ICTs can be conceptualized as including availability, affordability, and skills needed to use a set of ICTs (Gerster and Zimmermann 2003), as well as being affected by structural and gendered social norms governing mobility, reproductive and productive roles within households, and the use of space and time, which structures access (Kleine 2010). Gendered digital divides, particularly in the Global South, structurally limit women's opportunities to harness ICTs' emancipatory potentials because of marginalization in terms of access to time, resources, education, and mobility, as well as technophobia, safety, religious and cultural constraints, socio-economic status, and age—in addition to perceived relevance of technology to women's lives (see, e.g., Dodson, Sterling, and Bennett 2013; Hafkin and Taggart 2001; Huyer and Carr 2002; Odame 2013). In Kleine's (2013) choice framework, which maps development processes, she lists eleven different resources: educational, social, financial, material, geographic, natural, cultural, and psychological, as well as health, time, and information. Access to these resources affects women's choices to use ICTs and is often gendered. ICTs can also be used as a tool to control, harass, and oppress women, and many women experience severe constraints in using ICTs effectively as a result of cultural norms and power hierarchies (Buskens and Webb 2009; 2014).

Telecenters emerged as an approach to close global digital divides (Heeks 2008) and can form empowering community spaces, particularly for women, as well as provide social benefits that do not have to be ICT related (Kleine 2013; Madon et al. 2009; Wheeler 2007). Nonetheless, telecenters are gendered and socially coded spaces that impose practical and social constraints on women's usage, linked to norms on the use of space and time (Kleine 2010; 2013). While the long-term future of many public access points is in doubt because of the increasing availability of mobile Internet, research suggests that private usage may complement, rather than replace, communal use of ICTs (Donner 2015).

Women, ICTs, and Entrepreneurship

Increasing numbers of development interventions are focused on women's entrepreneurship development (WED). WED is commonly promoted as a key to unlocking job creation and employment opportunities, as well as to drive innovation in contexts with a high proportion of youth coupled with scarce formal and public employment opportunities (e.g., UNCTAD 2011; 2014). In government plans in donor and recipient countries, self-employment is often a key pillar in the strategies to replace lost jobs. The entrepreneurship discourse is also frequently mobilized in programs targeting the unemployed in traditional donor countries such as the UK.

Only limited critical analysis has been applied to identifying how far entrepreneurship and economic empowerment of women *leads to* and *supports more equal* gender relations in society overall; instead, these outcomes are often directly or indirectly equated and assumed. Links running in the other direction, however, emphasizing the economic benefits of a focus on women, can draw on a growing body of evidence that identifies wider benefits of businesses run and owned by women, such as higher rates of repayment of microloans, less risk taking, different management styles, and inherent insight into overlooked consumer segments (Guihuan 2005). Women are often found in small and medium enterprises (SMEs)—and typically within female (often less profitable) sectors—which inherently face structural challenges related to the size and nature of the business, gendered social hierarchies governing ownership of land, access to collateral, and gendered division of labor (King, Sintes, and Alemu 2012).

Women become entrepreneurs for various reasons, and not all of them do so voluntarily. Some may have no job opportunities or have lost their employment, in some cases as part of the public-sector cuts following neoliberal structural adjustment policies.

For women in the Global South, embedded in discourses of entrepreneurship is the invitation to reimagine themselves as neoliberal subjects. On the one hand, such discourses mirror neoliberalism's individualist bias, locating choice at the individual level, thus helping women escape the trap of having choices made for them (and in their name) at the household or community level, often by more powerful, usually male actors. On the other hand, this individualist bias in the neoliberal entrepreneurship discourse moves explanations for inequality away from structural factors and places it on the shoulders of the individual. Women are reframed from "beneficiaries with social rights to clients with responsibilities to themselves and their families." This discursive move devolves "responsibility for securing economic opportunity to individuals acting as responsible agents for their own well-being" (Rankin 2001, 20). Such a framing subjugates divergent, more collectivist social norms and allows the state, and funders, to shed responsibility: the poor are seen to be poor because of individual failure—the woman is poor because she is not entrepreneurial enough.

Widening the Discourse of Female Entrepreneurship

Within the ICT4D discourse, women are often framed as either consumers of ICTs whose relative lack of access represents a lost revenue opportunity for ICT-related goods and services (the emerging and untapped market; e.g., GSMA 2015) or as budding entrepreneurs who just need to be equipped with ICTs to be successful (UNCTAD 2014). Both opportunities exist to a point, but overemphasizing them carries the risk of underplaying broader structural and cultural factors.

UNCTAD (2014)—among others—ascribes women's entrepreneurship an important role in increasing gender equality, creating employment, improving economic growth, and ultimately reducing poverty levels at individual, household, and community levels. Meena and Rusimbi (2009) celebrate the individual success story of Bahati, who worked her way up from dressmaking to hairdressing, saving money to buy a mobile phone, which then enabled her to connect with clients and expand her business. Buskens (2010) agrees that Bahati, with her drive, discipline, and dedication, is an ideal candidate for ICT4D projects focusing on female entrepreneurs, and that women like her may need little more than access to ICTs to then run with the opportunities provided. Yet, Buskens asks critically how many women like Bahati there are. What, one might add, happens to the women who are not like her? From a broader perspective, one might also want to ask how "local gender ideologies treat the individual woman entrepreneur who begins to think in terms of private profit as an end in itself" (Rankin 2001, 21), and the implications this has on the overall well-being of the woman who has "succeeded" as an ICT-aided neoliberal entrepreneur but who may sense a conflict with local gender norms.

Further questions that are also rarely asked in WED literature include: Do all women *want* to become entrepreneurs? When these programs use the framing of the female entrepreneur, are the targeted women presented with a choice that includes alternatives to their reinvention as atomized individuals seeking to maximize profit? What price do women have to pay to live up to the imagined entrepreneurial selves they are expected to be within the neoliberal paradigm?

Case Studies

Moving now to our comparative case studies, we bring together data about and direct experiences of women in two projects, one in Zanzibar (Tanzania) and one in Chile. In Zanzibar, the wider research took place in a rural village in an area of great natural beauty and little tourism, where the economy is centered on farming, fishing, and small-scale businesses in the informal sector. Challenges for women include a high rate of gender-based violence, forced early marriages, and gender-based disadvantage in education and political participation. Zanzibar has low ICT penetration, a low ICT skills base, and a lack of reliable electricity and connectivity. The local center for ICT and vocational training is the main access point for ICT and Internet use. The center ran a twelve-month training program focused on English and ICT and entrepreneurship skills, in which twelve women were enrolled. Nine participating women between seventeen and twenty-seven years old and their three teachers were interviewed with a translator, with interviews lasting between forty and one hundred minutes.

In Chile, eight women entrepreneurs were interviewed as part of a wider research project on ICT and development in rural Chile (see Kleine 2013). These women had engaged with the local telecenter and free IT training offered there to different degrees. Some of them had been involved in an EU-funded competition celebrating local small and micro-entrepreneurs while others had been enrolled in "micro-entrepreneur" groups in order to be eligible for government grants. Members of these groups were then taking the IT courses. Interviews lasted between thirty and ninety minutes.

Data from these interviews enable exploring the discourse of female entrepreneurship and how it has been negotiated, assimilated, or resisted by these women. All interviews were translated, transcribed, and coded in open coding mode. Interview data were triangulated with observation and participant observation, focus group data, and (in Tanzania only) participatory action research data. The following sections draw on the conceptual language of the choice framework (Kleine 2013) to sketch out the resources women had access to and the structural factors they navigated.

Findings from Tanzania

The center for ICT and vocational training is located on the main road, about an hour's walking distance from the outskirts of the village. The center was the women's sole access point for familiarizing themselves with computers, since none of the women reported having access to computers outside the center (although approximately half of the women owned simple or feature phones). The distance to the surrounding villages in the community was an obstacle, in terms of both time and energy. As one woman remarked, "Women are immobile. Men have bicycles."

Because women lacked the material resource of a bicycle, they had to find more time and energy (affecting and affected by health and psychological resources) than men to get to the center. Because of informal social and cultural norms and fear of violent crime, women generally did not move outside after sunset, thus further reducing the times when they could access the center.

For the women that participated in the course, based on the main motivations they expressed in interviews, the goal was to get employment via IT skills, rather than acquiring IT literacy as a goal in itself. One trainer explained, "The overall goal of women is to learn in order to get certificate, and to be able to be self-employed." Yet, apart from one graduate of the center's IT training, who went on to be employed by the government, no respondent could provide any examples of the IT skills acquired having led to employment.

As is common in ICT4D projects (see Kleine, Hollow, and Poveda Villalba 2014), the center tended to work with the easier-to-reach groups in the community: men and those women with relatively fewer domestic care responsibilities. Despite their relative privilege compared to other women, these female participants reported being constrained in their participation by their care duties within their households. As one teacher put it, "The burden of domestic work most women have at home. ... I have never seen a man stay at home because their mother is sick. The men come to the extra courses, but the women often have to prioritize differently."

Eleven of the twelve female participants were unmarried and had no children. This pattern was also observed among women taking general IT and English classes at the center. Thus, the course was successful in reaching women who were relatively more privileged regarding their care responsibilities and available time. As women are not a homogeneous category, but vary according to social, economic, cultural, and material resources, courses targeting women should reflect on *which* women they reach.

Aspiring to Be Entrepreneurs?   Of the nine women interviewed, all but two women were involved in small-scale, informal income-generating activities. The women were engaged in sectors that were associated with women (and were often less profitable) and that were dependent on women's cultural resources (indigenous cultural knowledge and traditional skills), such as tailoring, lending or embroidering traditional clothing, handicraft, and basket weaving. When asked by the interviewer about their occupations, however, they identified themselves as students at the center, not as entrepreneurs. The women saw their entrepreneurial activities as a necessity to sustain themselves and meet family and household financial expectations. Except for one woman who aimed to expand her business, the women did not aspire to become or remain entrepreneurs, but neither could they identify many other viable alternatives to increase their economic independence or meet their current financial needs.

Staff working at the center had absorbed the entrepreneurship discourse: "They benefit because their life to be better. They get entrepreneur and they can sell something and they make life to be better because they get money."

Among the training staff, entrepreneurship was equated with success, reduced poverty, self-reliance, and economic empowerment. These assumptions were not shared by the interviewed women participating in the course. When asked in interviews, the women said they dreamed about becoming teachers or health professionals, working in TV or radio, or going to university. At the heart of many of the hoped-for trajectories women described in the interviews lay the ability to secure employment offering financial security, which some of them imagined would increase their autonomy and decision-making power. They also imagined it would free them from their current domestic roles and responsibilities, as well as from expectations of financial contributions they currently experienced under their (male-) dependent domestic situations, living with parents or other relatives. The interests expressed by the women in interviews reached beyond altruistic support for the household and dependents. Instead, it was common for women to "seek employment for other forms of income generation to liberate themselves from their families, to pursue alternative futures" (Pearson 2005, 182).

Despite the rhetoric of entrepreneurship, the program missed key practical opportunities in its design. For example, the women were not linked to a successful local cooperative, where eighty producers were creating doormats and bags; were not introduced to the opportunities and risks of microloan schemes in the community; and were not offered linkages between their newly acquired IT skills and their current businesses. According to the women, however, the course had facilitated discussions around the nature and implications of gendered subordination in the community, which the women perceived had led to an increased awareness about gendered structures and their own roles within these structures. The women identified possible transformative trajectories stemming from broader gendered interests and needs, within this element of the course (including reducing gender-based violence, enabling women's access to education, and challenging uneven domestic responsibilities). Nonetheless, the course did not link these trajectories with the ICT (or entrepreneurship) skills the women attained in other segments of the course.

The group of women took part in a participatory video workshop, organized by volunteers and center staff. Through discussions and participatory video work, issues such as gender-based violence and the structural discrimination the women experienced came to the fore. In response, one of the high-level board members at the center proposed that after the women had done so well with the video workshop, now professional actors could be hired to improve the groups' acting skills, and then they could open a company to capitalize on making commercial movies as well as filming weddings and other celebrations in the community. This is a telling example of how neoliberal ideas of commercializing creative expression, and indeed the primacy of economic over political or social empowerment, had been internalized by local decision makers. A potentially transformative trajectory of the project where deeper social issues could be discussed and potentially addressed was quickly bent back and rerouted to fit a more conformist agenda of neoliberal selfhood, in which every aspect of the center was focused on economic empowerment.

Barriers to Women's Enterprises Some of the women emphasized that they struggled, as seamstresses, embroiderers, and handicraft producers, with access to credit and to markets, especially when customers placed orders but did not pay (lack of financial resources). Despite all participants having individualized course components with a personal tutor, all but one did not see a connection between the ICT skills they had learned at the center and their existing business activity.

The exception was Fatuma (not her real name), one of the participating women, who had attended sewing training in the closest regional town (expansion of educational resources), had borrowed her grandmother's sewing machine (expansion of material resources), and had asked customers to bring their own fabric, since she had no access to funds to pay for materials in advance (lack of financial resources). She had furthermore worked out a way to receive payments from customers via mobile money and was looking up new designs online to improve her tailoring business. Fatuma's actions display a high level of drive, innovativeness, self-motivation, and risk taking. The hard work, discipline, dedication, and possible sacrifices made by women like Fatuma need to be recognized. Are the enterprising qualities Fatuma shows proof of something that can be expected of all women? Are women like Fatuma the norm or an exception? Research by Chew, Ilavarasan, and Levy (2013) found that individual motivation and entrepreneurial expectations correlated with women entrepreneurs' use of mobile phones to support their businesses, suggesting that the degree of motivation, and indeed the psychological resources and time resources (see

Kleine 2013) that underpin it, are not universal. This indicates that we need to look beyond factors of access and use, and acknowledge that too many policies assume that *all* women have the characteristics of the imagined neoliberal entrepreneurial subject illustrated in inspiring yet isolated success cases such as Fatuma's.

Defining Women's Needs and Interests The course, including its focus on entrepreneurship, originated from an initial situation analysis based solely on solicited views and opinions of board members of the main NGOs in the area (led largely by middle-aged men). It is hard to know whether committees with more women and/or youth representation in them would have come to a different set of priorities. Nevertheless, the actual process illustrated how women's needs and interests are too often co-opted and defined by others (men). The male-led local organizations were keen to promote women-focused projects and programs that aligned smoothly with donor priorities. At the same time, they were reluctant to invite women into the decision-making structures of the organizations. Shortly after the fieldwork was carried out, one of the organizations reappointed all men to the board and governing positions within the organization.

As illustrated here, women's supposedly objective and pressing needs and interests are depicted as homogeneous for "women" as a group. Some women may well agree with and assimilate them, but when formulated by representatives of the status quo (men but also some women), these broad interests are unlikely to include transformative interests that "enhance women's position overall" (Molyneux 2000, 79). As opposed to the NGO board members that set out the priorities for the course, the women, when asked in interviews, did not identify entrepreneurship as a preferred choice if other options of courses had been available. Instead they sought a means to secure their livelihoods, and while they aspired to other jobs, including salaried employment, it was entrepreneurship that was discursively framed as the only alternative on offer. Simultaneously, the conformist strategies adopted within the program restricted the women to their current sphere, which was socially and culturally coded as female. Key economic actors within the community discouraged them from participating in and influencing the direction of the project. An imagined trajectory of female economic empowerment was pushed, overlooking the fact that economic empowerment, in itself, cannot change gender subordination, as "the subordination of women is not caused by poverty" (Jackson 2005, 60).

Findings from Chile

In a very different context, rural Chile, the observations, interviews, and focus groups yielded similar patterns of gendered norms on the use of space. Because of unevenly distributed material resources and gendered social norms, bicycles were used only by men and young children, while women walked or sometimes rode on the back of a bicycle steered by a male relative or friend, or in the few cars, which were invariably driven by men. In the town of Algun, however, the public telecenter was located in the library, near the central square, and thus was easily accessible for women who lived in the small town. For the indigenous women living in the extensive rural area surrounding the town, the energy and time needed to access the telecenter was still significant. Similar to the case from Tanzania, the Chilean telecenter was arguably more accessible to the relatively more privileged women.

While the telecenter held no entrepreneurship classes specifically for women, the government agencies promoting entrepreneurship classes expected a significant number of participants to be women. Indeed, they felt under pressure to "count enough women." For instance, one all-male entrepreneur group (carpenters), who wanted to improve their chances of receiving a government grant for machinery, co-opted a woman into the group for strategic reasons, even though her link to carpentry was tenuous. This same group of carpenters was taking the IT course for entrepreneurs, partly motivated by an interest in learning about IT and partly to improve group members' chances of receiving government grants if they fit the image of dynamic self-improving entrepreneurial selves. In the case of the woman co-opted into the group, it was her elderly father who was a carpenter; however, carpentry was actually marginal to her livelihood strategies. Nevertheless, her presence allowed the group (and the government agency offering the course) to display having female (and younger) "entrepreneurs" in the course.

Some WED work witnessed in the town encouraged risk taking among otherwise rather cautious women. An NGO encouraged one woman to take out a large loan to set up a cookery school when she did not have a clear business plan. After she had taken out the loan and set up the school, it soon became clear that the intended participants for the cookery courses, women in the community, did not have control over the household budget and thus could not pay course fees. Instead, their husbands saw these cookery courses as relating to a "natural skill" of women and their responsibility in any case, not an additional household cost item. Thus, the cookery school was not financially sustainable, and the woman experienced deep financial and psychological distress over repaying the loan. Having been encouraged to be a risk-taking entrepreneurial neoliberal self, she had followed this lead and was left in a difficult situation when the enterprise began to fail.

Absorbing the Female Entrepreneurship Discourse Another local NGO was funded by EU money to organize an entrepreneurship competition designed to recognize micro- and small-business entrepreneurs in this rural and income-poor region of Chile. The winner was Ana Melihuen (not her real name), an indigenous (Mapuche) woman in her fifties. She used to be a teacher of the Mapuche language in schools, and she had lost this regular employment in the public sector, which she very much regretted. This forced her into being "entrepreneurial" to keep together her family clan of thirteen people, which she de facto headed. So, in a community with high unemployment and few prospects, she successfully applied for a state grant for the materials to build a *ruka*, a traditional large straw and wood longhouse, which was to be the cornerstone of her ethnotourism business. Drawing on her existing material resources (land) and the social resources she was able to mobilize (free labor from her family clan), as well as cultural resources (indigenous cultural knowledge and traditional skills), she was able to set up the ruka. She was now able to charge tourists an entrance fee and offer storytelling and lectures on the Mapuche language, cooking, weaving, and musical instruments for a fee. She explained in an interview the rationale for her business:

Ana: Because maybe because of the culture of my ancestors we lost many things. They lost their culture, their jewellery, their land; because of their ignorance. [...]
Author 2:   Can one mix the Mapuche [indigenous] culture with a logic of business?
Ana: Yes, because I sell my culture. I am not going to give out information just like that, I can't. (Kleine 2007, 198)

Ana had previously shared her indigenous knowledge with the next generation as a state-funded public service in the schools, in line with the more collectivist ethos of Mapuche tradition. Her new situation in the neoliberal marketplace was one where she felt she needed, as an individual entrepreneur, to "sell her culture" for a fee. Ana estimated that she needed US$195 a month to provide for her household, but she charged only 500 Chilean pesos (about $1) as an entrance fee, which meant she needed 195 visitors per month to cover the household income from this activity. Visitors were rare, but she did not have to pay back the state grant, and now she had won a prize of around US$2,000 in the business competition. "I have never in my life taken something that was not mine. This is why the competition that I won makes me fret. How can I invest my money so that they can say: 'This is what she is going to do'?"(ID21, F2).

Ana wanted to take the free IT course at the local telecenter "because if I want to be a micro-entrepreneur, I need something to communicate with" (ID21, F2).

In one reading of Ana's story, she was liberated to explore her own "inner female entrepreneur," which in some sense she very impressively did. Though her business was not yet profitable, at the same time it had already absorbed a large government subsidy. In another reading, Ana had been employed by the state as a teacher and public servant to share her knowledge of the Mapuche language "for free at the point of use," as a public good. When laid off, she had realized that in the neoliberal market logic, her traditional indigenous knowledge could be appropriated by an individual, privatized, and commodified, which is what she then set out to do, thus conforming to the expectations of neoliberal selfhood. Such an approach seemingly facilitated "a convergence between the interests of women and the promotion of economic liberalization" (Baden and Goetz 2005, 24). Like Fatuma in Tanzania, Ana in Chile subsequently became a poster child for female entrepreneurship. Her success, however, was de-rooted from women's self-defined and diverse interests (Jackson and Pearson 2005) in that she enjoyed sharing her culture and needed to earn a livelihood, but she would have preferred to be able to pass on her cultural knowledge regardless of whether people were able to pay, as she had done as a teacher in a public school. Ana was certainly entrepreneurial, but given the choice, she might have used this entrepreneurial energy as a creative teacher or as a social entrepreneur instead.

Before moving on, it is worth asking whether the extraordinary time and energy that Ana put into the ruka project (psychological resources) and the support she mobilized from her family (social resources) could be replicated by other teachers who had been laid off in rounds of government cuts. Ana had extraordinary levels of initiative and creativity, as well as her powerful position as the de facto head of her family clan—factors that can hardly be generalized to all the women looking for work in rural Chile.

Conclusion

In this chapter, we argue that in principle, ICTs can support women's empowerment in conformist, reformist, and transformist/transformative ways. WED discourse is in many ways conformist and reformist, aiming to integrate women into a neoliberal economic paradigm of development. We set out to interrogate this literature, based on evidence from two case study locations, focusing on two main points: First, what assumptions are being made about the ability of all women to be included and to benefit from such programs? Second, to be included and to benefit, what logics and framings must women subject themselves to or be subjected to?

Some ICT4D interventions are conformist and framed within the neoliberal strand of development discourse characterized by modernization thinking and econocentric discourse. The related "entrepreneurship" activities we witnessed in both Tanzania and Chile were not particularly sophisticated in and of themselves, although they were woven into women's lives in complex ways. Most of the women interviewed would not have chosen to be self-employed if given a choice but saw entrepreneurship, understood in the narrow sense of self-employment, as preferable to not having any income. They became flexible neoliberal selves out of necessity, when many of them craved more stability. The majority were not aspiring entrepreneurs in the way the discourse imagined them, although many were entrepreneurial in the wider sense of being innovative and open to new opportunities. The trainers, however, had fully assimilated the more narrow discourse of economic entrepreneurship and in many cases equated it with women's empowerment.

We encountered situations where the majority of the women were not aspiring to be entrepreneurs, yet even here we were able to identify heartening stories of individual women who, thanks to a quite extraordinary level of energy, creativity, and drive, were able to make the most of the opportunities arising. The ICT4D literature is heavily skewed toward these success stories.

Nonetheless, we need to critically ask how meaningful it is to endlessly repeat these success stories and whether a focus on such individual achievements distracts from the very real structural challenges these women face. Change, or even empowerment, needs to be acknowledged as a process that is long term and nonlinear. It requires, as one of the first steps, an analysis that goes beyond factors that enable or constrain women from becoming entrepreneurs toward addressing entrenched gendered power relations. Structural disadvantage cannot be changed by simplistic inputs of material resources (including ICTs) but needs to engage in systemic and long-term transformative processes, which by necessity includes engaging with gendered structures. Indeed, if the powerful and fashionable discourse of female entrepreneurship is uncritically adopted, such rhetoric risks leaving the door open to a logic of accusing all the other women of "just not being entrepreneurial enough." Women would be sorted into the more and the less "deserving poor," in a discursive move that shifts the responsibility for inequality onto the individual. Instead, we should be looking at structural barriers that hold women back, such as unequal access to the law, capital, autonomous time, education, and mobility. Since these structural inequalities can only be challenged collectively, celebrating the heroic individualism of female entrepreneurs can be distracting and potentially debilitating.

Further, we need to ask at what price women are offered the chance to successfully integrate into the hegemonic economic system. As women in the Global South are invited to become neoliberal subjects, they are expected to conform to the commodification of their world, including their cultural heritage, and to relate to others in terms of maximizing profit. While the care and community cohesion work they do remains unremunerated, with their entrepreneurial activity, they are invited to participate in a vision of a society that, in Oscar Wilde's words, "knows the price of everything and the value of nothing."

The framing of women as neoliberal subjects is shaping and narrowing the vision of women's digital inclusion in much of the mainstream ICT4D discourse. This risks many missed opportunities, since ICTs can be enabling for women not only to empower themselves economically, but also to connect and communicate across physical and social boundaries, to craft new identities for themselves, and to start questioning and challenging the environments they live in (Buskens 2010). There is potential for transformative use of ICTs. ICT4D does not have to be a neoliberal discourse; however, where ICT courses are combined with female entrepreneurship training, our evidence shows it can be just that. At present, the conformist trajectory of many women's entrepreneurship projects may or may not enrich women economically, but if it does not move beyond conforming, it risks impoverishing them in other ways.

Notes

1. The notion of "development" itself is normative and highly contested. For a summary of critiques of current development institutions and discourses, see, for instance, Kothari (2005).

2. The Chilean study was published as a book, *Technologies of Choice: ICTs, Development and the Capabilities Approach* (Kleine 2013), and the Tanzanian study is an unpublished master's thesis: "Moving Beyond 'Counting Women' in ICT4D: ICTs, Practical and Transformational Gender Interests and Female Entrepreneurship in Rural Zanzibar" (McCarrick 2014).

3. "Millennium Development Goals and Beyond 2015," United Nations, accessed October 29, 2018, http://www.un.org/millenniumgoals/; "Sustainable Development Goals," United Nations, accessed October 29, 2018, http://www.un.org/sustainabledevelopment/sustainable-development-goals/.

References

Baden, Sally, and Marie A. Goetz. 2005. Who Needs [Sex] When You Can Have [Gender]: Conflicting Discourses on Gender at Beijing. In *Feminist Visions of Development: Gender Analysis and Policy*, edited by Cecile Jackson and Ruth Pearson, 19–39. New York: Routledge.

Benería, Lourdes. 2003. *Gender, Development and Globalization: Economics as if All People Mattered*. New York: Routledge.

Benton, Ted. 1981. Objective Interests and the Sociology of Power. *Sociology* 15 (2): 161–184.

Boserup, Ester. 1970. *Woman's Role in Economic Development*. London: George, Allen, and Unwin.

Buskens, Ineke. 2010. Notes from the Field: Agency and Reflexivity in ICT4D Research: Questioning Women's Options, Poverty, and Human Development. *Information Technologies and International Development* 6:19–24.

Buskens, Ineke. 2015. Gender and ICT4D. In *International Encyclopedia of Digital Communication and Society*, edited by Robin Mansell, Ang Peng Hwa, Charles Steinfield, Shenja van der Graaf, Pieter Ballon, Aphra Kerr, James D. Ivory, Sandra Braman, Dorothea Kleine, and David J. Grimshaw, 1–11. Chichester, UK: Wiley-Blackwell.

Buskens, Ineke, and Anne Webb. 2009. *African Women and ICTs: Investigating Technology, Gender and Empowerment*. London: Zed Books; Ottawa: IDRC.

Buskens, Ineke, and Anne Webb. 2014. *Women and ICT in Africa and the Middle East: Changing Selves, Changing Societies*. London: Zed Books.

Callinicos, Alex. 1987. *Making History: Agency, Structure, and Change in Social Theory*. Cambridge: Polity.

Chew, Han Ei, Vigneswara Ilavarasan, and Mark R. Levy. 2013. When There Is a Will, There Might Be a Way: The Economic Impact of Mobile Phones and Entrepreneurial Motivation on Female-Owned Microenterprises. Paper presented at the Sixth International Conference on Information and Communication Technologies and Development, Cape Town, South Africa, December 7–10, 2013.

Dodson, L., S. R. Sterling, and J. K. Bennett. 2013. Minding the Gaps: Cultural, Technical and Gender-Based Barriers to Mobile Use in Oral-Language Berber Communities in Morocco. In *Proceedings of the Sixth International Conference on Information and Communication Technologies and Development: Full Papers*, vol. 1, 79–88. New York: ACM.

Donner, Jonathan. 2015. *After Access: Inclusion, Development, and a More Mobile Internet*. Cambridge, MA: MIT Press.

England, Kim, and Kevin Ward. 2007. *Neoliberalization: States, Networks, Peoples*. Chichester, UK: Wiley-Blackwell.

Gerster, Richard, and Sonja Zimmermann. 2003. *Information and Communication Technologies (ICTs) for Poverty Reduction?* Richterswil, Switzerland: SDC and Gerster Consulting. http://www.gersterconsulting.ch/docs/ict_for_poverty_reduction.pdf.

Gonick, Marnina. 2006. Between "Girl Power" and "Reviving Ophelia": Constituting the Neoliberal Girl Subject. *NWSA Journal* 18 (2): 1–23.

GSMA. 2013. *Unlocking the Potential: Women and Mobile Financial Services in Emerging Markets*. https://www.gsma.com/mobilefordevelopment/wp-content/uploads/2013/02/GSMA-mWomen-Visa_Unlocking-the-Potential_Feb-2013.pdf.

GSMA. 2015. *Bridging the Gender Gap: Mobile Access and Usage in Low- and Middle-Income Countries*. https://www.gsma.com/mobilefordevelopment/wp-content/uploads/2016/02/Connected-Women-Gender-Gap.pdf.

Guihuan, Li. 2005. The Effect of ICT on Women's Enterprise Creation: A Practical Example from China. In *Gender and ICTs for Development: A Global Sourcebook*, 25–31. Oxford: Oxfam GB.

Hafkin, Nancy, and Nancy Taggart. 2001. *Gender, Information Technology and Developing Countries: An Analytic Study*. Washington, DC: AED. http://www.mujeresenred.net/zonaTIC/IMG/pdf/Gender_Book_NoPhotos.pdf.

Heeks, Richard. 2008. ICT4D 2.0: The Next Phase of Applying ICT for International Development. *Computer* 41 (6): 26–33.

Hindess, B. 1982. Power, Interests and the Outcomes of Struggle. *Sociology* 16 (4): 498–511.

Huyer, Sophia, and Marilyn Carr. 2002. Information and Communication Technologies: A Priority for Women. *Gender, Technology and Development* 6 (1): 85–100.

Intel. 2013. *Women and the Web: Bridging the Internet Gap and Creating New Global Opportunities in Low and Middle-Income Countries*. Santa Clara, CA: Intel. https://www.intel.com/content/www/us/en/technology-in-education/women-in-the-web.html.

Jackson, Cecile. 2005. Rescuing Gender from the Poverty Trap. In *Feminist Visions of Development: Gender Analysis and Policy*, edited by Cecile Jackson and Ruth Pearson, 39–64. London: Routledge.

Jackson, Cecile, and Ruth Pearson, eds. 2005. *Feminist Visions of Development: Gender Analysis and Policy*. London: Routledge.

Jonasdottir, Anna G. 1988. On the Concept of Interests, Women's Interests and the Limitations of Interest Theory. In *The Political Interests of Gender*, edited by Kathleen B. Jones and Anna G. Jonasdottir, 33–65. London: Sage.

Kabeer, Naila. 1994. *Reversed Realities: Gender Hierarchies in Development Thought*. London: Verso.

King, Sally, Hugo Sintes, and Maria Alemu. 2012. Beyond Participation: Making Enterprise Development Really Work for Women. *Gender and Development* 10 (1): 129–144.

Kleine, D. 2007. Empowerment and the Limits of Choice: Microentrepreneurs, Information and Communication Technologies and State Policies in Chile. PhD diss., London School of Economics and Political Science. https://pure.royalholloway.ac.uk/portal/en/publications/empowerment-and-the-limits-of-choice-microentrepreneurs-information-and-communicaton-technologies-and-state-policies-in-chile(d6057d58-08b2-43a4-98c1-d4b30d684f85).html.

Kleine, D. 2010. ICT4WHAT?—Using the Choice Framework to Operationalise the Capability Approach to Development. *Journal of International Development* 22 (5): 674–692.

Kleine, D. 2013. *Technologies of Choice? ICTs, Development and the Capabilities Approach*. Cambridge, MA: MIT Press.

Kleine, Dorothea, David Hollow, and Sammia Poveda Villalba. 2014. *Children, ICTs and Development—Capturing the Potential, Meeting the Challenges*. Florence: UNICEF Office of Research.

Kothari, U. 2005. *A Radical History of Development Studies: Institutions, Individuals and Ideologies*. London: Zed Books.

Lemke, Thomas. 2001. "The Birth of Bio-Politics": Michel Foucault's Lecture at the Collége de France on Neo-Liberal Governmentality. *Economy and Society* 30 (2): 190–207.

Madon, Shirin, Nicolau Rinhard, Dewald Roode, and Geoffrey Walsham. 2009. Digital Inclusion Projects in Developing Countries: Processes of Institutionalization. *Information Technology for Development* 15 (2): 95–107.

McCarrick, H. 2014. Moving Beyond "Counting Women" in ICT4D: ICTs, Practical and Transformational Gender Interests and Female Entrepreneurship in Rural Zanzibar. Master's thesis. Royal Holloway University of London.

Meena, R., and M. Rusimbi. 2009. Our Journey to Empowerment: The Role of ICTs. In *African Women and ICTs: Investigating Technology, Gender and Empowerment*, edited by Ineke Buskes and Anne Webb, 193–205. London: Zed Books; Ottawa: IDRC.

Molyneux, Maxine. 1985. Mobilisation Without Emancipation? Women's Interests, the State and Revolution in Nicaragua. *Feminist Studies* 11 (2): 227–254.

Molyneux, Maxine. 2000. Analysing Women's Movements. In *Feminist Visions of Development: Gender Analysis and Policy*, edited by Cecile Jackson and Ruth Pearson, 65–88. London: Routledge. (First appeared in 1988 in *Development and Change* 29:2.)

Murphy, James T., and Pádraig Carmody. 2015. *Africa's Information Revolution: Technical Regimes and Production Networks in South Africa and Tanzania*. Chichester, UK: Wiley-Blackwell.

Odame, H. Hambly. 2013. Gender and ICTs for Development: Setting the Context. In *Gender and ICTs for Development: A Global Sourcebook*, 13–24. Oxford: Oxfam GB. http://www.bibalex.org/Search4Dev/files/281644/113589.pdf.

Pearson, R. 2005. "Nimble Fingers" Revisited: Reflections on Women and Third World Industrialization in the Late Twentieth Century. In *Feminist Visions of Development: Gender Analysis and Policy*, edited by Cecile Jackson and Ruth Pearson, 171–188. London: Routledge.

Rankin, Katherine N. 2001. Governing Development: Neoliberalism, Microcredit and Rational Economic Woman. *Economy and Society* 30 (1): 18–37.

Ren, Hai. 2005. Modes of Governance in Neo-liberal Capitalism: An Introduction. *Rhizomes: Cultural Studies in Emerging Knowledge* 11:1–9.

Sen, Amartya. 1999. *Development as Freedom*. New York: Oxford University Press.

Sennett, Richard. 2006. *The Culture of the New Capitalism*. New Haven, CT: Yale University Press.

Turken, Salman, Hilde E. Nafstad, Rolv M. Blankar, and Katrina Roen. 2015. Making Sense of Neoliberal Subjectivity: A Discourse Analysis of Media Language on Self-Development. *Globalizations* 13 (1): 32–46.

UNCTAD. 2011. *Information Economy Report 2011: ICTs as an Enabler for Private Sector Development*. New York: United Nations. http://unctad.org/en/PublicationsLibrary/ier2011_en.pdf.

UNCTAD. 2014. *Empowering Women Entrepreneurs through Information and Communication Technologies: A Practical Guide*. New York: United Nations. http://unctad.org/en/PublicationsLibrary/dtlstict2013d2_en.pdf.

Walkerdine, Valerie. 2003. Reclassifying Upward Mobility: Femininity and the Neoliberal Subject. *Gender and Education* 15 (3): 237–248.

Walkerdine, Valerie. 2006. Workers in the New Economy: Transformation as Border Crossing. *Ethos: Journal for the Society for Psychological Anthropology* 34 (1): 10–41.

Wheeler, Deborah L. 2007. Empowerment Zones? Women, Internet Cafés, and Life Transformations in Egypt. *Information Technologies and International Development* 4 (2): 89–104.

5 "Let the Private Sector Take Care of This": The Philanthro-Capitalism of Digital Humanitarianism

Ryan Burns

Introduction

At the heart of this chapter lies the imperative to understand a confluence that is centrally shaping digital economies at the global margins.[1] On the one hand, crowdsourcing, social media, and mass collaboration are increasingly affecting the humanitarian enterprise through *digital humanitarianism* (Meier 2015); on the other, private for-profit companies are becoming more involved in humanitarianism under the banner of *philanthro-capitalism*, commonly labeled "corporate responsibility" (Bishop and Green 2008). These two distinct shifts together are having profound effects on humanitarian knowledge, aid allocation, and humanitarianism's raison d'être. In fact, while much has been written about each of these developments individually, here I argue that they are integral components of the same process, which will significantly influence how humanitarianism is conducted in the twenty-first century.[2]

Following the historical arc of "disruptive" technologies, we can see that for more than a decade, technologists have been developing digital spatial technologies that they hope will "revolutionize" humanitarianism (Meier 2012).[3] Many claim these "liberation technologies," as they are often called (Meier 2015), can increase democratic decision making, citizen empowerment, and civic engagement, effectively dislodging humanitarianism from its established modus operandi. This trend seeks to accentuate more "voices" by crowdsourcing knowledge and recruiting labor in platforms like OpenStreetMap, Ushahidi, Tomnod, and the Standby Task Force, and by scraping social media resources like Twitter, Facebook, and Instagram. The evidence for digital humanitarianism's impact is mixed (Brandusescu, Sieber, and Jochems 2015; Read, Taithe, and Mac Ginty 2016). But, more importantly, digital humanitarianism has been shown to be a fundamentally social and political enterprise rather than a technical advance (Burns 2015b; Duffield 2016; Finn and Oreglia 2016). Thus, because the development, use, effects, and assumptions of digital humanitarianism are inherently situated within sociospatial contexts, those contexts themselves necessitate critical scrutiny (Burns 2014; Crawford and Finn 2015).

Concurrently, private for-profit businesses have become more intimately involved in philanthropy and humanitarianism. Most paradigmatically, this shift is characterized by private companies positioning philanthropy at the center of their business model, such that the act of philanthropic giving generates profit for the company. In this new age of philanthro-capitalism, companies accumulate capital by leveraging moral economies (Fridell and Konings 2013). The business models of shoe company TOMS and Starbucks' Ethos Water exemplify this shift, in that for the former, for every pair of shoes someone purchases, a second pair is donated to "a person in need";[4] and for the latter, US$0.05 is donated to charity for every bottled water purchase. Slavoj Žižek (2010, n.p.) has argued that far from this shift being a political-economic exception, "charity is no longer an idiosyncrasy of some good guys here and there, but the basic constituent of our economy."

In this chapter, I argue that these phenomena signal broader shifting relationships between the state and the private sector, enabled through digital humanitarian technologies. These shifts increasingly inculcate private for-profit logics, rationalities, and imperatives into humanitarianism, aid relief, and most broadly the public sector. More specifically, I argue that in the context of increasing austerity and the drive to "do more with less," humanitarian organizations see digital spatial technologies as an innovation that enables their continued functioning. In the process of adopting digital humanitarianism, the project of humanitarianism becomes more capitalist. I substantiate this argumentation by drawing on ethnographic research conducted in 2012–2013 with a public policy research institution involved in efforts to proliferate digital humanitarianism in the public sector.

I begin by contextualizing the emergence of digital humanitarianism and philanthro-capitalism within existing research. Following a brief description of the broader research project from which this chapter draws, I develop my argument along two lines. First, I demonstrate that for public sector humanitarian agencies, digital humanitarianism represents an "innovation" that assuages the pressures created by neoliberal reforms. Second, these incursions of the private sector into humanitarianism can be seen as philanthro-capitalism, wherein for-profit institutions and their charitable arms accumulate capital by developing digital humanitarian technologies and data. I conclude by arguing that private for-profit businesses benefit from these new configurations at the expense of those who produce and process digital humanitarian data.

The Social Origins of Digital Humanitarianism

Many histories of digital humanitarianism begin with the use of Ushahidi, Mission 4636, and OpenStreetMap in the response to the 2010 earthquake outside Port-au-Prince, Haiti (Sandvik et al. 2014; see, for example, Meier 2015). In the earthquake's aftermath, Mission 4636 collected SMS messages sent to a dedicated number; these messages typically requested resources, aid, or assistance (Burns 2015b). Geographically disparate volunteers translated, georeferenced, categorized, and amended these messages through the Ushahidi interface (Meier and Munro 2010). Many digital humanitarian organizations have since been established (and some terminated), including the Standby Task Force—loosely coordinated individuals who are tasked with data collection, processing, and mapping by formal humanitarian organizations—and the Digital Humanitarian Network—a liaison between the different digital humanitarian organizations working on a project (Crowley and Chan 2011). Increasingly spotlighted in contemporary digital humanitarian debates are the roles and purposes of unmanned aerial vehicles (UAVs, commonly "drones"; Sandvik and Lohne 2014), and the insights that may be generated by analyzing social media (Robinson, Maddock, and Starbird 2015). In contrast with this origin story, however, many technologies and constituent phenomena associated with digital humanitarianism were in use prior to 2010, emerging from complex histories surrounding data, software, hardware, and peer production (Roche, Propeck-Zimmermann, and Mericskay 2011; Barnes 2013).

Indeed, many have argued that these contexts indicate that digital humanitarianism should be conceptualized as not merely a technological advance, but instead an assemblage of knowledges, social relations, and political imperatives.⁵ Some research is now beginning to theorize digital humanitarianism as such, in contrast with early research, which largely sought to characterize the field and identify pressing technical issues such as privacy, data quality, and intellectual property (Goodchild and Glennon 2010; Liu and Palen 2010; Burns and Shanley 2013). I have argued (Burns 2015b) that digital humanitarianism has not ushered in a fundamental "revolution" but is instead a set of shifted practices toward data collection and processing, which make epistemological claims regarding what can and cannot be known about crises. These technologies themselves did not emerge out of a teleological, apolitical "progress" but instead are the outcomes of complex sets of negotiations around how knowledge and needs can be captured as data and represented cartographically (Burns 2014). Building on much broader conversations, here social relations are embedded within, and are in turned shaped by, data models, software code, hardware, protocols, and infrastructural knowledge (Graham 2005; Kitchin and Dodge 2012; Dalton and Thatcher 2014; Kitchin 2014). For humanitarianism, then, the use of digital technologies plays a significant role in shaping how crises, individuals, and knowledge come to be known, and by extension, the forms of social and political action appropriate for addressing them (Jacobsen 2015; Finn and Oreglia 2016).⁶

Within this set of discussions, digital humanitarianism's implications within political economy are under-researched. This is despite much research showing that digital and web-based spatial technologies profoundly affect urban consumerism (Graham 2010; Thatcher 2013), constitute markets and protectionist capitalist practices (Leszczynski 2012; Dalton 2015), and become sites of capital accumulation (Cupples 2015; Thatcher, O'Sullivan, and Mahmoudi 2016). In fact, in recent years, private businesses have begun playing an increasingly important role in digital humanitarianism, raising important questions that have yet to be addressed in research. Duffield (2016) has argued that the affirmatory politics of digital humanitarianism—"celebrat[ing] the restorative powers of smart technologies and fast machine thinking" (149)—negates any potential critique of existing capitalism. That is, the celebratory discourses surrounding digital humanitarianism contrast with late-modern capitalist trends toward precarity, crisis, and neoliberalization. For Duffield, however, the connections between digital humanitarianism and these economic turns are merely coincidental rather than causally related. I now turn to these emergent trends within capitalism.

No Free Gift: Philanthro-capitalism and Humanitarianism

The current political-economic moment in the Global North is often characterized by neoliberal reforms. These typically entail what Peck and Tickell (2002) call the dual "roll-back" and "roll-out" of the public sector, releasing roles and responsibilities to the realm of capital accumulation while developing the frameworks within which this capital can be readily accumulated. Neoliberalism as a hybrid political and economic project is guided by logics emerging from the private for-profit sector within capitalist economies, primarily driven by a reliance on free market principles and the posited superiority of the "competitive individual" (Hall, Massey, and Rustin 2013, 9). Yet it "looks" quite different in different venues, geographies, and technological moments, as well as with different actors and institutional relationships involved. The goal of research, then, is to explain the forms neoliberalism takes across different contexts (Larner 2003; Peck 2006).

Neoliberalization has produced humanitarian contexts in which new labor practices must emerge. For humanitarian practitioners, an emerging idea is that new forms of labor are able to mitigate the harmful effects of capitalism (Roy 2010). Within humanitarianism, neoliberal reforms have resulted in a climate of austerity, an increased role of the private and nongovernmental (NGO) sectors, and the incursion of capitalist rationalities (Roberts 2014; Mitchell and Sparke 2016). That is, private sector involvement in humanitarianism incorporates and normalizes metrics stemming from the drive for capital accumulation, such as "return on investment," "poorly performing countries," and "freedom of choice" (Carbonnier 2006; Mitchell 2016). This has had the twofold effect of major humanitarian institutions facing decreased budgets concomitant to an increased role of private sector contractors in humanitarian work, in what Norris (2012, n.p.) calls the "development-industrial complex."

Within this context private companies have begun making philanthropy and humanitarianism central to their business models and strategies. This marriage of profit motives and charity, which many call philanthro-capitalism, is enabling capital accumulation within humanitarianism; indeed, it is transforming the act of humanitarian assistance into the very means of accumulation (Morvaridi 2012; Fridell and Konings 2013). Philanthro-capitalism does this through a moral economy: it promises that by buying a product, the consumer is "buying into" a good philanthropic cause (Žižek 2009; Mitchell 2016).[7] The consumer, presumably, then wishes to purchase these products to help a disadvantaged community somewhere in the world. More indirectly, celebrities and icons of "success" (e.g., Bono and the Bill & Melinda Gates Foundation) are increasing collective influence over philanthropy and humanitarianism through actively promoting private sector companies and contractors in humanitarianism and philanthropy (Jenkins 2011; Hay 2013; Mitchell and Sparke 2016). This changes the terms of the debates around humanitarianism, deepening market-based rationalities and normalizing large-scale political-economic trends toward austerity (Adams 2013; Loewenstein 2015). It is in this sense that McGoey says "there's no such thing as a free gift" in philanthropy (2015).

Still, researchers have largely theorized this emergent shift toward philanthro-capitalism in humanitarianism independently of digital spatial technologies. While understanding that spatial technologies are embedded within, are coconstitutive, and compose entire economies, researchers have not to date drawn lines between philanthro-capitalism's emergence and the technologies through which this emergence is enabled. Roy (2010), for example, accounts for smartphone apps that streamline microfinance delivery while producing capitalist market subjects, but for her the app is coincidental to the microfinance industry. Elwood (2015), on the other hand, demonstrates how the spatial technologies employed by Kiva, a microfinance organization, produce a visual and affective way of understanding poverty, global political economy, and subject positionalities. Likewise, Maurer (2015) has shown that mobile payment platforms and apps have been a necessary condition for the proliferation of many capitalist markets in the Global South, and they have "captivated industry and philanthropic attention" (2015, 127).[8]

These two bodies of research—digital humanitarianism and philanthro-capitalism—ask us to look more deeply at the ways spatial technologies are situated within, and enable, political-economic reforms. If philanthro-capitalism has emerged in the context of neoliberal reforms toward humanitarianism, and digital humanitarian technologies enroll the new labor practices propagated by these same processes, analytically tying the two together can offer insights into the ways in which emergent technologies shape economies at the global margins.

The Extended Case Method

Below I draw on a one-year research project that employed the *extended case method* (Burawoy 1998). The extended case method is a theory-driven inductive analytic framework for generating theoretical propositions from evidence. Over the course of the year, I examined the institutional relations and imperatives underlying digital humanitarian technology adoption, usage, and development. My particular use of the extended case method entailed a combination of ethnography, in-depth semistructured interviews, and archival analysis. The ethnography combined work with a public policy research institute at the forefront of debates about public sector use of digital humanitarian technologies, and participant observation with digital humanitarian communities. Within this ethnographic work, I hosted and attended numerous conferences, workshops, and digital humanitarian "deployments," with the intention of understanding the complexities of public sector digital humanitarian adoption. These were attended also by high-level managers at international and domestic humanitarian agencies, digital humanitarian communities, academic researchers, and policymakers. This ethnographic work led to interviews with many of these leaders in the field, as well as archival analysis on social media, after-action reports, blog posts, crowdsourced maps, and policy white papers.

I transcribed and coded these data to identify patterns and trends around the interinstitutional pressures and opportunities for adopting digital humanitarian technologies. This involved discourse analysis with attention to how leaders in the field understand the roles and relationships between humanitarianism, the private sector, and spatial technologies. All quoted interviewees in this chapter have been pseudonymized, with all identifying information removed. Although the claims I cover emerge from the particular case I investigated, this case provides a unique window into how digital humanitarianism works and how formal humanitarianism is changing. Thus, despite my relatively small sample size, the principles I elucidate have strong theoretical purchase beyond my individual case.

Austerity and Innovation

The current "humanitarian moment" (Fassin 2012) is characterized by the incursion of capitalism into humanitarianism. The roles, rationalities, and imperatives of humanitarianism are being reconfigured to align more closely with those of the private for-profit sector. Digital humanitarianism plays a significant role in this reconfiguration. First, for the public sector, digital humanitarianism is the "innovation" that allows continued operation in the context of increased drive for efficiency, austerity, and decreased expenditures. This innovation constitutes a new wave of neoliberalization of humanitarian aid, in which, *via digital humanitarian technologies*, humanitarianism becomes a new site for capital accumulation. Second, for digital humanitarianism, the new roles of the private sector take the form of philanthro-capitalism, wherein private businesses and their philanthropic/charitable extensions generate profit through developing the technologies used in digital humanitarian operations. Importantly, by relying on unquestioned assumptions about the inherent "good" of philanthropy and humanitarianism, this private sector involvement is depoliticized, meaning it obscures the trade-offs and consequences of digital humanitarianism and removes them from the realm of legitimate critique.

For the formal humanitarian sector, digital humanitarianism is an innovation that allows continued operations in the context of precarious formal funding sources. This "innovation" comprises the new techniques, approaches, technologies, and procedures that reconfigure how the public sector relates to the private sector. Under new regimes of private sector rationalities and cutbacks, those who manage humanitarian aid are increasingly feeling pressure to more efficiently and wisely allocate their (labor, funding, technology) resources. Lauren, who works for a major US-based development agency, and who prior to our interview had recently conducted a digital humanitarian crowdsourcing project, characterized this pressure: "Without crowdsourcing we didn't have the resources or the time to [process data] ourselves, so that's why we really needed to rely on the public. And I think having crowdsourcing as an option for government agencies, especially in this financial time when you see sequestration happening and people having to do more with less, that we have no choice but to rely on and really engage the public."

As Lauren points out, this process is emblematic of broader public sector trends toward austerity and retrenchment. By her way of thinking (in a sentiment reflected broadly across my interviews with the public sector), the "public" constitutes a pool of reserve labor that can be mobilized via digital humanitarianism to meet increasing pressures for decreased resource expenditure. This at once creates a pool of reserve labor and makes that labor contingent on humanitarian project funding and initiatives. Importantly, Lauren is cognizant of these pressures and the search for new reserves of labor that they elicit. In this way, the statement should be understood as pointing to shifted data collection and production practices brought into being by emergent institutional imperatives.

Digital humanitarians promote the message that their mass collaboration tools *immediately* improve efficiency, presenting this narrative in most of their interactions with the public and with the formal sector. Jasmine, a leader of one of the largest digital humanitarian communities, provided a comprehensive depiction of this narrative. Asked what benefits her organization gives humanitarian agencies, she said, "Part of it is just pure manpower. … If we can get 200 people working … across time zones … when Geneva is asleep and everybody else is still working away, by the time that they wake up they see that a massive amount of work has been done overnight. And it gives them a sort of 24/7 workforce. … So I think that's not something that a lot of organizations would typically have in-house, is a breadth of really strong technical people that can work across time zones."

In this hypothetical scenario, the data producers and data-processing capacity are expanded beyond the state and formal humanitarian agencies. Typical of contemporary conceptualizations of neoliberalism, formal institutions delegate data production and processing responsibilities to digital humanitarian organizations and, by extension, to the large numbers of contingent laborers who contribute to these projects. Her account stands at odds with the previous quotation. Lauren accounts for the new practices and institutional changes necessary to adopt digital humanitarianism, whereas Jasmine assumes immediate improvement. Jasmine, in her attempts to market digital humanitarianism as the innovation needed by the public sector, offers idealistic descriptions of the technologies and communities. In both cases, the interviewees have identified digital humanitarianism as an innovation that allows humanitarian agencies to resolve their broader drive toward efficiency.

At the public 2012 "Connecting Grassroots to Government" workshop at the Woodrow Wilson International Center for Scholars, Eric Rasmussen (CEO of the private business Infinitum Humanitarian Systems and adjunct associate professor of medicine at the University of Washington) stated, "Robert Kirkpatrick, who's now at the U.N. Global Pulse program, used to be at Microsoft. And he used to argue that—in these discussions, please *let the private sector take care of this*. We will address this problem for you, we will take the research, we will commercialize it, and we'll sell it back to you for cheap. Everybody will be happy." (Woodrow Wilson Center 2012b, n.p., emphasis mine.)

Rob Munro, founder and CEO of the private company Idibon, followed this up immediately: "I'll second what Eric Rasmussen said about letting the private sector take care of this. Natural language processing and machine learning is just kind of that level of complexity beyond what you would get in most very good engineers who are working with NGOs. It's something that you want to give to the private sector" (Woodrow Wilson Center 2012b, n.p.).

These two quotations reveal that private sector involvement is seen to "make sense" by appealing to saving resources, as well as relying on notions of what the public sector is able to deliver. The emergence of such new data sources intertwines technical power limitations with technical expertise and time pressures. Both Rasmussen and Munro owned private businesses, Infinitum Humanitarian Systems and the now-defunct Idibon, respectively, that stood to benefit from the shift they encouraged in those quotes. By "letting the private sector take care of" digital humanitarian technology development, Rasmussen and Munro were both likely to see financial gain. Nonetheless, private sector companies become involved in humanitarianism because it ultimately—in their conception—leads to a greater "good," while generating new sites for capital accumulation. To do so, humanitarianism shifts to align itself more closely with private sector rationalities, including liberalizing market logics, decreasing public dependence on the formal/public sector, and adopting the techniques and language (e.g., investments, profit, "best practices") of the private sector.

Digital humanitarianism is a channel through which the private sector enters humanitarianism. Private sector businesses such as Esri, Google.org,

DigitalGlobe, and TechChange have all developed the tools and languages for digital humanitarians through, for example, the Google Crisis Response Team, DigitalGlobe's donations of imagery to OpenStreetMap, and TechChange's educational offerings in technology for emergency response (see figure 5.1).[9] These companies have been making inroads into digital humanitarianism in venues such as the International Conference of Crisis Mappers (ICCM), public policy workshops, and Esri's Disaster Response Program. For digital humanitarianism, the private sector enters directly by developing the tools, technologies, and data-sharing agreements for such encroachment, and indirectly through prioritizing logics such as profit and efficiency.

Philanthro-capitalism and Digital Humanitarianism

Private sector incursions into humanitarianism via digital technologies take the form of philanthro-capitalism, which here refers to private sector companies intervening in humanitarianism ultimately to accumulate capital. It appeals to contemporary economic ideology—generally that of neoliberalism—as well as digital humanitarianism's innovative nature. These factors depoliticize digital humanitarianism, setting its trade-offs and consequences outside the realm of legitimate critique.

In recent years, private businesses have begun investing heavily in charity and philanthropy, often incorporating these missions directly into their product marketing. This "enlightened capitalism" (Essex 2013, 152) situates the private sector as an important actor in humanitarian interventions. In this new configuration, the private sector provides financial support, project management, and services provision. It prioritizes economic and operational efficiency, often invoking the private sector to reach that goal. For example, in January 2016, UNICEF launched a "venture fund" to provide financial resources to "open source startups" that "brings together models of financing and methodologies used by venture capital funds" (Acharya 2016, n.p.; see figure 5.2).

Digital humanitarianism is a practice of philanthro-capitalism in that private for-profit businesses are involved in such philanthropic causes as a means of accumulating capital, and they do this both through developing the digital technologies for use in crises and by serving as a source for the logics and rationalities under which digital humanitarianism

Figure 5.1
Amazon has partnered with the Standby Task Force to mobilize a moral economy that encourages consumers to buy from their site. *Source*: Author.

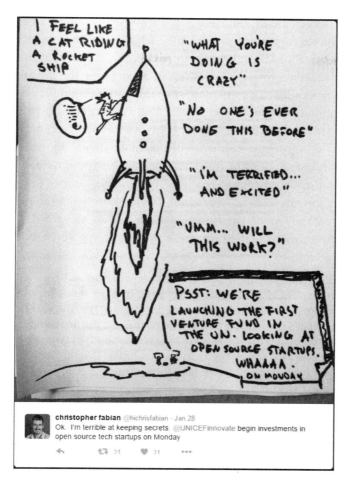

Figure 5.2
In 2016, UNICEF began its venture fund, which draws on ideas from venture capital finance to fund "open source startups." *Source*: Author.

operates. These processes are depoliticized for at least three reasons. First, as philanthro-capitalist practice, it invokes the ideological commitments of the contemporary neoliberal milieu. This includes the notion that the public sector should serve economic functions in areas where the private sector is known to fail; capital accumulation is the normative economic status, and the exception is the public sector (Peck and Tickell 2002). Digital humanitarians and the broader public implicitly invoke this new "common sense" when welcoming private sector involvement in digital humanitarianism. Digital humanitarianism thus runs parallel to the dominant political-economic discourses of "good practice." For instance, at the "Connecting Grassroots to Government" workshop, David Kaufman, a senior-level administrator at the Federal Emergency Management Agency (FEMA), conveyed the degree to which these practices are firmly entrenched in humanitarian imaginaries:

Right, so take feeding people. Government doesn't feed people in this country, even in ... like military and prisons and school systems, we still contract almost all that out. ... So we have private sector representation inside our [disaster] operation center now, started with the retail sector, the big box stores, and now the financial services sector. ... [W]hat we care about is the ability to see in real time how [private companies] are. ... So that we aren't setting up shop ... in the same places that they're open for business, that we're putting our efforts in places where there's a gap, and vice versa. (Woodrow Wilson Center 2012a, n.p.)

Kaufman here naturalizes the private sector's prioritization in emergency response. According to his statement, when private companies operate in a particular space, the formal response community should be removed. Using the analogy of "feeding people," Kaufman claims that the public sector is far less involved in state operations than most people assume, and that this current state is balanced. Rather than critiquing the political-economic negotiations behind this privileging of the private sector, Kaufman accepts its current status and implies it is the only possible structure. Naturalizing in this way takes trade-offs and consequences as necessary costs to be paid in a perfect political-economic organization. Kaufman further naturalizes the role of the volunteer and nonprofit sectors, again removing responsibility from the public sector and laying the groundwork for a role of digital humanitarians.

The second mechanism through which philanthro-capitalism depoliticizes digital humanitarianism is in invoking commonplace conceptions of, as four interviewees explained, "the good." That is, digital humanitarians view humanitarian practices as beyond the realm of critique because they are unequivocally altruistic things to do. Critiquing digital humanitarianism is often seen to be useless at best (by not contributing to applied practices) and perverse at worst. By extension, private sector businesses are guarded from critique by participating in these practices that exemplify "caring" standpoints on global issues.[10] Remaining unanswered by the businesses and by academic research is the question of why the private sector would involve itself in philanthropy, particularly around development and humanitarianism. A prominent reason for doing so is that such activities not only provide new spaces for capital accumulation, but also bolster companies' "images" and thus rely on an affective economy. By exuding a caring, cosmopolitan corporate persona, these companies are able to persuade consumers to purchase their products, while cultivating in the consumers feelings of altruism and global citizenry. Consumption here appends an affective stimulus onto an exchange, such that consumers purchase not only a commodity but also an affective experience.

Third, digital humanitarianism is depoliticized because it is seen as an innovation that improves humanitarian response. Digital humanitarians have successfully marketed their work as innovative and potentially revolutionary, a trope taken on by some in the formal humanitarian sector as well. This view of digital humanitarianism as an innovation contributing to altruistic ends is another justification for it to be beyond the realm of critique. While such "innovations" most often occur outside the context of an emergency, emergency imaginaries influence how people conceptualize critique in relation to innovation. To explain, Scarry (2011) argues that, despite high levels of thinking that happen in emergencies, they are seen to necessitate postponing thinking in favor of acting.

Philanthro-capitalism within digital humanitarianism produces inequalities (i.e., whose knowledge is seen to matter, which crises and needs are addressed, etc.) as a byproduct of the different logics through which privatized humanitarianism operates. Capitalist enterprises are driven by the imperative to accumulate capital, which privileges logics such as speed, production/consumption, competition, and privatization of public assets (Harvey 1982). Publicly traded companies are accountable to shareholders to increase profit. In contrast, humanitarianism's primary motive is claimed to be to decrease suffering (Weizman 2012). Weizman and Manfredi (2013) argue that this motivation privileges "saving lives," human rights, and mitigating violence. Private sector incursion into digital humanitarianism shifts imperatives from decreasing suffering to accumulating capital, constituting a new form of neoliberalized humanitarianism.

Conclusion

In this chapter, I have built on conversations around philanthro-capitalism and digital humanitarianism to argue that they are dual constituencies of the same broad processes. I argue that philanthro-capitalism signals shifting relationships between the state and the private sector, enabled by digital humanitarian technologies. Humanitarian agencies see digital humanitarian technology, and its underpinning philanthro-capitalist drives, as the innovation that allows humanitarian agencies to continue operations in the climate of neoliberal reforms. This has the effect of drawing on and fostering the rationalities and logics of capitalist imperatives. Philanthro-capitalism helps depoliticize digital humanitarianism because it appeals to an inherent "good" of charitable work, without attention to its consequences and drawbacks.

These processes have strong implications for economies at the global margins. What I have outlined above is the process of knowledge becoming captured, enclosed, and commodified to identify new sources of capital accumulation (Perelman 2004). More precisely, digital humanitarianism appropriates affected communities' knowledge as data in its new practices and approaches. In so doing, while affected communities might experience some benefit from strengthened assistance, the ultimate benefactors are private for-profit companies. These companies bolster their image while accumulating capital and improving their symbolic societal power (McGoey 2015). Because capitalist processes produce poverty and thus the conditions for philanthropy itself, philanthropy has been called a "protective layer for capitalism" (Roelofs 1995, 16). These processes have implications for the types of crises recognized and addressed, yet more research is needed to see what the precise effects are.

Most importantly, the broad contours of these processes shape how crises, people, and knowledge are conceptualized, represented, and captured as data. They produce new struggles over the purposes to which digital humanitarianism will be put as well as over control of the resulting platforms and datasets. I have charted the sociotechnical terrain of these struggles as they present in political-economic reforms. Digital humanitarianism facilitates the emergence of philanthro-capitalism and, by extension, new waves of neoliberalism.

Notes

1. Sections of this manuscript have been adapted from my unpublished doctoral dissertation (Burns 2015a).

2. In this chapter, I often use the term "humanitarianism" to describe processes, institutions, and imperatives many might associate more with emergency management, disaster relief, and international development. While each of these can be distinguished in concept, practice, and associated research agenda, they are fundamentally "deeply interlocked regimes of knowledge, power, and morality" (Burns 2015a, 25). Each can be understood as a set of sociopolitical projects aimed at establishing and normalizing relations that prioritize power and knowledge from the Global North, while invoking a moral economy of intervention (Calhoun 2004; Lawson 2007; Fassin 2012; Weizman 2012).

3. Digital spatial technologies are those that use geographic location, such as Google Maps, geolocated Tweets, and Facebook check-ins.

4. "Improving Lives," TOMS website, accessed October 29, 2018, http://www.toms.ca/improving-lives.

5. By "assemblage" here I mean that technologies always entail social bits: they are designed to address social problems, they produce knowledge to be used socially, and so on. This is a very important distinction from understanding technology as merely a "tool." Tools are politically and socially neutral on their own but are used for social and political ends; sociotechnological assemblages contain political and social imperatives and potentials. They have some agency.

6. More broadly, knowledge produced about emergencies is itself contested, limited, and proliferated by powerful people and institutions and is thus socially constitutive (Calhoun 2004). Digital technologies are not deterministic in their capture and representation of emergency-related knowledge; they may have unintended impacts, be repurposed, and be uniquely adopted in context (Pinch and Bijker 1987).

7. This tendency to turn crisis into capital accumulation opportunities is not entirely new: Klein (2007) argued that the "shock" of crisis is leveraged as a political tool to push privatization. The new phenomenon here is how humanitarian crises and emergencies have become *central* to business models, simultaneously drawing on the moral economy of "helping others" with a purchase.

8. See also Leszczynski (2012).

9. Google.org is, importantly, Google's *for-profit* philanthropic wing. Google.org also manages a nonprofit organization, the Google Foundation, as part of its portfolio (Boss 2010), which does not seem to have changed with Google's recent placement under the Alphabet structure.

10. In other words, paradoxically, digital humanitarian philanthro-capitalism is depoliticized in the process of private businesses deciding on a particular political standpoint. These standpoints usually are popular and avoid contentious topics.

References

Acharya, Sarmistha. 2016. Unicef to Invest in Technology Startups to Help Children. International Business Times. February 1. http://www.ibtimes.co.uk/unicef-invest-technology-start-ups-help-children-1541146.

Adams, Vincanne. 2013. *Markets of Sorrow, Labors of Faith: New Orleans in the Wake of Katrina*. Durham, NC: Duke University Press.

Barnes, Trevor. 2013. Big Data, Little History. *Dialogues in Human Geography* 3:297–302.

Bishop, M., and M. Green. 2008. *Philanthrocapitalism: How the Rich Can Save the World*. New York: Bloomsbury Press.

Boss, Suzie. 2010. Do No Evil. *Stanford Social Innovation Review*. http://ssir.org/articles/entry/do_no_evil.

Brandusescu, Anna, Renee Sieber, and Sylvie Jochems. 2015. Confronting the Hype: The Use of Crisis Mapping for Community Development. *Convergence* 22 (6): 616–632. First published online May 18, 2015. doi:10.1177/1354856515584320.

Burawoy, Michael. 1998. The Extended Case Method. *Sociological Theory* 16 (1): 4–33.

Burns, Ryan. 2014. Moments of Closure in the Knowledge Politics of Digital Humanitarianism. *Geoforum* 53:51–62.

Burns, Ryan. 2015a. Digital Humanitarianism and the Geospatial Web: Emerging Modes of Mapping and the Transformation of Humanitarian Practices. PhD diss., University of Washington, Seattle. https://digital.lib.washington.edu:443/researchworks/handle/1773/33947.

Burns, Ryan. 2015b. Rethinking Big Data in Digital Humanitarianism: Practices, Epistemologies, and Social Relations. *GeoJournal* 80 (4): 477–490.

Burns, Ryan, and Lea Shanley. 2013. *Connecting Grassroots to Government for Disaster Management: Workshop Report*. Washington, DC: Woodrow Wilson International Center for Scholars; http://www.scribd.com/doc/165813847/Connecting-Grassroots-to-Government-for-Disaster-Management-Workshop-Summary.

Calhoun, Craig. 2004. A World of Emergencies: Fear, Intervention, and the Limits of the Cosmopolitan Order. *Canadian Review of Sociology and Anthropology* 41 (4): 373–395.

Carbonnier, Gilles. 2006. Privatisation and Outsourcing in Wartime: The Humanitarian Challenges. *Disasters* 30 (4): 402–416.

Crawford, Kate, and Megan Finn. 2015. The Limits of Crisis Data: Analytical and Ethical Challenges of Using Social and Mobile Data to Understand Disasters. *GeoJournal* 80 (4): 491–502.

Crowley, John, and Jennifer Chan. 2011. *Disaster Relief 2.0: The Future of Information Sharing in Humanitarian Emergencies*. UN Foundation and Vodafone Foundation Technology Partnership. http://www.globalproblems-globalsolutions-files.org/gpgs_files/pdf/2011/DisasterResponse.pdf.

Cupples, Julie. 2015. Coloniality, Masculinity, and Big Data Economies. *Geography/development/culture/media* (blog). May 11, 2015. https://juliecupples.wordpress.com/2015/05/11/coloniality-masculinity-and-big-data-economies/.

Dalton, Craig. 2015. For Fun and Profit: The Limits and Possibilities of Google-Maps-Based Geoweb Applications. *Environment & Planning A* 47 (5): 1029–1046. doi:10.1177/0308518X15592302.

Dalton, Craig, and Jim Thatcher. 2014. What Does a Critical Data Studies Look Like, and Why Do We Care? Seven Points for a Critical Approach to "Big Data." *Society and Space Open Site*. http://societyandspace.org/2014/05/12/what-does-a-critical-data-studies-look-like-and-why-do-we-care-craig-dalton-and-jim-thatcher/.

Duffield, Mark. 2016. The Resilience of the Ruins: Towards a Critique of Digital Humanitarianism. *Resilience* 4 (3): 147–165.

Elwood, Sarah. 2015. Still Deconstructing the Map: Microfinance Mapping and the Visual Politics of Intimate Abstraction. *Cartographica: The International Journal for Geographic Information and Geovisualization* 50 (1): 45–49. doi:10.3138/carto.50.1.09.

Essex, Jamey. 2013. *Development, Security, and Aid*. Athens: University of Georgia Press.

Fassin, Didier. 2012. *Humanitarian Reason: A Moral History of the Present*. Berkeley: University of California Press.

Finn, Megan, and Elisa Oreglia. 2016. A Fundamentally Confused Document: Situation Reports and the Work of Producing Humanitarian Information. In *Proceedings of the 19th ACM Conference on Computer-Supported Cooperative Work and Social Computing*, 1349–1362. New York: ACM. http://www.ercolino.eu/docs/Oreglia_Pub_Fundamentally%20Confused%20Document%202016_AD.pdf.

Fridell, G., and M. Konings. 2013. Introduction: Neoliberal Capitalism as the Age of Icons. In *Age of Icons: Exploring Philanthrocapitalism in the Contemporary World*, edited by G. Fridell and M. Konings, 3–25. Toronto: University of Toronto Press.

Goodchild, Michael, and J. Alan Glennon. 2010. Crowdsourcing Geographic Information for Disaster Response: A Research Frontier. *International Journal of Digital Earth* 3 (3): 231–241.

Graham, Mark. 2010. Neogeography and the Palimpsests of Place: Web 2.0 and the Construction of a Virtual Earth. *Tijdschrift voor Economische en Sociale Geografie* 101 (4): 422–436.

Graham, Stephen. 2005. Software-Sorted Geographies. *Progress in Human Geography* 29 (5): 562.

Hall, Stuart, Doreen Massey, and M. Rustin. 2013. After Neoliberalism: Analysing the Present. In *After Neoliberalism? The Kilburn Manifesto*, edited by Stuart Hall, Doreen Massey and M. Rustin, 3–19. London: Soundings. https://www.lwbooks.co.uk/sites/default/files/00_manifestoframingstatement.pdf.

Harvey, David. 1982. *The Limits to Capital*. Chicago: University of Chicago Press.

Hay, Iain. 2013. Questioning Generosity in the Golden Age of Philanthropy Towards Critical Geographies of Super-Philanthropy. *Progress in Human Geography* 38 (5): 635–653. First published online September 2, 2013. doi:10.1177/0309132513500893.

Jacobsen, Katja. 2015. *The Politics of Humanitarian Technology: Good Intentions, Unintended Consequences and Insecurity*. London: Routledge.

Jenkins, G. 2011. Who's Afraid of Philanthrocapitalism? *Case Western Reserve Law Review* 61 (3): 1–69.

Kitchin, Rob. 2014. Big Data, New Epistemologies and Paradigm Shifts. *Big Data and Society* 1:1–12.

Kitchin, Rob, and Martin Dodge. 2012. *Code/Space: Software and Everyday Life*. Cambridge, MA: MIT Press.

Klein, Naomi. 2007. *The Shock Doctrine: The Rise of Disaster Capitalism*. New York: Picador.

Larner, W. 2003. Neoliberalism? *Environment and Planning D: Society & Space* 21 (5): 509–512.

Lawson, Victoria. 2007. *Making Development Geography*. Oxford: Oxford University Press.

Leszczynski, Agnieszka. 2012. Situating the Geoweb in Political Economy. *Progress in Human Geography* 36 (1): 72–89.

Liu, Sophia B., and Leysia Palen. 2010. The New Cartographers: Crisis Map Mashups and the Emergence of Neogeographic Practice. *Cartography and Geographic Information Science* 37 (1): 69–90.

Loewenstein, Antony. 2015. *Disaster Capitalism: Making a Killing Out of Catastrophe*. London: Verso.

Maurer, Bill. 2015. Data-Mining for Development? Poverty, Payment, and Platform. In *Territories of Poverty: Rethinking North and South*, edited by Ananya Roy and Emma Shaw Crane, 126–143. Athens: University of Georgia Press.

McGoey, Linsey. 2015. *No Such Thing as a Free Gift: The Gates Foundation and the Price of Philanthropy*. New York: Verso.

Meier, Patrick. 2012. Crisis Mapping in Action: How Open Source Software and Global Volunteer Networks Are Changing the World, One Map at a Time. *Journal of Map & Geography Libraries* 8 (2): 89–100.

Meier, Patrick. 2015. *Digital Humanitarians: How Big Data Is Changing the Face of Humanitarian Response*. Boca Raton, FL: CRC Press.

Meier, Patrick, and Robert Munro. 2010. The Unprecedented Role of SMS in Disaster Response: Learning from Haiti. *SAIS Review* 30 (2): 91–103.

Mitchell, Katharyne. 2016. "Factivism": A New Configuration of Humanitarian Reason. *Geopolitics* 22 (1): 110–128. doi:10.1080/14650045.2016.1185606.

Mitchell, Katharyne, and Matthew Sparke. 2016. The New Washington Consensus: Millennial Philanthropy and the Making of Global Market Subjects. *Antipode* 48 (3): 724–749.

Morvaridi, B. 2012. Capitalist Philanthropy and Hegemonic Partnerships. *Third World Quarterly* 33 (7): 1191–1210.

Norris, John. 2012. Hired Gun Fight. *Foreign Policy*, July 18, 2012. http://foreignpolicy.com/2012/07/18/hired-gun-fight/.

Peck, Jamie. 2006. Response: Countering Neoliberalism. *Urban Geography* 27 (8): 729–733.

Peck, Jamie, and Adam Tickell. 2002. Neoliberalizing Space. *Antipode* 34 (3): 380–404.

Perelman, Michael. 2004. *Steal This Idea: Intellectual Property and the Corporate Confiscation of Creativity*. New York: Palgrave Macmillan.

Pinch, Trevor, and Wiebe Bijker. 1987. The Social Construction of Facts and Artifacts: Or How the Sociology of Science and the Sociology of Technology Might Benefit Each Other. In *The Social Construction of Technological Systems: New Directions in the Sociology and History of Technology*, edited by Wiebe Bijker, Thomas Hughes, and Trevor Pinch, 11–44. Cambridge, MA: MIT Press.

Read, Róisín, Bertrand Taithe, and Roger Mac Ginty. 2016. Data Hubris? Humanitarian Information Systems and the Mirage of Technology. *Third World Quarterly* 37 (8): 1314–1331. doi:10.1080/01436597.2015.1136208.

Roberts, Susan. 2014. Development Capital: USAID and the Rise of Development Contractors. *Annals of the Association of American Geographers* 104 (5): 1030–1051.

Robinson, John, Jim Maddock, and Kate Starbird. 2015. Examining the Role of Human and Technical Infrastructure during Emergency Response. In *Proceedings of the ISCRAM 2015 Conference*, edited by Leysia Palen, Monika Büscher, T. Comes, and A. Hughes, 311–320. Cham: Springer.

Roche, Stéphane, E. Propeck-Zimmermann, and B. Mericskay. 2011. GeoWeb and Crisis Management: Issues and Perspectives of Volunteered Geographic Information. *GeoJournal* 78:1–20.

Roelofs, J. 1995. The Third Sector as a Protective Layer for Capitalism. *Monthly Review* 47 (4): 16–25.

Roy, Ananya. 2010. *Poverty Capital*. London: Taylor and Francis.

Sandvik, Kristin, M. Jumbert, J. Karlsrud, and M. Kaufmann. 2014. Humanitarian Technology: A Critical Research Agenda. *International Review of the Red Cross* 86 (893): 219–242.

Sandvik, Kristin, and Kjersti Lohne. 2014. The Rise of the Humanitarian Drone: Giving Content to an Emerging Concept. *Millennium: Journal of International Studies* 43 (1): 145–164.

Scarry, Elaine. 2011. *Thinking in an Emergency*. New York: Norton.

Thatcher, Jim. 2013. Avoiding the Ghetto through Hope and Fear: An Analysis of Immanent Technology Using Ideal Types. *GeoJournal* 78 (6): 967–980.

Thatcher, Jim, David O'Sullivan, and Dillon Mahmoudi. 2016. Data Colonialism through Accumulation by Dispossession: New Metaphors for Daily Data. *Environment and Planning D: Society & Space* 34 (6): 990–1006. doi:10.1177/0263775816633195.

Weizman, Eyal. 2012. *The Least of All Possible Evils: Humanitarian Violence from Arendt to Gaza*. New York: Verso Books.

Weizman, Eyal, and Zachary Manfredi. 2013. "From Figure to Ground": A Conversation with Eyal Weizman on the Politics of the Humanitarian Present. Qui Parle. *Critical Humanities and Social Sciences* 22 (1): 167–192.

Woodrow Wilson Center. 2012a. Legal and Policy Issues. Video of panel uploaded by the Woodrow Wilson Center, October 24, 2012. https://www.youtube.com/watch?v=apEMNJFnBEM&feature=youtu.be.

Woodrow Wilson Center. 2012b. Research Challenges. Video of panel uploaded by the Woodrow Wilson Center, November 2, 2012. https://www.youtube.com/watch?v=uTIT3mkQhew&feature=youtu.be.

Žižek, Slavoj. 2009. *First as Tragedy, Then as Farce*. New York: Verso.

Žižek, Slavoj. 2010. First as Tragedy, Then as Farce. Video of talk uploaded by The RSA, March 10, 2010. http://www.youtube.com/watch?v=cvakA-DF6Hc&feature=youtube_gdata_player.

6 The Digitalization of Anti-poverty Programs: Aadhaar and the Reform of Social Protection in India

Silvia Masiero

Introduction

The topic of digital governance is central to the study of information and communication technologies (ICTs) for development. Over the last two decades, the potential for ICTs in governance has been widely recognized. Heeks (2001), for instance, noted that e-governance can be viewed as the "ICT-enabled route to good governance," and new technologies have been seen as practical tools to tackle institutional frailty on a world scale. In the early days of digital governance, a "tool-and-effect" logic led the debate, seeing technology as the heart of the solution to problems of effectiveness and accountability (World Bank 1999; UNDP 2001). This is especially important in developing nations, where state failure may yield severe consequences for the lives of the poor and vulnerable (Corbridge et al. 2005).

However, a tool-and-effect logic has proved unable to satisfactorily describe the effects of digitalization on the governance of development projects.[1] Over the last ten to fifteen years, the literature on failure in ICT for development (ICT4D) projects has been growing, posing theoretical problems with which research in the field has engaged directly. In particular, scholars have questioned the tool-and-effect logics of the early days. Contemporary work tends to present a different set of ICT4D hypotheses, in which technology is only as good as the policy decisions it embodies. The idea of technology as a "carrier of policy" is at the heart of this line of reasoning, stating that ICTs embody the decisions of governors, who inform the design of artifacts to advance their own agendas (Cordella and Iannacci 2010).

So devised, the vision of technology as a "carrier of policy" has loomed large in ICT4D. But to what extent does this paradigm capture the theoretical link between ICTs and governance? Compared to a "deterministic tool" view, it certainly seems to provide a more faithful representation of reality. But two key issues indicate the need for further research on the topic. First, the logic of technology as a "carrier of policy" describes how artifacts are informed by ideas of the people who govern them but says little about the reverse process, that is, how technology may in turn reshape the policies it is supposed to advance. Second, the effects of this mutual shaping on socio-economic development are unclear. As such, if we are to take Walsham's (2012) invitation to engage in "making a better world" with ICTs, the feedback of ICT-based policy reform on development should be openly taken into account.

This chapter advances the view that technology not only *carries* anti-poverty policy but actively *reshapes* it in ways that affect development trajectories. To explore this hypothesis, I present a study of the ongoing transition to digitality in the Public Distribution System (PDS), the main food security program in India, under the aegis of the Unique ID project (Aadhaar), a central government scheme enabling the biometric identification of all citizens enrolled. The Aadhaar project aims to solve identification issues, by endowing each enrolled resident with a twelve-digit number and registered biometric details (fingerprints and an iris scan). A recent economic survey (Government of India 2015) openly recommends incorporating Aadhaar in the reform of the main social safety schemes in the country. In this chapter, I explore the rationale and dynamics of reform, based on empirics from a state (Kerala) that conducted a pilot project of biometric transformation of the PDS.

Studying Aadhaar's role in social protection, I discover a set of mechanisms through which biometric technology affects its underlying policies, in turn reshaping the development trajectories in which it participates. At the micro level, the system leads to monitoring a specific part of the PDS supply chain, specifically *ration dealers* (last-mile retailers), on whom much of the corruption in the program is blamed. At the macro level, Indian states can appropriate the same technology, aimed at reshaping the PDS with a cash transfer program, to protect the program in its existing form and to implement reforms to minimize leakage and corruption. Aadhaar's infrastructure hence envisages a new policy direction for the poor and marginalized, oriented to direct cash transfers as opposed to food subsidies, which states can partially renegotiate according to local visions and priorities.

In this chapter, I first explore the logic of digital governance as a "carrier of policy," and its implications for the study of ICT adoption and reform of social protection. I then articulate the idea that digitality may instead "reshape" policy in a deeper way, illustrated through the example of the PDS reforms advanced by Aadhaar in Kerala and in India at large. The chapter then elucidates the micro- and macro-level effects of the adoption of biometrics on anti-poverty policy, detailing its influence on the development trajectories of its adopters. On this basis, I explore the effects of digitality on users' entitlements, arguing that these are imbued with ambiguity and may in fact lead to further exclusion rather than empowerment.

Theoretical Perspective: Technology as a Shaper of Policy

The tool-and-effect logic that previously dominated the discourse on ICTs and poverty alleviation entailed little problematizing of context and conditions. E-governance found its rationale in the improvement of effectiveness and accountability of social safety nets: this deterministic logic leads, for instance, to the identification of best practice for ICTs to improve social security mechanisms. This was a "tool view" of technology for poverty reduction, depicting technology in terms of the objectives that its material features were meant to pursue (Orlikowski and Iacono 2001).

Yet, this logic soon proved suboptimal to account for the use of technology in anti-poverty programs. A "tool view" ended up neglecting relevant mechanisms instead of unpacking them, preventing the observer from making sense of the conditions under which ICT-based intervention could actually work. As noted by Richard Heeks, most projects in this area result in failure, and taking stock of this is important for designing any constructive intervention. Heeks's explanation of failure points to gaps between reality and the perceptions held by designers, which prevent them from providing interventions tailored to the actual needs and requirements of beneficiaries (Heeks 2003).

As theorized by Wanda Orlikowski and Suzanne C. Iacono (2001), moving beyond a tool-and-effect logic leads to a vision in which technology is *embedded* in its context of action and emerges from it while influencing its dynamics. The notion of social embeddedness, widely affirmed in the discipline of information studies, is itself predicated on an ensemble view: technology is not simply a dependent variable but generates feedback mechanisms that shape the dynamics around it. Over the last decade, social embeddedness has gained substantial ground in ICT4D, countering the notion of "technology transfer" in this domain. According to a socially embedded vision, technology is not necessarily *transferred* to the developing world but conceived according to locally relevant needs (Avgerou 2008).

In the domain of anti-poverty policy, a theory of technology as socially embedded needs to take into consideration social policy principles (which are seldom considered openly in ICT4D). Two particular social policy principles matter highly to the construction of a theoretical perspective on this topic.

First, social safety schemes are informed by an intrinsic *rationale*, that is, the core objective for which they were conceived. For example, a food security program finds its rationale in guaranteeing a people's right to adequate nutrition, and a workfare scheme has the purpose of ensuring employment in exchange for a wage. The rationale of a program informs how the scheme is designed and what policy mechanisms it involves. As a result, the general consensus among designers of pro-poor technologies is that IT artifacts should be designed in continuity with their rationale.

Second, social safety nets are the expression of underlying *political* programs, put forward by their policymakers (Cordella and Iannacci 2010). Social policy design is a political exercise: different views will lead, for example, to more or less narrow targeting of schemes or to different propensities to rights-based approaches. Furthermore, social safety nets are often at the core of electoral promises: their making is an integral part of electoral competition and mobilizes existing party interests around them (Mooij 1999). The discourse that views technology as a "carrier of policy" hence depicts IT artifacts as the material embodiments of the policy agendas they put forward.

The logic of technology as a "carrier of policy" has become deeply entrenched in ICT4D. But the field has evolved rapidly over the last decade, leading to questions of the extent to which this logic accounts for intertwining technology and policymaking. On the one hand, it is reasonable to assume that technology embodies the objectives of the policymakers behind it; yet, on the other hand, this logic can conceal the feedback effect that ICT systems may yield on the policies themselves. The information systems literature reminds us that technology invites human action along specific guidelines. Jannis Kallinikos (2011), for instance, notes that contemporary politics are "governed" through technology, and overlooking the feedback effects of ICTs would mean bracketing an important part of the picture.

In a world in which technology is ever more entrenched in economic development and poverty reduction, the consequences of technology uptake on the "substantial unfreedoms" that affect poorer people's lives are paramount (Sen 2001, xii). It is hence important to consider the idea that technology, beyond carrying policy, may act as a *shaper* that directs it toward specific routes and objectives. On this basis, the hypothesis examined in this chapter is that ICT adoption may influence the course of social protection reform in a developing nation. To study this, I needed to look at the digital infrastructures that are entrenched in the making of social welfare schemes. This led specifically to a focus on the adoption of biometrics in the Public Distribution System (PDS), the largest food security net in India.

Aadhaar and the Indian Food Security System

I began collecting data on computerization of the state-level PDS in Kerala in 2011, when the digitalization of the program was in its early stages. Since then, I have conducted multiple rounds of fieldwork to monitor the system's evolution from back-end automatization to front-end and subsequently biometric recognition of users. This chapter draws on interviews and observations in the ration shops, telecenters, and administrative offices (known as Taluk Supply Offices) adopting the digital PDS. It uses a narrative analysis of the contents of interviews as the main tool to reconstruct causal processes (Riessman 2008). Following the case study method, I further triangulated primary data with statistics, press releases, and government documents on the PDS and its digital transformation.

The purpose of the PDS is to provide primary necessity goods (mainly rice, wheat, sugar, and kerosene) at subsidized prices to households below the poverty line, thereby improving their nutritional levels and welfare. Rather than relying on imports, the PDS uses internal redistribution of commodities: food grains are reallocated from surplus states to food-deficit ones through a centralized redistribution system. The reallocation mechanism is governed by the Food Corporation of India (FCI), a government agency that buys goods from private producers at the minimum support price and redistributes them to states based on the theoretical requirement.

Figure 6.1 illustrates the PDS supply chain. First, goods procured from the FCI and private producers are distributed at the district level through authorized wholesale dealers. These are then lifted from the wholesale points by the ration dealers who own the fair-price shops, known as *ration shops* because goods are rationed monthly. Finally, beneficiaries buy PDS commodities from the ration dealers at subsidized prices, which make the goods affordable to the poor and vulnerable.

Launched in 1965, the PDS was initially predicated on equal entitlements for all citizens, based on the principle of a universal right to food (Mooij 1999). During the severe balance of payment crisis that affected India in the 1990s, however, international funding institutions strongly criticized the scheme's leakage to the nonpoor, as well as the high subsidy cost, which was estimated to be around 0.5 percent of the country's GDP in 1990–1991 (Ahluwalia 1993). This led to the shift to a targeted system in 1997, in which entitlement is based on poverty status and aimed specifically at the households classified as below the poverty line. While actually reducing the subsidy cost (Umali-Deininger and Deininger 2001), the policy shift also had several unintended effects, of which the Kerala case is paradigmatic.

Kerala originally operated what was widely recognized as the most effective PDS in India, serving 97 percent of the state's population (George 1979) and thus having a significant impact on beneficiaries' nutritional status. Rice and wheat produced in Kerala account for only 15 percent of the state's total consumption of food grains, which makes a well-functioning

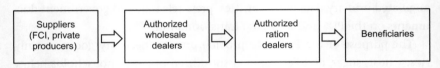

Figure 6.1
India's Public Distribution System (PDS) supply chain. *Source*: Author.

The Digitalization of Anti-poverty Programs

Figure 6.2
Ration shop, Trivandrum Central, Kerala. *Source*: Author.

PDS crucial to the state's nutritional security. The Keralite system, and the operational efficiency that characterized it, were based centrally on the universality of PDS and on its capacity to serve almost the whole population through ration shops (figure 6.2).

But the move to a targeted system caused deep changes in the scheme, reducing the supply of PDS goods to Kerala to less than 10 percent of the pretargeting amount (Swaminathan 2002). Users classified as above the poverty line were left with only very limited subsidies, hence their leaving the system en masse between 1997 and 2001.[2] Many ration shops in slum areas (figure 6.3), finding themselves with considerably fewer customers, became unviable and closed down, which also caused a wave of suicides among ration dealers (Suchitra 2004). In addition, since Kerala's poverty incidence was estimated at only 25 percent, many citizens found themselves excluded from the food security net on which they had long relied, and this put numerous households in a dire situation of exposure to food insecurity (Swaminathan 2008).

Figure 6.3
Chenkalchoola, slum area, Trivandrum, Kerala. *Source*: Author.

What targeting has left in Kerala is a collapsed system, ridden with issues that were not present at the time of universal coverage. The main one is leakage, meaning the loss of PDS goods along the supply chain due to recurrent illegal sales of these goods on the private market (Khera 2011). This phenomenon, known as the *rice mafia* (as rice is the staple commodity in the PDS), endangers the program's capacity to serve the poor, especially as illegal sales have persisted despite many reform efforts over the years. Diversion of goods from the PDS to the market is spurred by the price difference of commodities, which are more profitably sold on the nonsubsidized market system than on the heavily subsidized PDS.

In Kerala, the digitalization of PDS, delegated to the state-level section of India's National Informatics Centre based in Trivandrum (the state's capital), has been at the core of the program's reconstruction. Computerization was conceived specifically to monitor the PDS supply chain, thus allowing any diversion of goods to be detected. Government officials, and the National Informatics Centre staff that I followed during implementation, were consistently clear about the antileakage nature of the effort, and the IT-based PDS was indeed built with the explicitly declared objective of combating the rice mafia (Masiero 2015). To achieve this purpose, the program was reconstructed in two phases, a first one in which infrastructure was built and a second one, still ongoing, in which it is being augmented with biometric recognition of users.

In the first phase, the National Informatics Centre developed a suite of software articulated in three modules—a front-end one involving citizens, and two back-end ones used by staff at administrative offices. The front-end interface consists of a Ration Card Management System, a workflow-based application for ration card requests. Once received through the registration counter, applications are verified by rationing inspectors, who then submit them to the local Taluk Supply Office for approval. The two administrative functions cover the district-level allocation of goods and the legal inspection of ration shops, with details recorded in a database to flag suspicious activity by ration dealers. By digitizing these key functions of the PDS, the state aimed to ensure their safe and constant monitoring.

The second phase involves integrating the PDS with a form of biometric control, to make sure that access to subsidized commodities is reserved for genuinely entitled users. A pilot project run in six ration shops in Trivandrum enabled biometric recognition through Aadhaar, the unique identification system rolled out by the Unique Identity Authority of India since 2009. While the legal system is currently being adapted to accommodate the adoption of biometrics in nationwide welfare schemes, pilot projects based on Aadhaar are already being run across the nation, with mixed results in terms of access (Bhatti, Drèze, and Khera 2016). In Kerala, a set of ration shops were endowed with point-of-sale machines, performing Aadhaar identification on users and allowing or denying transactions on that basis.

The idea of linking Aadhaar to welfare schemes is rooted in its capability to combat diversion (Government of India 2015). A frequent practice of ration dealers is that of attaching bogus ration cards to their shops to mask illegal sales on the market. Biometric recognition, by linking each user to their own biometric details, ensures that sales are made only to real beneficiaries, as all transactions need to be matched by a valid number. Point-of-sale machines, connected to the central Aadhaar database, confirm the user's identity and entitlement to access the PDS.

More recently, the National Democratic Alliance (NDA) central government voiced the intention of replacing the PDS with a cash transfer program—removing the entire system of subsidies and simply paying people to buy food in an open market—and has already envisaged operational measures to make the transition possible. To do so, Aadhaar is being combined with a financial inclusion program (Jan Dhan Yojana) and mobile technologies. Jan Dhan Yojana, literally meaning "people's money plan," aims to provide each household with a bank account, to which direct benefits can be transferred.[3] The combination of Jan Dhan Yojana, Aadhaar, and mobile payments, known as the "JAM trinity," is being planned with the explicit purpose of rebuilding the nation's anti-poverty system, converting existing subsidies into a direct transfer to those below the poverty line. This new intentionality is taking shape in the present debate and needs to be considered in analyzing the link between digitality and policymaking.

Analysis: Two Perspectives on Aadhaar

As mentioned, digitality in the PDS can be observed through two complementary paradigms, seeing technology as a *carrier* or as a *shaper* of social policy. The former paradigm details the rationale and policy agenda embodied in Aadhaar's inscription in the PDS. The latter explains the feedback effects of Aadhaar's adoption on food security policy, demonstrating its actual and potential effects on the development trajectory in which it participates.

Technology as a Carrier of Policy

The rationale for using Aadhaar in the PDS is stated clearly by policymakers and lies in a problem-solution nexus between the root cause of PDS leakage and the biometric technology devised to fight it. Government staff have been explicit about the presence of leakage, which is uniformly seen as due to the ubiquity of the rice mafia within the state. While losses in transportation and storage are acknowledged, they are almost never cited as a reason for leakage: this is blamed on the black market networks on which rice is sold for much higher prices.

Although PDS supply chains can be long, attribution of blame falls almost entirely on the ration dealers, who are frequently reported to turn down beneficiaries with the excuse of having "run out" of food grains or even keeping their village shops closed for many days. A senior official in Trivandrum clarified the problem: "In Kerala there are a lot of bogus cards, which copy existing ones or make up households that do not exist. ... It is the ration dealers that fabricate bogus cards, not the customers, so they can pretend having sold goods to the BPL, when instead they have sold them elsewhere" (rationing inspector, Trivandrum).

If the problem is depicted as such, the solution is embodied in the integration of biometric recognition in the PDS and is grounded in a link between technology and the nature of diversion. This link is in turn articulated into three mechanisms, all embedded in the functioning of the Aadhaar system.

The first mechanism ensures that users are securely identified because the main means for ration dealers to divert food grains is based on bogus cards, registered in the shops to mask illegal sales to the black market. By linking each card to the owner's data, biometrics will make such identity fraud virtually impossible, since point-of-sale machines require recognition of customers' fingerprints to allow transactions. As reported by a senior member of staff at the Kerala State IT Mission (KSITM), "Ration cards have barcodes. … Often they are copied, and ration dealers claim false sales as a result. With Aadhaar, there is no risk of this, because control will be biometric."

In parallel, the second mechanism ensures that *all* transactions are led through point-of-sale machines, so that commodities are not diverted to nonentitled users. The monthly bill, summing up all transactions conducted through Aadhaar, is the basis for the subsequent month's allocation and needs to be presented by ration dealers to receive their monthly stock of commodities from the wholesale point. This leaves no room for "inventing" sales, as noted by another senior official at KSITM: "The system will reveal what goods are sold, and to whom. Ration dealers … claim that stocks have finished and sell them on the market. But now, the system will be able to track exactly who buys what."

A third mechanism aims to reduce the incentive of ration dealers to cheat their customers by pretending to have run out of goods or selling them at higher prices. At present, all households are registered with a ration shop and cannot decide to opt out of it in case of suspected deception by the owner. But Aadhaar-based biometric authentication can be performed anywhere, so that citizens can access the system from any ration shop. As a National Informatics Centre officer said, "Ration dealers will be unable to count on their usual customers, because people will be able to buy [PDS] food from everywhere. … They [ration dealers] won't be able to compete if they continue their cheating."

The two elements characterizing the logic of technology as a carrier of policy are evident in these narratives. The rationale is to combat leakage in the program by targeting the last mile, represented by transactions in the ration shops. The policy agenda behind this rationale embodies specific assumptions and priorities: it depicts the ration dealers as the guilty party and informs a system that monitors them specifically, as opposed to monitoring the supply chain as a whole.

Technology as a Shaper of Policy

We have seen how policy decisions are embodied in ICT infrastructures, but a holistic perspective on the implications of this for marginalized communities should include a focus on the reverse mechanism, namely the feedback effect of digitalization on the food security policies in which it participates. I observe this at the micro (intrastate) and macro (national) level of operation.

At the micro level, I have described the action of Aadhaar-enabled monitoring of ration dealers. But at least three issues arise in the logic underlying its implementation. First, the technology is predicated on the ration dealers' guilt in terms of diversion; however, recipients often claim that agents *before* ration shops, along the PDS supply chain, are also guilty. In particular, citizens are concerned with a border mafia that subtracts commodities before they even reach the shops, as reported by a PDS user: "A lot of goods are stolen at the border. ... The goods that get to the shop, and are not sold somewhere else, are just a small share. It is easy to say, the ration dealers are causing the system not to work, but we should talk more about goods being stolen during transportation" (Ayesha, PDS user).[4]

These concerns are augmented by the discussion often found in the media of border mafia and diversion happening before the ration shops. Early stage diversion is reported by many ration dealers, who claim to be unable to procure the amount of food grains they need to serve all users: "When I go to the godown [for the monthly collection of food grains], I never get enough rice for all households registered in the shop. So in the first ten days of the month, a lot of people come to the shop. ... After that if there are more people, I may have run out" (Chaitram, ration dealer).

While it is hard to verify these assertions, recent field observation revealed that the back-end modules of the e-PDS are still in an early stage of development, and in several administrative offices, they are not at all used or known by the staff. Investing in back-end modules would probably be comparatively less popular with citizens, as these modules do not transform the ration shop directly—hence they do not affect people's *direct* access to the system. Still, the current strategy trades visibility for effectiveness, as it focuses on the last mile alone rather than on holistic monitoring of the PDS.

The second issue with Aadhaar implementation is that it doesn't take into account the *root cause* of corruption. Ration dealers' narratives establish a close relation between market diversion and the reasons for it. The main reason they identify is the shift to a targeted PDS: this has put the shops at serious risk of unviability, taking away a large share of their customers and the financial sustainability that was previously assured. As noted, this happened after the system classified most Keralan households as above the poverty line because of relatively low estimates of poverty incidence. Given the low subsidies, these households have massively moved to the private market, leaving ration dealers with a limited customer base, as a ration dealer explained: "We get a commission on the goods we sell, but that is not enough to make a living. If we were allowed to sell other goods, other than just PDS, that would help a lot, but the government has not yet authorized us to do so. This is why many shops close down, and people do not trust us anymore" (Pratap, ration dealer).

The need to resort to market diversion to survive strongly emerges from the narratives of PDS actors, including the Taluk supply officers. The loss of customers that followed the move to a targeted system still conditions ration dealers' activity, and narratives of PDS beneficiaries cluster strongly around this point: leading a ration shop business, and preserving its viability, means indulging in a certain amount of exchange with the private market. This is the only way for many ration dealers to survive the threat of unviability that led to the closure of many ration shops.

Aadhaar controls ration dealers' behavior and detects illegal transactions, or at least it is designed to do so. Yet, its limitations are the core of the matter: as the system is constructed, it does not act on the root cause of the problem, namely the perverse consequences of targeting (Masiero and Prakash 2015). With Aadhaar, the invitation for shop owners to comply with the law will be technologically enforced, but the system does not offer them an alternative such as a credit concession or other means to make a living without resorting to market diversion. In this way, technology comes across as both partial and mistargeted: partial because it targets only the ration dealers and not the other actors in the supply chain, and mistargeted because it addresses the effect of the problem and not its root cause.

Third, Aadhaar's infrastructure enforces another policy decision, that of defending and sustaining a narrowly targeted PDS. Kerala has indeed opted for a targeted system with narrow criteria for below-poverty-line status and reservation of food subsidies. Advocates of the opposite policy (a universal PDS) remark that the right to food is universal, and that abolishing the duality of prices would also reduce incentives to market diversion (Sen and Himanshu 2011). Nonetheless, two subsequent state governments preferred a targeted PDS, in which the entitlements of the poor would be preserved.

Kerala's choice of narrow targeting emerges in the design of the biometric PDS. In targeted systems, determination of beneficiaries can incur two types of errors: inclusion errors, incorporating nonentitled citizens, and exclusion errors, rejecting genuinely entitled ones. The biometric PDS is designed to prevent the inclusion error, but no mechanism has been put into place for the opposite problem, even though exclusion errors do occur, as explained by a community volunteer: "In Karimadom [Trivandrum's main slum], still a lot of families have pink cards [above poverty line]. People in abject poverty still fail to be recognized as poor. ... Aadhaar will make it worse, because it will add one more layer of exclusion" (Vijaya, community volunteer).

Technology designed for the Kerala PDS focuses on last-mile monitoring and targets the system narrowly to the poor, leading to the feedback effect of technology on policymaking: by putting responsibilities on the shoulders of specific actors, and enforcing a system that tackles the inclusion error as its top priority, Aadhaar's infrastructure directly participates in the development trajectory of the state. It leads to a more selective PDS and shapes monitoring mechanisms in a way that implicitly blames the ration dealers for corruption. At the micro level of everyday access, technology already works as an active shaper rather than a passive carrier of policy decisions.

At the macro level, the policy choices of the central government should be considered. The last economic survey presents an open argument for dismantling the PDS (Government of India 2015), claiming that moving to cash transfers would eliminate distortions and minimize opportunities for leakage. But beneficiaries, ration dealers, and civic activists interviewed across Kerala present a more complex picture, influenced by the effect that a transition to cash transfers would have on their entitlements: "Aadhaar is not at all inclusive; in fact it excludes the poor ... because it excludes all those that are not registered, and many poorer citizens are among these" (Julian, PDS user); "Aadhaar will make cash transfers mandatory for everyone. This is very difficult for the poor, who may have never used a bank at all" (Swetha, right-to-food activist); "The real problem with Aadhaar ... will be in the long term. Since people can use every ration shop, ration dealers will not know how many customers they have, and so they will not be able to require the right amount [of food grains]. The only way is giving licenses to grocery shops. ... The PDS will disappear, and people will have to use the market" (Rajesh, right-to-food activist).

Exclusion of the poor, lack of protection from inflation, and vulnerable groups' unfamiliarity with banking systems are the main practical concerns surrounding a potential shift to cash transfers. These need to be added to political understandings of the shift, based on the unwanted involvement of the market in a system that was isolated from it since the beginning. Crucially, political readings of the move to cash transfers have been found recurrently among respondents and are by no means confined to activists and volunteers. As long-term work in the field has revealed, users with a range of political affiliations are concerned about the effects that the demise of the PDS would have on their entitlements.

In September 2013, an order by the Supreme Court of India forbade states from making enrollment in social programs conditional on Aadhaar registration. In March 2016, however, the central government passed the Targeted Delivery of Financial and Other Subsidies, Benefits and Services bill, most commonly known as the Aadhaar bill, providing legal backing to the use of Aadhaar in social welfare schemes. Crucially, the order was passed as a money bill, which allowed it to be enforced without needing approval from the upper house of Parliament, a move that generated vivid criticism from the public.[5] Legitimizing Aadhaar-enabled schemes, the bill paves the way for the transition to cash transfers, which is supposed to mark the beginning of an effective anti-poverty system, hence "wiping every tear from every eye" (Government of India 2015, 52).

Such a rosy picture, however, is problematized by ground reports from states that have already adopted Aadhaar-based recognition of PDS users.

In Rajasthan, where Aadhaar has been extended to all ration shops, civic unrest recently culminated in a group of construction workers smashing a biometric machine, having lost their food entitlements after Aadhaar identification was made mandatory for PDS (Yadav 2016). Reports of malfunctioning machines pervade the states that have adopted the system (Bhatti, Drèze, and Khera 2016), turning a technical problem into a political one: the very same machines that should empower recipients end up disempowering them, denying access to the entitlements that the technology should guarantee. The short-term issues implied by the advent of digitality coexist, in the long run, with fixations on transition to a market-led system of direct benefits.

Surveys of citizens, conducted across India, generally reveal strong preference for the current PDS over a hypothetical move to cash transfers (Aggarwal 2011; Khera 2014; Puri 2012). What is striking in Kerala is that with the adoption of biometric technologies, the state government is building infrastructures to reduce leakage, hence attempting to improve the program without disrupting it. This entails a strong position taken by the state government: technology can be used to *protect* the system, rather than to dismantle it in favor of cash transfers, as the central government would recommend.

Conclusion

Aadhaar's inclusion in the Indian PDS enables exploration of the diverse feedback effects that digital infrastructures yield on anti-poverty policies. At the micro level, technology shapes the functioning of the PDS: closely monitoring ration dealers and enforcing a targeted system. At the macro level, ICT infrastructures sustain the decisions of policymakers on the program, tailoring it toward cash transfers on a national scale. Aadhaar's infrastructure hence envisages a new direction for social policy, which can be partially renegotiated at the state level, as the government of Kerala is doing to reinforce existing anti-poverty mechanisms.

Drawing on the data collected, the broader argument I make here is twofold. First, the role of technology goes beyond that of simply "carrying" policy as digitalization reshapes the anti-poverty system in ways that affect existing development trajectories. This is important when considering digital economies at the global margins as it reveals that ICT infrastructures may have long-term effects on the governance of social safety nets and their beneficiaries.

Second, digitality has ambiguous consequences on users' entitlements under anti-poverty schemes. This problematizes a techno-rational logic, which uniformly advocates the benefits of digitalization. As observed here, transition to digitality may result in exclusionary outcomes, which further marginalize the poor and vulnerable instead of leading to empowerment, such as how the Rajasthani construction workers were marginalized. Aadhaar is used to combat the erroneous inclusion of beneficiaries without tackling exclusion and to support a program of cash transfers conditional on biometric registration. The effects of digitalization should thus be monitored in ways that problematize the implications of intervention and minimize the risk of illicit exclusion.

Walsham's invitation to contribute to "making a better world" with ICTs implies a focus on how development theories inform practice on a global scale (Walsham 2012). If technology adoption yields effects on policymaking, it is crucial to manage the governance of development accordingly. For social safety schemes, this means achieving open integration between ICTs and policy design: the Indian government has devised the JAM trinity as an integral part of its social policy, rather than as a self-standing technological intervention. As integration is achieved, a focus on guaranteeing access to benefits is a prerequisite for the effectiveness of ICT-enabled social safety schemes.

At present, digital technologies are being adopted in anti-poverty programs around the world. These technologies are fundamental in reconstructing development policies and play a major role in adapting them to the needs of beneficiaries within social welfare schemes. In doing so, however, digitalization runs the risk of resulting in the perverse effects of exclusion and disempowerment. It is important that designers be mindful of these effects, and protect the entitlements on which the livelihoods of marginalized communities are predicated.

Notes

1. The notion of digitalization is used here as "a sociotechnical process of applying digitizing techniques to broader social and institutional contexts that render digital technologies infrastructural" (Tilson, Lyytinen, and Sorensen 2010, 749).

2. Per Reetika Khera's estimates, statewide purchases of food grains from the PDS dropped from 4.64 tons in 1997 to 1.71 in 2001, largely because many above-poverty-line households abandoned the system (Khera 2011).

3. According to the World Bank's Global Findex database, currently 53 percent of Indians have a bank account, but the share drops to roughly 30 percent in rural areas (Tiwari 2016).

4. The term *border mafia* refers to diversion that occurs at the border between two states, when commodities are redistributed across the country. In Kerala, a border mafia is widely reported to divert goods as they pass the borders with Tamil Nadu and Karnataka.

5. As noted by Tiwari (2016), the ruling party of India (Bharatiya Janata Party, BJP) holds the majority of seats in the lower house of Parliament (Lok Sabha), but not in the upper house (Rajya Sabha), which was bypassed by framing the order as a money bill.

References

Aggarwal, Ankita. 2011. The PDS in Rural Orissa: Against the Grain. *Economic and Political Weekly* 46 (36): 21–23.

Ahluwalia, Deepak. 1993. Public Distribution of Food in India: Coverage, Targeting and Leakages. *Food Policy* 18 (1): 33–54.

Avgerou, Chrisanthi. 2008. Information Systems in Developing Countries: A Critical Research Review. *Journal of Information Technology* 2 (2): 133–146.

Bhatti, Bharat, Jean Drèze, and Reetika Khera. 2016. Experiments with Aadhaar. *Hindu*, June 20, 2016.

Corbridge, Stuart, Glyn Williams, Manoj Srivastava, and René Véron. 2005. *Seeing the State: Governance and Governmentality in India*. Cambridge: Cambridge University Press.

Cordella, Antonio, and Federico Iannacci. 2010. Information Systems in the Public Sector: The E-government Enactment Framework. *Journal of Strategic Information Systems* 19 (1): 52–66.

George, P. S. 1979. *Public Distribution of Foodgrains in Kerala: Income Distribution Implications and Effectiveness*. Washington, DC: IFPRI.

Government of India. 2015. *Economic Survey*. New Delhi: Department of Economic Affairs, Ministry of Finance.

Heeks, Richard. 2001. Understanding E-Governance for Development. Working Papers Series 11. Institute for Development Policy and Management, University of Manchester.

Heeks, Richard. 2003. *Most E-government for Development Projects Fail: How Can Risks Be Reduced?* Manchester: Institute for Development Policy and Management, University of Manchester.

Kallinikos, Jannis. 2011. *Governing through Technology: Information Artefacts and Social Practice*. London: Palgrave MacMillan.

Khera, Reetika. 2011. Trends in Diversion of PDS Grain. *Economic and Political Weekly* 46 (21): 106–114.

Khera, Reetika. 2014. Cash vs. In-Kind Transfers: Indian Data Meets Theory. *Food Policy* 46:116–128.

Masiero, Silvia. 2015. Redesigning the Indian Food Security System through E-Governance: The Case of Kerala. *World Development* 67:126–137.

Masiero, Silvia, and Amit Prakash. 2015. The Politics of Anti-Poverty Artefacts: Lessons from the Computerization of the Food Security System in Karnataka. In *Proceedings of the Seventh International Conference on Information and Communication Technologies and Development*. New York: ACM.

Mooij, Jos. 1999. Food Policy in India: The Importance of Electoral Politics in Policy Implementation. *Journal of International Development* 1 (4): 625–636.

Orlikowski, Wanda, and Suzanne C. Iacono. 2001. Research Commentary: Desperately Seeking the "IT" in IT Research—a Call to Theorizing the IT Artifact. *Information Systems Research* 12 (2): 121–134.

Puri, Raghav. 2012. Reforming the Public Distribution System: Lessons from Chhattisgarh. *Economic and Political Weekly* 4 (5): 21–23.

Riessman, Catherine K. 2008. *Narrative Methods for the Human Sciences*. Thousand Oaks, CA: Sage.

Sen, Amartya. 2001. *Development as Freedom*. London: Sage.

Sen, A., and A. S. Himanshu. 2011. Why Not a Universal Food Security Legislation? *Economic and Political Weekly* 46 (12): 38–47.

Suchitra, M. 2004. Undermining a Fine Public Distribution System in Kerala. *India Together*, January 1, 2004. http://www.indiatogether.org/2004/jan/pov-keralapds.htm.

Swaminathan, Madhura. 2002. Excluding the Needy: The Public Provisioning of Food in India. *Social Scientist* 30 (3): 34–58.

Swaminathan, Madhura. 2008. *Programmes to Protect the Hungry: Lessons from India*. New Delhi: United Nations Department of Economic and Social Analysis.

Tilson, David, Kalle Lyytinen, and Carten Sorensen. 2010. Digital Infrastructures: The Missing IS Research Agenda. *Information Systems Research* 21 (4): 748–759.

Tiwari, Pragya. 2016. "I Am Not Aware that This Has Been Thought Through": Jean Drèze Speaks to Pragya Tiwari about the New Aadhaar Act. *South Asia @ LSE*, London School of Economics. July 30, 2016. http://blogs.lse.ac.uk/southasia/2016/07/28/even-if-aadhaar-is-deemed-inevitable-there-are-many-ways-of-using-it-some-more-helpful-or-harmful-than-others-jean-dreze/.

Umali-Deininger, Dina, and Klaus Deininger. 2001. Towards Greater Food Security for India's Poor: Balancing Government Intervention and Private Competition. *Agricultural Economics* 25 (2–3): 321–335.

UNDP. 2001. *Making New Technologies Work for Human Development.* United Nations Human Development Report. New York: Oxford University Press.

Walsham, Geoff. 2012. Are We Making a Better World with ICTs? Reflections on a Future Agenda for the IS Field. *Journal of Information Technology* 27 (2): 87–93.

World Bank. 1999. *World Development Report 1998/99: Knowledge for Development.* Washington, DC: World Bank.

Yadav, Anumeha. 2016. In Rajasthan, There Is "Unrest at the Ration Shop" Because of Error-Ridden Aadhaar. *Scroll.in.* April 2, 2016. https://scroll.in.

7 The Myth of Market Price Information: Mobile Phones and the Application of Economic Knowledge in ICTD

Jenna Burrell and Elisa Oreglia

Introduction

The mobile phone as a platform for the dissemination of information, particularly market prices, has become shorthand for the transformative possibilities of *information* in general for low-income rural populations in the Global South. This new variant of economic development thinking after the decline of capital fundamentalism, defined by Peter Evans as the assumption that "the problem of underdevelopment was primarily about increasing poor countries' stock of capital" (Evans 2005, 91), places mobile phones and other network infrastructures in a critical role, hastening an end to a state of presumed "information scarcity" in remote regions. Such thinking is gaining influence in the domains of development policy and practice. Thus, we seek to gain specificity about the role of information in the emerging field of information and communication technologies and development (ICTD), which brings together academic researchers and practitioners.

Our critique focuses on "market price" as a particular type of information within agriculture and natural resources work. We consider the translation of "market prices" from neoclassical economic model to ICTD truism, and then to application in technological system building. "Information" in this process of translation is reified: it comes to be understood as a real and separable substance and is treated as existing in the world in the same way as the isolated variable in the economic model. It is imagined as unproblematically extractable, especially from the relationships between actors who exchange it. Yet, information is also understood to escape conventional material constraints. It may traverse digital networks at the speed of light and be reproduced without cost. The characterization of information as a kind of substanceless substance, which offers a practically cost-free way to enhance the incomes of the poor (i.e., by recapturing profits lost to market inefficiencies), explains some of its appeal to narratives of poverty alleviation.

We give particular attention in our analysis to market information systems (MIS), which are designed to collect and distribute "market price information" impersonally. Recent evaluations show MIS as having a disappointing lack of impact (Fafchamps and Minten 2012; Camacho and Conover 2011). Such negative evaluations raise questions about how economic knowledge is incorporated into technological system building and what understanding of the decision-making processes of farmers they assume and embed. We arrive at a plausible explanation for farmers' lack of interest in MIS through methods that ascertain as directly as possible the decision-making practices of rural agriculturalists. Such an approach draws attention to counter-narratives that are unavailable from within the conceptual and epistemological frameworks of the econometric studies and economic models that have been most influential to the thinking on market prices (and their scarcity) in ICTD and in the broader field of international development work.

In describing the emergence of this particular bit of economic knowledge as a "myth," we are noting its circulation within elite technocratic circles and the way it is fueled by repetition and an increasing tone of factuality, conviction, and presumed breadth of applicability. An ongoing conversation at the intersection of economic sociology and science and technology studies (STS) questions the relationship between economics and economies, considers economics as *performed*, and attends to the way its theories and ideas are composed materially as well as the role of economic actors, ranging from experts to "nonexperts" (such as consumers) in this performance (Barry and Slater 2005; MacKenzie, Muniesa, and Siu 2007). The success of modern (and particularly mathematical) thinking in economics in promoting itself as a resource for powerful real-world solutions is evident in how noneconomists from other domains of expertise find economic models compelling and seek to extract from them what may be actionable. What are we then to make of the plausible charge that failures in application (such as in MIS) derive not from the inadequacy of the model but from the "misapplication" or "misinterpretation" of economic

The Myth of Market Price Information 175

knowledge? Could the models themselves be held accountable for misinterpretation in relation to economists' claims of real-world relevance and applicability? We examine this through the notion that the myth of market price information is a boundary object, which is by necessity translated and recast across fields and approaches.

To illustrate a counter-narrative to the myth, we draw from our qualitative research on trade, livelihoods, and mobile phones among low- to medium-income rural fishermen and fish traders in Uganda and farmers in northern rural China. In Uganda, the first author completed two periods of fieldwork in four villages (including two fishing villages) exploring general questions of mobile phone use in livelihood activities. Sites included a small and quite remote fishing village on Lake Kyoga, visited in November 2007, and a larger landing site where fish are sold locally as well as packaged in trucks for export, visited in July 2008. In China, the second author carried out fieldwork in three corn- and wheat-growing villages in the provinces of Shandong and Hebei in 2010 and in the summer of 2011. In both cases, the data were gathered through semistructured interviews covering the use of ICT and people's livelihoods, accompanied by participant observation and casual conversations with residents and traders. With such methods, we emphasize the meaning and motives that fishermen and farmers attach to their actions. What economists investigate as "mechanisms," sociologists and anthropologists refer to as "processes" or "practices," which emphasizes a stronger sense of agency in the work done by human actors (Cetina, Schatzki, and von Savigny 2005). For example, in interviews, when our informants had the opportunity to describe their key decision-making points, they consistently disclaimed any practice of acquiring market price information for the purpose of comparison between markets (by phone or other means), with a few rare exceptions. In Uganda, however, fishermen and fish traders still described the mobile phone as critical to their trade activities. In China, by contrast, farmers found little use for the mobile phone in agricultural activities, even though mobile phones were widely available and actively used for other purposes. We suggest that the contrasting behavior of these market actors is logical in the context of the available resources and pressures related to their socioeconomic circumstances and social settings of village life, both of which shape livelihood strategies.

Birth of a Myth: "Market Price Information" as Boundary Object in the ICTD Community

The emerging field of ICTD brings into contact various types of experts and professionals (from academia, research institutes, NGOs, aid agencies, and the commercial sector) with different forms of institutional backing and warrant for their work. They share an interest in understanding how digital technologies may help to realize development outcomes, by whatever definition one might attach to "development." They are a community in a rather broad sense, not contained by any one institution, and whose key contributors do not necessarily all identify as members. How, then, are the ideas that become common reference points understood and applied by such diverse players? Problems arise, in part, from the multiple challenges of this field, where members trained in different disciplinary traditions meet and attempt to draw on one another's efforts. One challenge has to do with disciplinary values, that is, what members of different fields consider to be priorities in the pursuit of knowledge and practice (Burrell and Toyama 2009). A second is validity, what members of different fields consider to be compelling evidence or a convincing argument. The third relates to communication and the terminology, case studies, publications, and so forth that become a kind of shorthand within a discipline, but that become distorted while moving between groups. By examining the particularly widespread notion that farmers use mobile phones to seek market price information, we seek to specify concretely some of the challenges that stem from the involvement of NGOs and commercial entities in ICTD, specifically in building market information systems, which often draw from academic research for their justification.

We suggest that "market price information" has come to serve as a *boundary object*, "both plastic enough to adapt to local needs and the constraints of the several parties employing them, yet robust enough to maintain a common identity across sites" (Star and Griesemer 1989, 393). Boundary objects have a common representation between diverse groups (such as economists and computer scientists or academics and practitioners) but are "weakly structured in common use" (393). Subgroups of the broader community develop a deeper understanding of a boundary object but "use the boundary objects in very different ways" (Bødker and Bannon 1997, 85), and "problems ... may occur in the subsequent interpretation of information by others where the origins of the information, in terms of the person or system that constructed it, or aspects of the context within which the information was produced, may not be available to other actors in the space" (85). When farmers are described by economists as *acquiring market prices via mobile phone*, this claim is embedded within assumptions that are clear, if implicit, to the members of the same epistemological tradition, who also understand the limitations of their models. Yet, members of other traditions (e.g., computer scientists, engineers, designers) interpret this claim according to their own ways of acting in the world and translate it into other shorthand (e.g., a set of user requirements) that is meaningful within *their* community but that does not capture enough of what is necessary to know about these farmers to make their solutions work in the way the economic model seems to suggest they should.

A Counter-Narrative

An alternative reading of the promise of better market price information can be found by focusing on how fishermen and farmers commonly describe how prices affect their decision making about trade, the role that mobile phones play (or do not) in acquiring price information, and how regulatory and political frameworks influence all business decisions.

Through our empirical work, we find four aspects of the "myth of market price information" that require reconsideration: (1) that information critical to decision making is scarce and actively sought after by farmers, fishermen, and small traders in rural settings; (2) that in their key decision-making practices, market price is the most critical piece of information; (3) that improvements in market functioning following the arrival of mobile phones necessarily stem from acquiring market price information; and (4) that the provisioning of market prices is the primary application of mobile phones in the context of rural trade activities.

Myth 1: Information on Prices Is Scarce

The broader narrative of information scarcity was not met with a similar account by the inhabitants of the rural regions we studied, who were generally not preoccupied with the search for a better price. Scarcity of information on prices is highly dependent on location—not all rural areas in developing countries experience such drought. In China, for example, the going prices for crops are widely known: "I know the prices of crops and all those agricultural news from television. Also, there is a government official who comes to the village and tells us; he is from the agricultural office in the town" (Male farmer, Shandong Province).

In the Chinese villages where we carried out fieldwork, information on prices, agricultural techniques, fertilizers, diseases, and new crops comes from sources such as television, radio, newspapers (for those few who read them), traders, neighbors, agricultural extension workers, the head of the village, and so forth. People find out prices from multiple sources and constantly double-check them in the course of casual conversations. Most of this information gathering and sharing is based not on written text but on oral exchanges in person among people who know each other. Prices, at least at this level of small trade, are inextricably embedded within relationships among people. Another older farmer received a daily weather forecast SMS on his mobile. He was aware of the opportunities offered by the Internet, but even more aware that a lot of information he could find online already reached him through the agricultural extension worker: "There is an agricultural extension worker, actually there is one in the county and one in the town, so we get the one from the town; he comes here to tell us about fertilizer or pesticide and all that. So we don't need to find out this information, because he tells us" (Male farmer, Shandong Province).

The Chinese agricultural extension worker brings not only information, but also "meta-information" that helps farmers place what he says in context. For example, this farmer pointed out that he knew the agricultural extension worker personally; therefore, he could evaluate the information he received. The information is certainly not all good, or impartial, or useful, but by knowing who the agricultural extension worker is, how he works within the community, and what kind of relationship he has with, for example, seed sellers, farmers know how to parse his advice and understand it in context. All this context is lost when the same information is detached from the information provider (Oreglia, Liu, and Zhao 2011).

Fieldwork in Uganda pointed to similar matters of relational context and a concern with information source. Fishermen in these villages also spoke of the material resources they needed to be able to act on better information. A focus group with members of a savings group in a remote fishing village on Lake Kyoga showed a relatively low level of interest in information relative to other needs and priorities. When participants were asked directly about what types of information they desired, they continually turned the discussion back around to assets and facilities that would improve their lives. This started with a problem of translation: there was no word in Luganda that directly translates to "information," so the word for "news" was used as the best substitute. When asked, "If you want information [news] on fishing issues, the information you will be interested in, what will it be talking about?" one fisherman responded, "The information [news] I would be interested in is that the government has put in place a good way of fishing, like giving people new fishing nets."

The fisherman, in light of government regulations against the use of traditional fishing nets because of overfishing, wanted "information" that the government would be giving away the legal (and expensive) nets. He did not truly desire information—he knew the rules and how they affected him—instead he sought tangible assets.

Adding further nuance to the issue of information in remote rural communities and its perceived scarcity, the village chairman (also a fisherman) expressed a desire not simply for information, but for "advice." Speaking now of the use of mosquito nets (or lack thereof) in the village to prevent malaria, the chairman commented, "You may be having the money [to buy a mosquito net], but if no one has encouraged or advised you to use the mosquito net, you may not bother" (Village chairman, Buyende district). What he drew attention to was the question of information *source* and of the quality of the relationship between what he envisioned as a kind of mentoring figure and the village community.

The examples of the Ugandan chairman's desire for "advice" and the Chinese farmer's reliance on the agricultural extension worker contrast with the impersonal nature of "information" as it is conceived in scholarship that explores its role as a catalyst for socioeconomic development. The general enthusiasm surrounding information as a developmental salve entails promoting a concept in sociocultural settings where it simply may not be as salient.

Myth 2: Market Prices Are the Most Critical Piece of Information in Trade-Related Decision Making

An abstracted view of the role of information in the market removes prices from the trade practices and relationships among trade partners in which this information is ordinarily embedded. Yet such relationships appear to be critical at the level of smallholder farmers and fishermen. Price is often an important factor in decision making, but as one of several variables embedded in specific local conditions. Existing business relationships, trust, attitude toward risk, and institutional rules and policies around the goods traded—these are all inputs for fishermen's and farmers' final decisions on whether or not to sell, whom to sell to, what species to fish and what crops to grow, and so on. Among our research participants, two factors took precedence over price in making sales decisions: long-term relationships with trade partners and individual attitudes toward risk.

Long-Term Relationships with Trade Partners

Among our research participants, an ability to act on information was often tied to who the source of this information was and the trust built over a history of interactions. In "new institutional economics," the need to trade with known and trusted trade partners is treated as an adaptation within a certain institutional context, specifically one lacking structures for effectively enforcing contracts (Fafchamps 2004). Granovetter (1985) suggests more broadly that markets everywhere are embedded in social relations between specific individuals and emphasizes that the "concrete personal relations and the obligations inherent in them" (488) are the basis of the trust that prevents market actors from malfeasance.

In Uganda, the mobile phone proved critical as a tool for building and maintaining a social network of "concrete personal relations" in an industry of remote and distributed suppliers and buyers. On the problem of unreliable trade partners, one middleman trader in the fish export business noted, "Some other people can lie to you that they will give you cash immediately. You bring the fish, but then when you bring it, they disappoint you" (Fish trader, Mukono district).

The significance of relationships was all the more evident on Lake Victoria, where fishermen were given credit from middlemen who bought their fish and transported it to nearby factories for export. Given these credit dependencies, fishermen (specifically those who had progressed in trade enough to own assets such as a boat or nets) sold exclusively to the middleman to whom they were indebted, removing the possibility of comparing and making decisions about whom to sell to based on the best price. Moreover, for the lowest-level fishermen, working exclusively on salary, checking prices could be a threat to their employment or even their freedom, as they own no assets and have no say over the sale of the fish catch. One such fisherman, who also worked as a porter, commented on market prices: "I leave it to the boss because if [I] am caught [checking prices] he would throw [me] in jail. It would clearly indicate that I clearly want to operate behind his back" (Male fisherman, Mukono district).

He referred to fishermen who, once outside the surveillance of their employer, will attempt to sell some portion of a fish catch. A woman who worked as a smoked fish seller in the fishing village on Lake Victoria noted that the mobile phone was most critical for capturing supply. It was essential that she maintain her availability so that if her supplier called she could be there immediately to buy his fish before another seller did: "I have been his customer for a long time. I have been dealing with him for three years now. ... I buy from him at good price. I don't disturb him" (Female smoked fish seller, Mukono district).

In other words, she made transactions with her supplier as seamless as possible, neither haggling over price nor calling other suppliers for price comparisons, and she offered this as an explanation for why her supplier treated her with preference over other sellers.

Attitude toward Risk

Among agriculturalists, traders, and retailers at the low-income end of the spectrum, income predictability (an expression of their conservative attitude toward risk) often took precedence over a short-term focus on maximizing profits. This was the case with the smoked fish seller mentioned above, who, since separating from her husband seven years prior, was her family's sole breadwinner. She was the one on whom her children (and specifically their education) were totally reliant. She was explicit about the purpose to which her profits were put: "I am gaining some money, which I use for the children's school fees."

In both sites we have seen varying degrees of willingness to take on risk, and to diverge from the patterns of others to realize a gain, often related especially to family composition and stage of life. Among the Chinese participants, who were mostly middle aged or elderly, farming served as a combination of income generation and social security. Not having state pensions, older farmers grow crops that can be both sold and eaten; their main concern is predictability. In emergencies, it is easier to rely on remittances from migrant children or to find a casual job nearby, as an elderly farmer explains:

> There isn't a big pressure to get a better income from the land, because almost everybody has income from work outside. I'd say for most families, half of the yearly income is from the land, half from other work. ... Also, my goal is not to grow my income or business, as long as things remain okay, that's all I need. The Internet is useful for young people who want to improve and grow their business, not for old people like me. My children are all grown up and have good jobs, so I don't need much and don't have lots of worries. Until two years ago I also went out to work, but now I don't. There's no need. (Male farmer, Shandong Province)

Prior to any decision making about prices, the Chinese farmers and the Ugandan fishermen had to decide what crop to plant or what species to fish. These decisions were made in anticipation of price, but also often in terms of how stable or predictable the price was likely to be. For the Ugandan fishermen, fishing the variety called mukene (minnows), as opposed to the larger Nile perch or Tilapia, for export meant staying closer to shore and facing less exposure and danger (from storms or pirates) out on open water. For the Chinese farmers, planting the same crops as their neighbors was another way of mitigating risk, as other farmers in the village provide a network of support for the individual. They shared their knowledge of farming, directly by giving suggestions, or indirectly by starting to do a specific task, such as using fertilizer in their field and thereby communicating to the others that it was time to do that work. They shared risks in the sense that if something happened to a crop, it was usually a common problem and it could be easier to come up with a common solution. The network of support represented by neighbors growing the same crop disappeared when a farmer decided to grow something different and therefore did not have anyone to consult in case of trouble. For farmers who depend entirely on their crop for food and income, such a risk could be potentially ruinous.

If the selling and buying activities of these farmers and fishermen are seen as one discrete decision point, they might seem illogical. Nonetheless, the coherence of their reasoning is apparent in the broader context of life events and opportunities that unveil over the course of a longer period, and that are shaped by past experiences and current conditions of both the individual and the community.

Myth 3: Improvements in Market Efficiency Realized by the Mobile Phone Stem from the Better Circulation of Market Prices

A constant refrain among rural mobile phone users is how, by using the mobile phone, they avoid wasted trips. But the information necessary to avoid such wasted trips, and waste in general, was not specifically market prices. A relatively affluent fisherman working on Lake Victoria noted the value of his mobile phone for calling and requesting ice for preserving fish. He would call any of his contacts at the landing site and have them send out ice to him on the next boat. Ice, storms, and equipment failure were all unpredictable factors. The trader from Mukono district first mentioned above, who bought fish for export, spoke of a recent incident where just such a series of factors were in play and a shipment of fish was saved from being dumped by the use of the mobile phone: "After the coming of the phone, I remember one time the engine failed when we were supposed to arrive here at 4:00 p.m., and if we didn't get in contact with people here, the truck would leave us. So we had to inform them about our problem and assure them that we were coming, and we arrive at almost 10:00 a.m. because of engine failure and the storm. But because we had informed them, they were here waiting for us. So the phone helped us so much."

This is not simply information exchange but coordination work, specifically work to synchronize buyers and sellers in time and space. Information of various sorts is part of this work, but the broader practice of coordination does not readily conform to the reification of information-as-extractable-good. In Uganda, the information being passed around had to do with quantities of fish, availability of supplies (ice, fuel), location of vehicles and people, estimated time of arrival, sufficiency of cash for making payments, and so forth. Along the way, reputational information was not necessarily explicitly communicated but was nonetheless acquired through the process of arranging these transactions. This is reflected in the fish export trader's comment above: "people can lie to you that they will give you cash immediately. You bring the fish, but then when you bring it, they disappoint you." Thus, the reputation of one who came reliably with cash as they had promised would be enhanced.

Similarly, the head of one of the Chinese villages had a contact at a wheat mill whom he would call at harvest time to negotiate the sale of wheat on behalf of most of the villagers. The price was usually slightly higher than what traders offered, and farmers trusted the head of the village to negotiate a good deal for everybody because of his personal relationship with the mill buyer. The phone facilitated a relationship and the practical coordination of it, both elements that had been in place before the arrival of any kind of telephony.

Myth 4: Obtaining Market Prices Is the Most Valued Application of the Mobile Phone in Trade

Apart from coordination work, fishermen found the mobile phone useful—indeed, in some cases essential—for its most basic functionality: connecting two individuals across sometimes vast distances for synchronous speech-based communication. The phone can help establish and maintain one's reputation as a market actor, as noted above. Phone calls picked up immediately or made to communicate the status of a shipment contribute to one's reputation just as successful face-to-face transactions do. For some, having a phone was considered absolutely critical to being able to participate in trade at all, as the smoked fish seller notes: "If you do not have a phone, you can't get these kinds of jobs." Phone calls did not simply transfer information, but also communicated requests or commands—to "send ice" or to "meet the boat at a particular time and place," or commitments such as, "I will come with cash." These phone calls were speech acts that had some force. Looking at communicated speech in this way, it is helpful to distinguish between locutionary and illocutionary acts of speech. The former refers to what the speaker says specifically, the latter to the force of what is said and the intended effect on the listener, to drive the listener to specific actions (Austin 1960). Information communicated about price also entailed an indication (if not a firm commitment) that the buying party, by imparting a price, would be willing to buy at that price.

Uses of the mobile phone differed quite substantially among roles in the fish supply chain. For frontline fishermen in Uganda, who worked for salaries, by far the most critical use of the phone was to seek rescue when an engine died, a storm struck, or the boat was attacked by pirates, as other studies have also found (Abraham 2007; Sreekumar 2011). For middlemen in the fish supply chain, the phone could be useful as a tool for doing surveillance and monitoring at a distance. The fish export trader used his widely dispersed social network, a product of a lifetime living and working in the area, to keep track of his debtors. The phone was critical to this, as he noted: "When you come to me, I first find out who you are, your family, and about your work so even if he [the fisherman] got lost, I would locate him" (Fish trader, Mukono district). He called around to other villages to get reports of whether fish had been sold without his knowledge or to locate a fisherman who had disappeared.

This is not to say that "disembedded" information sources are never valuable. In rural China, by far the most successful use of mobile phones in farming has been the weather forecast delivered daily via SMS. The subscription back then was about RMB 3 per month (US$0.42), and many farmers had it, even those who had a hard time reading the screen or finding the message itself. The forecast helped to decrease short-term uncertainty and augmented existing sources. As the Shandong farmer cited above summarizes, "First I watch the national weather report on television; then I watch the local one; then I compare them with the weather forecast I get on my mobile. Then I analyze this information and come up with my forecast, and it's 70 percent reliable."

The weather forecast is something immediately actionable, which fits the farmer's existing routine (listening to news from multiple sources) and complements existing sources of information, which may be neither specific nor accurate enough.

This fourth and final "myth" about market price information illustrates how an investment in a particular scholarly conceptualization can obscure comprehension of the full range of ground-level priorities. Information sought out in our field sites covered an array of topics that went well beyond market prices to include status updates about shipments and transactions in process, information about trade partners that might reshape reputation assessments, and weather predictions. The phone was a platform for relational work, for communication, for sparking action. The information exchanged was inseparably intertwined with this work.

Conclusion: Information as Boundary Object in Market Information Services

The first generation of MIS for agriculture in developing countries began in the 1980s, but the appearance of affordable mobile phones in this century's first decade has given a new impetus to the development of agricultural applications, with a wide variety of digital services on offer (Aker 2011). While evidence of whether these systems work is mixed, what is constant is a generally poor rate of adoption and limited effect of SMS-based market price information services on market efficiency (Egg, Dembéle, and Diarra 2014; Fafchamps and Minten 2012; Camacho and Conover 2011). From the counter-narratives discussed above, we can see two explanations for this. First, "information" likely loses its usefulness once extracted from actual trade relationships and presented impersonally (i.e., as an SMS message), apart from any commitment from a buyer to pay a particular price. The way many scholars and ICTD practitioners represent mobile phone uses as *impersonal* information exchange is a consequence of abstraction: the extra details of the conversations are excluded from the economists' model in order to communicate insights parsimoniously, according to discipline-specific practices of knowledge building. In the subsequent application of such findings to build MIS, person-to-person phone calls in which *more* than just market prices were communicated became SMS messages, dispensing with the personal and business relationships between callers. Second, the encoding of market actors and their decision-making practices in system designs are based on epistemological assumptions (of the utility-maximizing market actor of neoclassical economics) that are supported by *indirect* evidence from econometric studies (e.g., of shifts in price in the market as opposed to direct observation of phone use). Even the indirect evidence, however, has been shown to apply only to middle-income or affluent agriculturalists (e.g., Jensen 2007). In the circulation of "market price information" as poverty reduction across different fields of practice in ICTD, these important distinctions have often been lost. Thus, in the wider circulation of "market price information" as a boundary object, and through the process of "deletion of modalities" (Woolgar and Latour 1986, 79), this model of decision making has come to characterize the category of "farmers" or "fishermen" as a whole, and is assumed to be inclusive of those at the lowest income levels as well.

In the ICTD field, both researchers and practitioners are generally interested in how users in lower socioeconomic strata might benefit *directly* from digital technologies. Our field data show that the smaller-scale market actors—low-income farmers and fishermen who own few or no assets—have less ability to act on better information about market price because of a reasonable reluctance to take on risk and their general lack of resources. Further exclusions follow from the shift to SMS-based market information services, which *introduce* literacy barriers that did not exist in voice-based modes of phone use, consequently blocking access to the least educated (and typically lower-income) groups who may be the ones purportedly targeted by such services.

Taking the perspective of small-scale agriculturalists on market prices, rather than an abstract view of economic principles, would likely result in different and varied types of MIS, based on the political economy and social organization of local markets. By examining the role market prices do and do not play in the decision making of rural agriculturalists, we have contributed a critical view on the rise of "information" in development policy and practice, how its relationship to the market is described, and consequently how its capacity for poverty alleviation is imagined and enacted, especially as embedded in the code and configurations of MIS. The very idea of "information" is of something that can circulate intact with its utility to end users unaltered. The declining expense of infrastructure building and the accompanying spread of mobile phones into rural and remote regions is considered an important step toward overcoming a state of information scarcity. Such regions are newly diagnosed with this affliction, and information comes to be positioned as a powerful potential salve for poverty. Our findings contribute similar insights toward an alternative understanding of the role of market prices and information delivery via mobile phone, even though the livelihood strategies and trade practices we observed were organized in different ways. In the Chinese field sites, the sharing of risk was a key consideration in deciding what to plant and in selling the resulting crops. Farmers generally did not seek a competitive edge by differentiating from other farmers, but rather followed along with their rural neighbors as a way of buffeting themselves against the vagaries of weather, crop pests, and the global economy. In the Ugandan fishing villages, the nature of fishing entailed travel onto the lake and away from the landing site for a few days at a time. This, as well as the perishability of the commodity, yielded a special emphasis on mobile phone use for contingency handling and for efficient coordination across time and space and between different roles in the fishing industry to supply ice and fuel, to call for rescue, and to predict arrival times.

By presenting a comparison of two different sites, we seek to map out the patterns and "missteps" in the way economic knowledge is extrapolated and materialized in applications such as MIS. In the notion of the boundary object, we indicate the impossibility of a perfect translation across disciplines and between model and practice. Economists might reasonably claim misinterpretation in how the notion of "market price information" is employed by noneconomists. Yet, at the same time this allows the field of economics to seize a victory of influence while sidestepping accountability. Additionally, where an attempt is made to apply the model and such an application fails (as is apparent in some evaluations of MIS), it is exceedingly difficult to arrive at an explanation based on the world selectively represented within the model.

If we take the perspective of prices, to the exclusion of all else, then as small-scale agriculturalists are put in a position to easily discover them, one would expect that they would get better prices and that general welfare (as measured by income) would increase. If this does not happen, innumerable additional elements in the context might explain why agriculturalists are still not getting better prices. Yet, a parsimonious economic model is compelling because of its simple clarity, which is accomplished by ignoring any elements that might be considered extraneous. As a result, the predominance of price, for example, is naturalized into the way the market works, and the model provides no way of arriving at any other intelligible counter-narrative. Furthermore, groups outside the circuit of this reified economic knowledge (such as the rural agriculturalists who fail to conform to the myth) come to appear irrational. And yet, the counter-narrative about market prices can be heard by involving these agriculturalists in conversations about how they make decisions. Thus, our critique is, fundamentally, also a call for methodological diversity both in ICTD and in development policy and practice. Narrow definitions of empiricism in influential strains of development economics prevent the methods we have used here from being routinely incorporated into how knowledge about poverty is generated. The result is an echo chamber that continues to reinforce a compelling myth—that farmers are using mobile phones to get market prices.

Acknowledgments

An extended version of this chapter first appeared as J. Burrell and E. Oreglia, "The Myth of Market Price Information: Mobile Phones and the Application of Economic Knowledge in ICT4D," *Economy and Society* 44, no. 2 (2015). It is reprinted in revised form by permission of the publisher, Taylor and Francis, https://www.tandofonline.com. This material is based on work supported by the National Science Foundation under Grant No. 1027310, titled "How Marginalized Populations Self-Organize with Digital Tools."

References

Abraham, Reuben. 2007. Mobile Phones and Economic Development: Evidence from the Fishing Industry in India. *Information Technologies and International Development* 4 (1): 5.

Aker, Jenny C. 2011. Dial "A" for Agriculture: A Review of Information and Communication Technologies for Agricultural Extension in Developing Countries. *Agricultural Economics* 42 (6): 631–647.

Austin, John Langshaw. 1960. *How to Do Things with Words*. Oxford: Oxford University Press.

Barry, Andrew, and Don Slater, eds. 2005. *The Technological Economy*. New York: Routledge.

Bødker, Susanne, and Liam J. Bannon. 1997. Constructing Common Information Spaces. In *Proceedings of ECSCW '97*, edited by John Hughes, Wolfgang Prinz, Tom Rodden, and Kjeld Schmidt, 81–96. Dordrecht: Kluwer Academic Publishers.

Burrell, Jenna, and Kentaro Toyama. 2009. What Constitutes Good ICTD Research? *Information Technologies and International Development* 5 (3): 82–94.

Camacho, Adriana, and Emily Conover. 2011. The Impact of Receiving SMS Price and Weather Information in Colombia's Agricultural Sector. IDB Working Paper Series 220. Inter-American Development Bank.

Cetina, Karin Knorr, Theodore R. Schatzki, and Eike von Savigny, eds. 2005. *The Practice Turn in Contemporary Theory*. New York: Routledge.

Egg, Johny, Nango Dembélé, and Salifou B. Diarra. 2014. La décentralisation des systèmes d'information de marché (SIM), une innovation pour répondre aux besoins des acteurs: Le cas de l'observatoire du marché agricole (OMA) au Mali. *Cahiers Agricultures* 23 (4–5): 288–294.

Evans, Peter. 2005. The Challenges of the Institutional Turn: New Interdisciplinary Opportunities in Development Theory. *The Economic Sociology of Capitalism*, edited by Victor Nee and Richard Swedberg, 90–116. Princeton, NJ: Princeton University Press.

Fafchamps, Marcel. 2004. *Market Institutions in Sub-Saharan Africa: Theory and Evidence*. Comparative Institutional Analysis Series 3. Cambridge, MA: MIT Press.

Fafchamps, Marcel, and Bart Minten. 2012. Impact of SMS-Based Agricultural Information on Indian Farmers. *World Bank Economic Review* 26 (3): 383–414.

Granovetter, Mark. 1985. Economic Action and Social Structure: The Problem of Embeddedness. *American Journal of Sociology* 91 (3): 481–510.

Jensen, Robert. 2007. The Digital Provide: Information (Technology), Market Performance, and Welfare in the South Indian Fisheries Sector. *Quarterly Journal of Economics* 122 (3): 879–924.

MacKenzie, Donald A., Fabian Muniesa, and Lucia Siu. 2007. *Do Economists Make Markets?: On the Performativity of Economics*. Princeton, NJ: Princeton University Press.

Oreglia, Elisa, Ying Liu, and Wei Zhao. 2011. Designing for Emerging Rural Users: Experiences from China. In *Proceedings of the SIGCHI Conference on Human Factors in Computing Systems*, 1433–1436. New York: ACM.

Sreekumar, T. T. 2011. Mobile Phones and the Cultural Ecology of Fishing in Kerala, India. *Information Society* 27 (3): 172–180.

Star, Susan Leigh, and James R. Griesemer. 1989. Institutional Ecology, Translations and Boundary Objects: Amateurs and Professionals in Berkeley's Museum of Vertebrate Zoology, 1907–39. *Social Studies of Science* 19 (3): 387–420.

Woolgar, Steve, and Bruno Latour. 1986. *Laboratory of Life: The Construction of Scientific Facts*. Princeton, NJ: Princeton University Press.

II Digital Production at Global Margins

8 Hope and Hype in Africa's Digital Economy: The Rise of Innovation Hubs

Nicolas Friederici

Introduction

I'm starting [my Kenya visit] at a place called iHub, where entrepreneurs can build and prototype their ideas. ... [A]cross the continent, things are really shifting. Things are moving from a resource-based economy ... to [an] entrepreneurial, knowledge-based economy. ... This is where the future is going to be built.
—Mark Zuckerberg, Facebook founder and CEO, quoted in Shapshak 2016, n.p.

Africa is experiencing a boom in technology entrepreneurship. High hopes have been invested in the continent's home-grown digital economy, envisioned to become an engine of rapid socioeconomic development and transformation. African governments are building entire cities for technology companies (Saraswati 2014; Vourlias 2015), larger and larger amounts of risk capital are being invested (Disrupt Africa 2016; VC4Africa 2014), and officials at the highest levels are celebrating and seeking affiliation with grassroots entrepreneurs (Hersman 2015; Wakoba 2014). Slogans like "Africa rising" (Economist 2011), "The Next Africa" (Bright and Hruby 2015a), and "Silicon Savannah" (Graham and Mann 2013) capture the sentiment that Africa is now a continent of economic opportunity and growth, driven by ubiquitous entrepreneurship, a growing middle class of consumers, a well-educated and driven economic elite, and improving Internet infrastructure.

In this chapter, I document and analyze a phenomenon that has been at the heart of the continent's technology entrepreneurship boom: the rise of innovation hubs. The number of hubs on African soil increased from a handful in 2010, to about 90 in 2013, to 117 in late 2015, and finally to 173 in June 2016 (BongoHive 2013; Firestone and Kelly 2016; Kelly 2014).

Regardless of the precise number, the World Bank was swift to conclude that hubs "drive economic growth in Africa" (Kelly and Firestone 2016, 1). Outgoing UN secretary general Ban Ki Moon addressed an audience at Nairobi's iHub, Africa's role model hub, with these words (Wakoba 2014, n.p.): "I believe I am seeing the future of Kenya. Technology can be used as a great power to change our life. Kenya is a thriving country. Being at the iHub has inspired me. You are the hope of Africa."

More specifically, hubs have been promoted as enablers of opportunity for grassroots digital entrepreneurs, allowing them to forge local clusters of software production.[1] Hubs have come to be seen as "one of the main sources of locally developed [mobile and software] applications," representing an "African digital renaissance [that] is increasingly home grown" (Kelly 2014, n.p.).

Notwithstanding this enthusiasm, in practice hubs are small organizations with a relatively simple functional setup. A hub usually consists of a Wi-Fi–connected space with hot desks and meeting rooms, used for laptop-based work or for training and mentorship sessions, networking events, presentations, and small innovation competitions like hackathons (see figures 8.1–8.3). Most hubs are located in major cities, and their funders and sponsors are quite varied, including development organizations, governments, and technology corporations, but also grassroots interest groups and philanthropic foundations (Friederici 2014).

Despite hubs' small physical size, proponents believe they are far "more than just a space to work" (e.g., Ofori 2016, n.p.). Agendas are usually rooted in lofty, far-reaching visions: hubs often (1) aspire to foster collaboration, openness, community, creativity, and diversity; (2) attempt to improve conditions in "tech communities" or "ecosystems" of innovation and entrepreneurship; (3) envision achieving positive social impact in addition to economic development; and (4) want to be an interface for diverse actors beyond entrepreneurs, including government representatives, investors, experts from the Global North, nongovernmental organizations (NGOs), training providers, artists, and many others (Gathege and Moraa 2013; Kelly and Firestone 2016; Toivonen and Friederici 2015).

Yet, not everyone has accepted these high hopes. In particular, hubs have been increasingly compared to or grouped with business incubators and accelerators (Baird, Bowles, and Lall 2013; Kelly and Firestone 2016).[2] Incubators are a longstanding form of entrepreneurship support, focusing

Hope and Hype in Africa's Digital Economy

Figure 8.1
The kLab space in Kigali, Rwanda. *Source*: kLab Rwanda.

Figure 8.2
Groups working in kLab space. *Source*: kLab Rwanda.

Figure 8.3
Event at kLab. *Source*: kLab Rwanda.

on the marginal increase in the performance (survival, revenues, capital raised, etc.) of incubatee companies (Amezcua et al. 2013; Sherman and Chappell 1998). But African hubs have rarely created wildly successful ventures, and consequently, some have considered them a hopeless endeavor (Essien 2015).

All in all, it is unquestionable *that* hubs have been extremely popular in Africa; yet, *why* this is remains unclear. Specifically, there is a paradox between, on the one hand, hubs' small physical size and simple functional setup and, on the other hand, the grand aspirations of transformative impact that have been attributed to them. Relatedly, there appears to be disagreement over what should be the yardstick for success: broader and vaguer goals (such as community development), or narrower and more specific ones (such as venture development).

The purpose of this chapter is thus threefold. First, I chart the diffusion history of African hubs. Second, I elicit the key expectations for hubs, held by different actor groups. Third, I ask why hubs have spread so quickly across Africa. My exploration in this chapter is part of a multiyear empirical study on African hubs and digital entrepreneurship.

Ultimately, this chapter warns against the supply-side focus and functionalism that is implicit in donors' and the media's accounts of hubs: the fact that hubs have diffused quickly does not say anything about the local demand for hubs, nor does it speak to hubs' impact or success. Instead, the diffusion of hubs appears to have been the result of a match between what hubs have been envisioned to do and contemporary paradigms of entrepreneurship- and technology-led economic development (Avgerou 2010; Friederici, Ojanperä, and Graham 2017; Steyaert and Katz 2004). For policy and practice, it will be necessary to move beyond the hub hype, and to think through limitations and negative side effects.

Method

I use discourse analysis as a suitable method to elicit visions and expectations that practitioners have had about African innovation hubs as an organizational form. Discourse analysis provides an approach to systematically examine "groups of statements that structure the way a thing is thought, and the way we act on the basis of that thinking" (Rose 2012, 190).

Since innovation hubs are not an established, clearly predefined type of organization (Toivonen and Friederici 2015), this chapter starts from a referential understanding (Ruef 1999), meaning that I simply examine those organizations in Africa that have been referred to as "hubs," "tech hubs," or "innovation hubs." Within Africa, iHub in Nairobi, founded in 2010, was the first widely recognized organization using the "hub" moniker in its name, and iHub soon became a role model for organizations across Africa (Gathege and Moraa 2013; Moraa 2012). The analysis thus first traces back how iHub's leaders envisioned it would work and then examines the content of visions of "hubs" as a wider organizational form that spread across Africa.

Methodologically, this means that I initially focus on meso-level discourse analysis, "sensitive to language use in context, but interested in finding broader patterns" (Alvesson and Kärreman 2000, 1133). Accordingly, I use extensive quotations from statements made by commentators and hub leaders to provide a sense of the typical narratives and imagery that have been employed and encountered by hub practitioners.

Thereby, the chapter moves from a referential understanding to a combined ideational and relational theory of hubs as an organizational form (Ruef 1999): it shows how the visions of one particular archetypical hub (iHub) have informed other practitioners' concept of hubs *in general*. In the process, the chapter ultimately condenses meso-level discourses into a macro-level Grand Discourse, "an assembly of discourses, ordered and presented as an integrated frame" (Alvesson and Kärreman 2000, 1133).[3] These Grand Discourses represent more universal expectations of African hubs.

Broad reviews of media articles and influential texts (such as accounts by known or powerful organizations and actors) are particularly suitable for this analysis (Rose 2012). Therefore, I identified reports about hubs, reviewing seven (Moraa 2012; Moraa and Mwangi 2012; Gathege and Moraa 2013; GIZ 2013; Koltai, Mallet, and Muspratt 2013; infoDev and CAD 2014; Kelly and Firestone 2016). In addition, I conducted an online keyword search on October 26, 2015, using the keywords "Africa hubs." The first 150 results generated 36 relevant media articles, covering mainstream business media (e.g., the *Economist*, *Wall Street Journal*, *BBC*), technology entrepreneurship blogs (e.g., *TechCrunch*, *Tech Cabal*, *VC4Africa*), and donor blogs (e.g., World Bank, The Rockefeller Foundation, British Council).[4] I reviewed additional blog posts (e.g., Hersman 2009), websites (e.g., AfriLabs 2016), and one master's thesis (Sanderson 2015) when necessary to fill gaps in the analysis.

Note that this chapter deals with representations of hub ideals and visions. These are related but not identical to on-the-ground realities or lived experiences (Alvesson and Kärreman 2000). I discuss potential differences and tensions between discourses and observable realities at the end of the chapter.

A History of African Hubs and Their Diffusion

iHub Nairobi, the African Role Model Hub

Outside of Africa, the first organization using the label "hub" was the Hub London, founded in 2005, later rebranded as Impact Hub.[5] Within Africa, iHub in Nairobi, founded in 2010, was the first widely recognized organization using the "hub" moniker in its name.[6] Erik Hersman (2009), the main founder of iHub, was influenced by coworking spaces he had visited in the United States and in the United Kingdom. He knew of but had not visited the Hub London when iHub was founded, and he maintains that use of the term "hub" was not directly inspired by it: "Names are interesting, because once you name something it becomes real. ... Why 'Hub'? Because it was supposed to be a hub of activity, something where people could dip in/out of and be the real nerve center for the tech community in Nairobi" (Hersman, personal communication, July 14, 2016).

The primary purpose of iHub was to connect individuals scattered across Nairobi, allowing them to collectively develop and implement ideas: "There was a clear gap in the market," Hersman remembers. While Nairobi did have a growing tech community, it was virtual. They needed a physical space in which to interact, collaborate, and gain more respect and attention from the outside world. "It all started with the idea of getting cool people into a cool place with the goal of having something cool happen" (Sanderson 2015, 5).

iHub focused explicitly on technology and the Kenyan and African context. Individual entrepreneurs and software developers in Nairobi and other African cities operate under harsh infrastructural and economic constraints. Traffic conditions and poor public transport during work hours inhibit on-time meetings and events. Fast Internet connections and office space are either entirely unavailable or prohibitively expensive, and homes without a generator regularly go without electricity for hours and sometimes days at a time. Entrepreneurs also have very limited personal financial resources; venture projects are often started with a few hundred rather than tens of thousands of pounds or dollars. For many, "iHub [was] an oasis of modern order in the otherwise so chaotic Kenyan capital Nairobi" (Lindijer 2013, n.p.).

The ambition behind iHub was fueled by the aspiration that Kenya and other African countries, following the arrival of broadband connectivity in 2010, could catch up with and become more closely connected to the Global North and the West (Graham 2015): Nairobi was reimagined as a "Silicon Savannah" (Graham and Mann 2013). iHub became the symbolic center of this transformation: "Africa always lagged behind; we always heard that we needed technological help from outside. But in the digital age, we are at par with other parts of the world. ... Before, we always heard that Africa had not gone through the industrialization age. I tell you, Africa skips the industrial revolution. We jump straight into the digital age" (iHub director quoted in Lindijer 2013, n.p.).

This focus had consequences for the types of innovations envisioned as coming out of iHub (locally adapted ones) and the types of people who were meant to create them (young Kenyan grassroots entrepreneurs): "Africa's innovative ideas are based on local needs, many of them stemming from budgetary constraints, others from cultural idiosyncrasies. People from the West often can't imagine or create the solutions needed in emerging markets, as they don't have the context and do not understand the 'mobile [technology] first' paradigm" (Hersman 2012a, 67).

iHub leaders claimed that the hub was a uniquely suitable answer to the challenges of young African innovators. They wanted iHub to be different from existing forms, such as coworking spaces and incubators, while also combining known elements: "We realized we needed something that was more of a hybrid than just the hot desking and paid space operations in [the UK and the US]. In our case we needed something that was part coworking, but also part community centre (like you would find in a university)" (Hersman, personal communication, July 14, 2016). Already in the initial launch announcement (Hersman 2010), iHub was thus referred to as "part open community workspace (co-working), part investor and VC [venture capital] hub and part incubator."

iHub's founders envisioned connections that would lead to entrepreneurship and innovation outcomes. This process was seen as indirect, and they acknowledged that the lion's share of affiliated startups had existed before (Moraa and Mwangi 2012). Thus, iHub leaders and the media used language to indicate a facilitative role of the hub, saying that startups "emerge from" iHub (Moraa and Mwangi 2012, 22) or that iHub "gives rise to" ventures (Hersman 2012b, n.p.).

"Community" was a core concept for iHub. The term indicated that members were not controlled by the hub management, that they shared values but were also responsible for shaping the organization, as Hersman explained: "The iHub is what we as a tech community make it. It is a blank canvas ... that needs some input from people within the community to design, and create a culture around" (2010, n.p.). Similarly, iHub's first manager recalled, "We were looking for techies who were doers and not talkers" (quoted in Moraa and Mwangi 2012, 7).

Actions taken by iHub staff were meant only to facilitate, and ultimately individual members made connections "serendipitously": "One of [the unique things about iHub] for sure is that we engineer serendipity. We do not push people in a certain direction; we allow them to try, fail, learn and revise their ideas, connect with others and explore their limitless potential. By putting a lot of smart people in one room, we have seen great innovations emerge" (iHub PR and communications manager, quoted in Wangari 2015, n.p.).

The idea was that "through iHub, the technology community, industry, academia, investors and venture capitalists [could] meet, share ideas and collaborate [and thus] transform their ideas into actions" (Moraa 2012, 9). The iHub "brand" was meant to provide "exposure" for innovators by pooling and providing access to opportunities, such as jobs, freelancer group contracts, or training (Moraa and Mwangi 2012). The hub's community was said to enable knowledge sharing and mentorship, thereby raising the skill levels of members, particularly university graduates (Moraa 2012; Moraa and Mwangi 2012). More broadly, iHub was meant to be a place that attracted visits from local technology businesspeople and representatives of international technology corporations, allowing young Kenyan innovators who regularly occupy the hub to make connections that they could not make otherwise (Hersman 2012a).

This technology- and entrepreneurship-specific notion of relevant connections generates a particular understanding of "ecosystems." iHub itself was understood as an ecosystem; yet, at the same time, it was also framed as embedded within a wider ecosystem: "You can think of iHub as an attempt at creating an innovation ecosystem in a box. The facility ... is run by the local startup community. ... The iHub organizers look for gaps in the local business ecosystem and try to fill them" (IBM 2013, 3).

iHub leaders made no sharp distinction between a hub-internal and a hub-external sphere. The "community" of iHub was imagined to blend in with the "Kenyan tech community," "tech ecosystem," "tech scene," or broadly the "tech environment." It was acknowledged that "the right growth environment for tech ... is nuanced, organic, and grows over time through the aggregated acts of individuals" (Hersman 2012a, 61). Yet, through the conceptual blending of connections made within the hub and those existing in its environment, iHub was envisioned to have positive effects on the local, national, and regional level (Hersman 2012a), for example: "The iHub catalyses the growth of the Kenyan tech community by connecting people, supporting startups, and surfacing information" (iHub

2016, n.p.); and "Your ideas and drive will make the iHub into the space to be in all of East Africa for tech-related activities" (Hersman 2010, n.p.).

In sum, the expectations around iHub nourished by founders and leaders centered on facilitating connections among like-minded, driven, and independent yet interdependent individuals in a city, leading to the formation of a community, which enabled a collective entrepreneurial process that generated innovations. iHub focused on Kenyan innovators and on locally adapted technological innovations. Based on the notion that iHub facilitated connections within a broader technology ecosystem, the hub was argued to "catalyze" the development of technology startups and ultimately make a significant contribution to economic development.

This vision proved hugely appealing for a range of actors seeking technology-driven economic development in Africa, and iHub quickly won prominence and acclaim. For instance, Fast Company (2014) named iHub as Africa's most innovative company in 2014, and the list of dignitaries and technology celebrities who have visited iHub has grown to include Uhuru Kenyatta (Kenya's president), Ban Ki Moon (outgoing UN general secretary), Mark Zuckerberg (CEO and founder of Facebook), Eric Schmidt (former Google CEO), and Vint Cerf (one of the inventors of the Internet).

The Spread of Hubs across Africa

The confluence of several events and dynamics allowed the spread across Africa of organizations similar to iHub, resulting in at least partial recognition of hubs as a distinct organizational form. iHub's widely perceived success had broadened the appeal of hubs beyond African technologists to reach development institutions, governments, and technology companies.

Hersman (2009) had already called on the "emergent, yet disconnected, technology community ... growing in many of the major African cities," to establish "tech coworking spaces" in 2009, one year before iHub opened. He was already using the term "hub" at this point to convey the envisioned interconnection function, while referring to "coworking spaces" in the title of his blog post and when pointing out organizations in the US and UK (including the Hub London).

In fact, several organizations following similar visions were founded shortly after iHub, in 2010 and 2011. This group includes Active Spaces

(Cameroon), kLab (Rwanda), KINU (Tanzania), Hive Colab (Uganda), BongoHive (Zambia), Nailab (Kenya), Banta Labs (Senegal), and the Co-Creation Hub or CcHUB (Nigeria). Several of them formed AfriLabs, a "network of tech innovation hubs in Africa" with the mission to "build the capacity of hubs, which support the growth of tech communities around them" (AfriLabs 2016, n.p.). AfriLabs was akin to a traditional business association, with independent member organizations. This effort was again promoted, if not led, by Hersman, with support from other bloggers and technologists from across Africa (Hersman 2011). "Hub" became the agreed upon terminology to refer to the class of organizations, even though "lab" features in the names of the association and some of its members (e.g., kLab in Rwanda).

A further milestone in instituting hubs as an organizational form was a study funded by British NGO Indigo Trust and conducted by iHub Research, the hub's think-tank-like branch (Gathege and Moraa 2013). The study helped establish "hubs" as a common label, variously using the terms "ICT Hubs," "Innovation Hubs," or just "hubs." It claimed that there was "growing evidence that the ideals of openness and collaboration are often in-built in the architecture of these hubs through their events and activities. … Innovation hubs can be most effective when they harness openness and community-driven approaches" (Gathege and Moraa 2013, 6).

Finally, hubs were affirmed as an important Africa-wide phenomenon through two widely noted stock-taking exercises. First, a crowdsourcing campaign by BongoHive (2013) provided a map and a count of hubs. Two World Bank specialists then built on BongoHive's work and the iHub Research study to publish curated maps in 2014, 2015, and 2016 (Firestone and Kelly 2016; Kelly 2014). The most recent map included 173 hubs (figure 8.4).

The following years were marked by intensifying and broadening interest in African hubs. The reports and online articles discussing them illustrate the attention paid by actors ranging from technology bloggers and mainstream business media (including the BBC and the *Economist*) to large development organizations (like the German Corporation for International Cooperation [GIZ] and the World Bank). Most of the sources refer to the BongoHive or World Bank maps. With core funding from GIZ, AfriLabs conducted three meetings of African hub leaders at the annual re:publica conference in Berlin from 2013 to 2015. The social development organizations

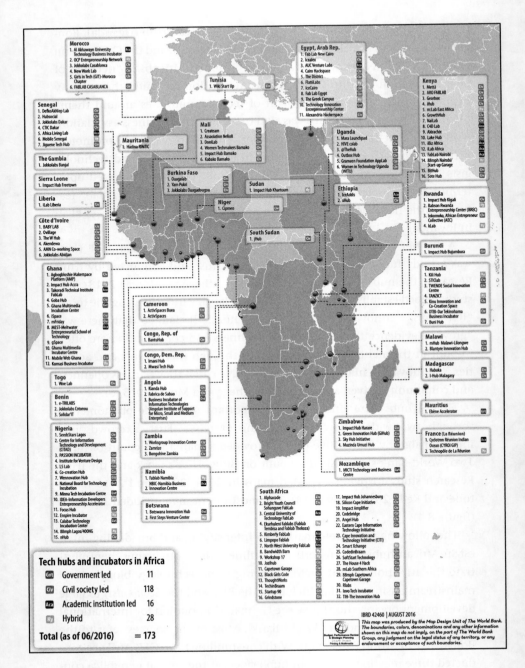

Figure 8.4
Technology innovation hubs in Africa mapped by the World Bank. *Source*: Firestone and Kelly (2016).

Hivos, Indigo Trust, and DEON initiated a fund for African hubs in 2014 (Treisman 2014).

In this context of increasing attention from media and development organizations, iHub's vision of being a positive contributor to local economic development became attributed to African hubs in general. For instance, a World Bank blog post (Kelly 2014, n.p.) describes hubs as "one of the main sources of locally developed [mobile and software] applications," and the same author later proclaimed that hubs "drive economic growth" in Africa (Kelly and Firestone 2016, 1). The notion was quickly adopted by other development agencies, including the International Telecommunication Union (Lamanauskas 2015). Christine Lagarde, head of the International Monetary Fund, following her tour of kLab in Rwanda, tweeted, "Innovative technology projects like kLab are vital to Rwanda's future" (Lagarde 2015). Not unlike discourses about the impact of Internet connectivity in general (Friederici, Ojanperä, and Graham 2017), such decontextualized, sweeping claims were made irrespective of missing evidence.

This background helps to understand popular expectations about how African hubs work as an organizational form. While the "hub" terminology was widely adopted in Africa after 2012, it also lost some of the specific meaning that the iHub founders had attached to it. Actors other than the immediate founders of the first generation of African hubs tended to rely on simpler, more heuristic mental representations of hubs, seemingly without fully recognizing the elaborate mental model of community-centric facilitation of connections that had originally been developed.

Network Infrastructures or Incubators? Two Expectations for Hubs

The Network Infrastructure Expectation

Several palpable features of hubs have been uncontested, such as physical layout and services. Hubs have consistently been understood to consist of a Wi-Fi–connected space with hot desks and meeting rooms, allowing for laptop-based work. Activities have been described as including events, presentations, small innovation competitions (like hackathons or pitching contests), group meetings on topics of interest, training (for instance, on coding), and mentorship sessions. It is also uncontested that hubs are generally meant to contribute to early stage innovation and entrepreneurship processes, and that they allow entrepreneurs to network with each other and with hub-external partner organizations.

Nonetheless, two divergent macro-level Grand Discourses about African hubs as an organizational form have developed out of the original iHub-specific meso-level discourses. The more optimistic of the two Discourses can be summed up as the *network infrastructure expectation*. It maintains the gist of iHub's original vision: hubs are assumed to be open, hosting and supporting diverse actors with a stake in technology entrepreneurship. Specifically, local entrepreneurs have the opportunity to make connections with each other and other actors, such as technology corporations, investors, and international visitors. Hubs are seen to be new, and unlike other known forms of organizations.

The network infrastructure expectation has been nurtured by representatives of African hubs (leaders, funders, and AfriLabs) and several journalists, who argue that hubs have an appeal as an organizational form that encourages innovation and entrepreneurship in Africa in unique and timely ways: "As these spaces become the nerve centres for the tech community, they also become critical international touch-points for those seeking to engage in technology and business in Africa. They are the new points of exchange for long-term expatriates and short-term visitors looking to identify trends, find local talent, and catch the African wave of innovation" (Kalan 2014, n.p.). Similarly, "These meeting places act as a bridge between technology start-ups, investors and academics, nurturing collaborations. Entrepreneurs experiment together and learn from one another, supported by the hubs' ICT infrastructure and mentoring programmes" (Oxford and Jeffries 2013, n.p.).

The hub metaphor is based on notions of interpersonal connectivity; hubs are seen as central nodes in wider networks, letting people interact who would not otherwise do so. Key expressions employed by proponents of this assumption are "networks," "openness," "knowledge sharing," and "collaboration": "Not only do [hubs] act as physical infrastructure, providing access to power, pipes (internet), and space, but they also offer network infrastructure, or access to the human, financial and social capital" (AfriLabs 2015, n.p.).

Yet, while this expectation maintains the general idea that hubs facilitate connections, it also tends to be more abstracted from the iHub founders' more specific visions. Namely, the originally detailed understanding of collaboration and coordination processes inside hubs has been lost. The role of a hub-internal and hub-shaping community is now emphasized less, or less elaborately. "Community" is still used as a concept, but it is often missing the original notion that community members are the ones owning and shaping the hub. Now the hub "creates" or "nurtures" community, not the other way around: "But the secret of [hubs'] success lies in the collaborative environment and close-knit communities *that they create*. These nurturing conditions promote a rapid exchange of ideas, skills sharing and problem solving which allows serendipity to happen" (Treisman 2015, n.p., emphasis added).

In other accounts, "community" is used exclusively in a wider sense, referring to the "tech community" of a city. The notion of a blended hub-internal and external community that was essential for iHub's vision is absent in the network infrastructure expectation, and instead the hub's facilitative function toward its environment is emphasized more (for instance, expressed as the "enabling environment" or "ecosystem"): "The opportunity is that hubs serve as *infrastructure* to support tech, entrepreneurship and innovation. This is because there are so many gaps in the enabling environment to be filled, or to be filled better, cheaper, or more easily" (Akinyemi 2015, n.p., emphasis in the original).

International connections to the Global North are heavily emphasized, as they are imagined to bring expertise and capital (Heuler 2015; Kalan 2014). AfriLabs goes so far as to argue that hubs have thereby become an infrastructure that embeds local entrepreneurship and innovation within a supranational or global "knowledge economy" or "digital economy" (AfriLabs 2015; 2016): "Our belief is that investing in these spaces will create an innovation infrastructure that will encourage the growth of Africa's knowledge economy by supporting the development of start-ups, technology, and innovation" (AfriLabs 2015, n.p.).

As continent-wide "innovation infrastructures," hubs are depicted as complementary to physical Internet and ICT infrastructure: "Across [Africa] new Silicon Savannahs are in the making and the components of a budding ... ecosystem are emerging. ... iHub-like innovation centers are becoming a mainstay of the continent's progressing ICT infrastructure ... these IT spaces are becoming central connect points for ideas, entrepreneurs, investment, and innovation across the continent" (Bright and Hruby 2015b, n.p.). Similarly, "Due to the increase in speed and affordability of connectivity, the sense of possibility the digital world has provided, and interest by investors, tech hubs, business incubators, hacker spaces, and maker spaces are providing places for the continent's young (and not so young) people to express their innate creativity and ambition, make up for disparity in educational opportunities, allow for the creation of wealth that is more sustainable than the resource extraction of the past, and help to produce products and services to address the needs of their communities" (Hopkins 2015, n.p.).

In short, according to the network infrastructure expectation, hubs work as interconnection nodes in geographically dispersed networks of people and organizations, connecting technology entrepreneurs in a given African city with each other, with important partners located in the city (such as mentors, investors, etc.), and with important partners normally located abroad, often in the Global North. Proponents of the network infrastructure expectation have called for patience in evaluating hubs, arguing that impacts are potentially transformative but will take time to materialize, given that effects are systemic and indirect in nature (Douglas 2013; Oluwagbemi 2015; Treisman 2015).

The Incubator Expectation
The second common assumption of hubs is much more simply structured and pessimistic. The *incubator expectation* is held mainly by actors who do not discern between incubators and hubs, or who dismiss hubs as a fad in development circles.

Lumping together organizations under the hub label appears to have been a consequence of the quickly rising interest in hubs since 2012. As the hub term became popular, the rich meaning developed in the iHub context got lost to the point where organizations that clearly differed from iHub were included in accounts about hubs (e.g., in maps, lists, and blog posts). This development can be traced back to iHub Research's comparative study on hubs (Gathege and Moraa 2013), which included the Meltwater School of Technology (MEST), even though MEST explicitly follows a traditional training and incubation approach (MEST 2016). It continued when BongoHive and the World Bank included several existing incubators and technology parks in maps of hubs in Africa.

Ironically, the terminological ambiguity allowed others to do the opposite, that is, to understand hubs narrowly as incubators. Accounts from

2014 and 2015 began to refer to "hubs, accelerators, and incubators," or simply to incubators, when discussing organizations that self-identified as hubs. In other words, many observers began simply to subsume hubs under the wider organizational form "business incubator."

Consequently, observers applied pre-existing notions about incubators to hubs, allowing the incubator expectation to develop. The perceived traditional goal of business incubators—increasing the survival and performance of young companies—was assumed to also be the goal of hubs. Implicitly, the incubator expectation is based on the rationale that the only essential innovation and entrepreneurship outcome is the number and performance of startups that an incubator produces, an assumption that is then transferred to hubs. For instance, Bright and Hruby (2015a, 177) estimate that 3,500 technology startups have been "forming out of [Africa's] innovation hubs," implying that hubs, like incubators, create or develop startups.

This reasoning, however, has opened hubs up for criticism. For instance, known Nigerian technology entrepreneur Mark Essien (2015, n.p.), in a widely noted blog post, makes critical statements generally about incubators but refers to a well-known (self-pronounced) hub to illustrate his argument: "Of the 9 biggest software startups in Nigeria, none was built by an incubator. … Of the 15 next biggest software startups … only one used to operate from CcHub. Incubators just don't work, otherwise they would have produced more successful startups in Nigeria. Even Kenya and Ghana that have a stronger incubator scene have produced nothing of note."

Although the skeptical element of the incubator expectation was not very prevalent in the textual material I analyzed for this chapter, similarly skeptical views have been reflected in African technology media (Masuku and Kalenga 2015) and appear to be widespread without being expressed. In some cases, hub leaders themselves began to revert to the incubator terminology (Mubaiwa 2015; Oluwagbemi 2015). For instance, in a response to Essien's provocative post, Michael Oluwagbemi (2015) defends his organization Wennovation Hub, outlining its goal to "build the ecosystem" rather than to support a few individual startups, but still refers to "incubators" throughout his account, even when referring to Wennovation Hub.

In sum, the incubator expectation does not retain much from context-specific iHub meso-level discourses. Specifically, hubs are not seen to be new or different kinds of organizations. Here, "hub" is simply a new label for organizations that support early stage technology entrepreneurs. Because such organizations have traditionally been referred to as incubators, the newly prevalent label "hub" is either abandoned or used synonymously with "incubator." The incubator expectation understands hubs as black boxes and is concerned only with their outcomes. Accordingly, hubs are measured against their results in terms of technology startup creation or development.

The network infrastructure and incubator expectations are not entirely contradictory. What distinguishes them is their implied level of optimism about hubs and the degree to which they retain elements from the more elaborate hub visions that iHub had established. While the network infrastructure expectation assumes that startups are created within wider "ecosystems," which the hub supports, the incubator expectation presumes hubs create ventures directly.

Discussion

Many innovation hubs have indeed been established within a short period, but the question of *why* this has happened is not trivial. The present findings show that any answer has to take into account the discursive landscape that emerged following the founding and acclaim of iHub.

Nairobi's role model hub promoted the idea that a single organization, physically embodied by not much more than a space with hot desks, could be a catalyst for the economic development of an entire low-income nation. iHub's leadership deftly connected its own context-dependent visions (meso-level discourses) with paradigmatic notions of entrepreneurial and technology-led development (Avgerou 2010; Friederici, Ojanperä, and Graham 2017; Steyaert and Katz 2004), forming a narrative with a wide appeal for a large and diverse set of audiences, including media, development organizations, and governments.

Namely, iHub promised to deliver idolized notions within contemporary international development, such as "community," "(mobile) technology," "grassroots/local," "youth," "openness," "diversity," "hacking/hackathons," "prototyping/experimenting," "startups," and, of course, "innovation" and "entrepreneurship." iHub's narrative was so compelling and the buy-in among media outlets and other discourse generators became so overwhelming that iHub's story spread swiftly across Africa and beyond. Consequently, dozens of organizations were created in iHub's image in a matter of just a few years. While it is hard to pinpoint exactly which discourses drove which hub implementations in what places, the diffusion history that I present in this chapter leaves hardly any doubt that aspirational images inspired by iHub played an important role.

Conversely, my findings offer little in support of the theory that hubs were created because they compensate for market failures, which both academic and practice-oriented literatures have offered as a rationale for the existence of incubators and accelerators (Baird, Bowles, and Lall 2013; Phan, Siegel, and Wright 2005). Similarly, the analysis does not provide backing for functionalist arguments ("the fact that hubs exist indicates that they are effectively fulfilling a need"), which are implicit in many practitioner accounts (Hopkins 2015; Firestone and Kelly 2016) and typical of top-down entrepreneurship interventions (Perren and Jennings 2005). At this point, no reliable evidence indicates that innovation hubs are effective at helping startups grow or at delivering development. Rather, hubs have spread quickly *despite* the absence of such evidence, not because their impact is unquestionable.

In fact, it is difficult to imagine what such evidence would look like, and an expectation of measurable socioeconomic development as a direct outcome of hubs' action appears unwarranted or nearly impossible to verify (Friederici 2014; Oluwagbemi 2015; Treisman 2015). To this day, whether and how iHub indeed generates the transformative impacts it has been imagined to produce remains unclear. For instance, commentators noted that only a few equity investments in technology startups have been made in Kenya because the hundreds of fledgling teams in the country lack the skills to develop market-ready businesses (Jorgic 2014)—a systemic barrier that iHub might not be able to do much about. The great divide between the aspirations of hubs and their verifiable outcomes highlights a basic but important risk of powerful discourses: inflated expectations toward a phenomenon and misguided allocation of resources and attention.

Similarly, the present findings show that, as hub discourses evolved over time, much of the context specificity and detail contained in iHub's original vision got lost. The network infrastructure expectation became pitted against the incubator expectation, without either side's proponents being aware of the situation. Different commentators started to debate the merits of hubs as an organizational form while using fundamentally different assumptions about how hubs work and what their goals should be. Ultimately, the broadening of hubs' appeal, and the transformation of meso-level discourses into Grand Discourses, meant that "hub" became an umbrella term, glancing over any given organization's structure and internal mechanics. While the term "hub" is today used everywhere in Africa, it has come to mean drastically different things for different people, which has led to poor communication among various stakeholder groups, and ultimately to confusion and frustration.

Finally, what is strikingly missing from both meso-level discourses and macro-level Discourses is the perspective of hubs' predestined users: grassroots, early stage technology entrepreneurs. It is unsurprising that hub leaders and funders have been vocal about hubs' promise, and it may also not be extraordinary that media and development organizations find the network infrastructure Discourse appealing. Yet, whether and how hubs can work like network infrastructures and thereby transform Africa's digital economies are empirical, not normative, questions. Any such empirical inquiry would have to start and end with entrepreneurs, as they are the ones ultimately using and (hopefully) benefiting from hubs. Such an empirical and participant-oriented grounding of hub discourses is strikingly absent from ongoing debates.

Conclusion and Outlook

This chapter provides a window into the productive power of hype and hope in Africa's digital economy. From about 80 hubs in 2013 (BongoHive 2013), the tally increased to 173 in 2016 (Firestone and Kelly 2016). Following this speedy diffusion, multiple development organizations, governments, and technology corporations have taken note of the phenomenon, while no effort has been made to rigorously and clearly discern opportunities and limits.

The question of why hubs have been established appears to be disconnected from questions of whether and how they actually work. The diffusion of innovation hubs happened when meso-level discourses developed from the specific case of Nairobi's iHub became disembedded from context and attributed to innovation hubs as a wider organizational form. Hubs spread regardless of the lack of evidence of positive effects, and confusion arose as to what hubs were trying to do, how they worked, and what their potentials and limitations might be.

These insights have several implications for scholars and practitioners of technology and entrepreneurship in Africa. First, the findings of this chapter tell a cautionary tale about excessive functionalism: the observation that hubs are implemented on a wide scale does not imply that they are operating successfully. Hubs have been carried by a widespread belief in grassroots technology entrepreneurship, and their diffusion was triggered by the rise of iHub as a continent-wide inspiration. Premature and acontextualized celebration of a new phenomenon as a panacea for development problems appears to be consistent across technology-related topics, as a similar mechanism has been found for sectors such as business process outsourcing (Graham 2015; Graham and Mann 2013) and Internet connectivity in general (Avgerou 2010; Friederici, Ojanperä, and Graham 2017).

Similarly, framings of hubs as direct and unavoidable consequences of Internet connectivity misses the variegated nature of discourses and processes of abstraction and decontextualization that goes into Discourses, which this chapter has highlighted. Much like technological determinism, such framings can lead to an underconceptualization of culture and power structures, ultimately resulting in a "discourse of inevitability" (Leonardi 2008). When broadband connectivity arrived in South Asia or Latin America, software developers and technology entrepreneurs also required physical places to work and collaborate, but they did not rely on innovation hubs as an organizational form in the same way as developers and entrepreneurs in Africa did. Unlike what a discourse of inevitability would suggest, actors with discursive power (mainly iHub's leadership, the AfriLabs association, development organizations, and an array of media outlets) worked together to promote hubs as a viable network infrastructure for Africa's digital economy. Hubs may well work as these actors imagine them to work, but it is also clear that their potential may be—deliberately or inadvertently—misconstrued or oversold.

As hubs continue to be set up, and as donors and governments invest their hopes in them, it will be important to see hub Discourses for what they are. The boosterish network infrastructure Discourse has had productive energy: far fewer hubs would exist today without the grand aspirations that developed on the back of the iHub success story. Moreover, the hype around hubs does not mean that they are harmful; hubs might indeed have some transformative effects that many envision them to have. Yet, Discourses have also generated simplistic and decontextualized expectations for hubs, which can have negative consequences within (local) realities of implementation.

These reflections point to two areas where we do not yet know enough about African hubs. First, a precise analysis of which discourses spread where and when could improve our understanding of why hubs have spread across Africa as fast as they have, and what the concrete consequences are of Discourses in local contexts. An institutional theory lens could help capture how hub models got transferred and translated to different contexts (Dalpiaz and Tracey 2013). Such a study could collect data on the timing and missions of hub implementations, as well as the continent-wide spread of hub Discourses, or the evolution of discourses into Discourses. This could shed light on how relational ontologies and discursive patterns that define and promote organizational forms are driven by local actors versus philanthropic and development organizations (Korff, Oberg, and Powell 2015; Ruef 1999), resulting in the diffusion of an organizational form irrespective of its effects or service offerings.

Second, what is missing from current debates is a better understanding of the organizational processes that hubs trigger. Development organizations and the media appear to have jumped to the conclusion that hubs are "the future of Africa" (Oxford and Jeffries 2013; Wakoba 2014), while others dismiss hubs as hopeless (Essien 2015). Strictly speaking, however, we can neither assess, nor should we assert, whether hubs work as long as we do not understand *what* hubs do and *how* they work. The contradictions and tensions between the network infrastructure and incubator expectation of hubs illustrate that these are far from easy questions. More in-depth empirical scholarship is needed to examine how hubs connect entrepreneurs with others, and how such coordinative processes play out within the complex and challenging realities of African cities. For hubs to lead to better outcomes, implementers and funders should move beyond the hype—acknowledging the indirect and indeterminate nature of hub outcomes and working toward a more grounded understanding of what hubs can and cannot do for African technology entrepreneurs.

Acknowledgments

This work was funded by the Clarendon Fund and the Skoll Centre for Social Entrepreneurship in Oxford. Additional support came from the Geonet project, established under the European Union's Seventh Framework Programme (FP/2007–2013)/ERC Grant Agreement no. 335716. Mark Graham (Oxford Internet Institute) and Marc Ventresca (Saïd Business School) supervised the underlying doctoral work. Tuukka Toivonen and other members of the Entrepreneurial Spaces and Collectivities group provided further valuable input at workshops in London and Oxford.

Notes

1. "Digital entrepreneurship" is used in this chapter as a more specific term than "technology entrepreneurship." Digital entrepreneurship refers to the creation of ventures and novel initiatives to market products and services where value creation is tied to software, mobile applications, or digital content.

2. A business incubator can be defined as an "organisation dedicated to the support of emerging ventures" (Bergek and Norrman 2008, 21), "providing tenant firms with a portfolio of new venture support infrastructure" (Mian, Lamine, and Fayolle 2016, 2) including office space and business assistance.

3. I adopt Alvesson and Kärreman's notation in this chapter, spelling meso-level discourses with a lowercase *d* and macro-level Discourses with uppercase *D*.

4. The full list of articles can be obtained from the author upon request.

5. This is not to say that traditional organizational forms, such as incubators and science parks, had never before used the term "hub" in brand names. For instance, in Africa, the science and technology parks Botswana Innovation Hub (Gaborone) and the Innovation Hub (Pretoria, South Africa) were both founded in 2002, before the Hub London. Here, however, the goal is to trace back the roots of hubs as a *type* of organization, which was later recognized as such by practitioners.

6. The official brand name of the organization is *iHub_, as a play on social media and coding culture. For the sake of simplicity, this chapter uses the common spelling without asterisk and underscore.

References

AfriLabs. 2015. AfriLabs. Accessed October 29, 2018. http://afrilabs.com.

AfriLabs. 2016. AfriLabs—the Network of African Innovation Hubs. Accessed October 29, 2018. http://afrilabs.com/copy-of-homepage/.

Akinyemi, Tayo. 2015. 11 Key Lessons for Innovation Hubs in Africa. *Venture Capital for Africa*. March 30, 2015. https://vc4a.com/blog/2015/03/30/11-key-lessons-innovation-tech-hubs-in-africa/.

Alvesson, Mats, and Dan Kärreman. 2000. Varieties of Discourse: On the Study of Organizations through Discourse Analysis. *Human Relations* 53 (9): 1125–1149. doi:10.1177/0018726700539002.

Amezcua, Alejandro S., Matthew G. Grimes, Steven W. Bradley, and Johan Wiklund. 2013. Organizational Sponsorship and Founding Environments: A Contingency View on the Survival of Business-Incubated Firms, 1994–2007. *Academy of Management Journal* 56 (6): 1628–1654. doi:10.5465/amj.2011.0652.

Avgerou, Chrisanthi. 2010. Discourses on ICT and Development. *Information Technologies and International Development* 6 (3): 1–18.

Baird, R., L. Bowles, and S. Lall. 2013. Bridging the "Pioneer Gap": The Role of Accelerators in Launching High-Impact Enterprises. Aspen Network of Development Entrepreneurs & Village Capital. https://www.aspeninstitute.org/publications/bridging-pioneer-gap-role-accelerators-launching-high-impact-enterprises/.

Bergek, Anna, and Charlotte Norrman. 2008. Incubator Best Practice: A Framework. *Technovation* 28 (1–2): 20–28. https://doi.org/10.1016/j.technovation.2007.07.008.

BongoHive. 2013. Hubs in Africa. Digital map. https://africahubs.crowdmap.com.

Bright, Jake, and Aubrey Hruby. 2015a. *The Next Africa: An Emerging Continent Becomes a Global Powerhouse*. New York: Thomas Dunne Books.

Bright, Jake, and Aubrey Hruby. 2015b. The Rise of Silicon Savannah and Africa's Tech Movement. *TechCrunch*. July 23, 2015. http://social.techcrunch.com/2015/07/23/the-rise-of-silicon-savannah-and-africas-tech-movement/.

Dalpiaz, Elena, and Paul Tracey. 2013. New Venture Creation and the Use of Cultural Resources: The Case of H-Farm. *Academy of Management Proceedings* 2013 (1): 10455. doi:10.5465/AMBPP.2013.10455abstract.

Disrupt Africa. 2016. *Funding Report 2015*. Cape Town, South Africa: Disrupt Africa. http://disrupt-africa.com/publications/.

Douglas, Kate. 2013. Impact of Africa's Innovation Hubs: Too Early to Call? *How We Made It in Africa*. December 17, 2013. http://www.howwemadeitinafrica.com/impact-of-africas-innovation-hubs-too-early-to-call/33491/.

The Economist. 2011. Africa Rising. *Economist*. December 3, 2011. http://www.economist.com/node/21541015.

Essien, Mark. 2015. Startup Incubators in Africa and Why They Don't Work. *Venture Capital for Africa*. April 21, 2015. https://vc4africa.biz/blog/2015/04/21/startup-incubators-in-africa-and-why-they-dont-work/.

Fast Company. 2014. The Top 10 Most Innovative Companies in Africa. *Fast Company*. April 2, 2014. http://www.fastcompany.com/3026686/most-innovative-companies-2014/the-top-10-most-innovative-companies-in-africa.

Firestone, Rachel, and Tim Kelly. 2016. The Importance of Mapping Tech Hubs in Africa, and Beyond. *Information and Communications for Development* (IC4D) (blog). World Bank. August 24, 2016. http://blogs.worldbank.org/ic4d/importance-mapping-tech-hubs-africa-and-beyond.

Friederici, Nicolas. 2014. *The Business Models of mLabs and mHubs: An Evaluation of infoDev's Mobile Innovation Support Pilots*. Washington, DC: infoDev, World Bank. http://www.infodev.org/mobilebusinessmodels.

Friederici, Nicolas, Sanna Ojanperä, and Mark Graham. 2017. The Impact of Connectivity in Africa: Grand Visions and the Mirage of Inclusive Digital Development. *Electronic Journal of Information Systems in Developing Countries* 79 (2): 1–20.

Gathege, Duncan, and Hilda Moraa. 2013. *Draft Report on Comparative Study on Innovation Hubs Across Africa*. Nairobi: iHub Research. Accessed October 12, 2018. https://docgo.net/philosophy-of-money.html?utm_source=draft-report-on-comparative-study-on-innovation-hubs-across-africa.

GIZ. 2013. *Technology Hubs—Creating Space for Change: Africa's Technology Innovation Hubs*. Deutsche Gesellschaft für Internationale Zusammenarbeit (GIZ) GmbH. http://10innovations.alumniportal.com/fileadmin/10innovations/dokumente/GIZ_10innovations_Technology-Hubs_Brochure.pdf.

Graham, Mark. 2015. Contradictory Connectivity: Spatial Imaginaries and Technomediated Positionalities in Kenya's Outsourcing Sector. *Environment & Planning A* 47 (4): 867–883. doi:10.1068/a140275p.

Graham, Mark, and Laura Mann. 2013. Imagining a Silicon Savannah? Technological and Conceptual Connectivity in Kenya's BPO and Software Development Sectors. *Electronic Journal of Information Systems in Developing Countries* 56 (1): 1–19.

Hersman, Erik. 2009. African Cities Need Tech Coworking Spaces. *WhiteAfrican* (blog), January 14, 2009. http://whiteafrican.com/2009/01/14/african-cities-need-tech-coworking-spaces/.

Hersman, Erik. 2010. iHub: Nairobi's Tech Innovation Hub Is Here! *WhiteAfrican* (blog), January 25, 2010. http://whiteafrican.com/2010/01/25/ihub-nairobis-tech-innovation-hub-is-here/.

Hersman, Erik. 2011. Afrilabs Provide a Model for African Innovation, Collaboration. *Memeburn*. February 8, 2011. http://memeburn.com/2011/02/afrilabs-provide-a-model-for-african-innovation-collaboration/.

Hersman, Erik. 2012a. Mobilizing Tech Entrepreneurs in Africa (Innovations Case Narrative: iHub). *Innovations: Technology, Governance, Globalization* 7 (4): 59–67. doi:10.1162/INOV_a_00152.

Hersman, Erik. 2012b. We Need More, Not Less. *WhiteAfrican* (blog). November 25, 2012. http://whiteafrican.com/2012/11/25/we-need-more-not-less/.

Hersman, Erik. 2015. A Busy Week for Tech Entrepreneurs in Kenya. *WhiteAfrican* (blog). August 5, 2015. http://whiteafrican.com/2015/08/05/a-busy-week-for-tech-entrepreneurs-in-kenya/.

Heuler, Hilary. 2015. Hub Life: Africa's Techies Have Found a Model to Bring Together Local Startups and Western Capital. *ZDNet*, January 6, 2015. http://www.zdnet.com/article/hub-life-africas-techies-have-found-a-model-to-bring-together-local-startups-and-western-capital/.

Hopkins, Curt. 2015. How Africa Grew More than 200 Local Tech Scenes. *Daily Dot*. October 5, 2015. http://www.dailydot.com/debug/africa-tech-hubs-hacker-spaces-incubators/.

IBM. 2013. Building Africa's Innovation Ecosystems. Point of View Essay. IBM Communications, Growth Markets. Armonk, NY: IBM Corporation. http://ihub.co.ke/ihubresearch/jb_BuildingAfricasInnovationEcosystemspdf2013-5-20-08-10-38.pdf.

iHub. 2016. The iHub—Technology Innovation Community. July 14, 2016. http://ihub.co.ke/about.

infoDev and CAD. 2014. *Do mLabs Make a Difference? A Holistic Outcome Assessment of infoDev's Mobile Entrepreneurship Enablers*. Washington, DC: infoDev, World Bank.

Jorgic, Drazen. 2014. Kenya's Technology Push Leaves Investors Cold. *Reuters*. December 31, 2014. http://www.reuters.com/article/kenya-tech-idUSL6N0UE15920141231.

Kalan, Jonathan. 2014. Inside East Africa's Technology Hubs. *BBC Future*. November 18, 2014. http://www.bbc.com/future/story/20121018-inside-africas-technology-hubs.

Kelly, Tim. 2014. Tech Hubs across Africa: Which Will Be the Legacy-Makers? *Information and Communications for Development (IC4D) Blog*. World Bank. April 30, 2014. http://blogs.worldbank.org/ic4d/tech-hubs-across-africa-which-will-be-legacy-makers.

Kelly, Tim, and Rachel Firestone. 2016. *How Tech Hubs Are Helping to Drive Economic Growth in Africa*. Background Paper for the *World Development Report 2016: Digital Dividends*. Washington, DC: World Bank. https://openknowledge.worldbank.org/bitstream/handle/10986/23645/WDR16-BP-How-Tech-Hubs-are-helping-to-Drive-Economic-Growth-in-Africa-Kelly-Firestone.pdf.

Koltai, Steven R., Victor K. Mallet, and Matthew Muspratt. 2013. *Ghana Entrepreneurship Ecosystem Analysis*. Report prepared for the UK Department of International Development. October 25, 2013. http://koltai.co/wp-content/uploads/2014/02/KolCo-Final-Report-DFID-Ghana-Entrepreneurship-Ecosystem-Analysis-FINAL-REDACTED1.pdf.

Korff, Valeska P., Achim Oberg, and Walter W. Powell. 2015. Interstitial Organizations as Conversational Bridges. *Bulletin of the American Society for Information Science and Technology* 41 (2): 34–38.

Lagarde, Christine (@Lagarde). 2015. Innovative Technology Projects like kLab Are Vital to Rwanda's Future. @klabrw. Twitter, January 28, 2015, 9:54 a.m. https://twitter.com/lagarde/status/560466067336990720.

Lamanauskas, Tomas. 2015. Why SMEs Are at the Heart of ITU Telecom World 2015. *ITU Telecom World*. October 7, 2015. http://telecomworld.itu.int/blog/why-smes-are-at-the-heart-of-itu-telecom-world-2015/.

Leonardi, Paul M. 2008. Indeterminacy and the Discourse of Inevitability in International Technology Management. *Academy of Management Review* 33 (4): 975–984. doi:10.5465/AMR.2008.34422017.

Lindijer, Koert. 2013. The Great Digital Leap Forward in Africa. *The Africanists*. July 14, 2013. http://theafricanists.info/the-great-digital-leap-forward-in-africa/.

Masuku, Andile, and Musa Kalenga. 2015. 04: Are Tech Hubs in Africa Effective? + The Week's Biggest News. *African Tech Roundup*, podcast, episode 4, 20:28. https://soundcloud.com/african-tech-round-up/atru004-are-tech-hubs-in-africa-effective-the-weeks-biggest-news.

MEST. 2016. About the Meltwater Entrepreneurial School of Technology (MEST). July 16, 2016. http://meltwater.org/about/.

Mian, Sarfraz A., Wadid Lamine, and Alain Fayolle. 2016. Technology Business Incubation: An Overview of the State of Knowledge. *Technovation* 50–51 (April): 1–12. https://doi.org/10.1016/j.technovation.2016.02.005.

Moraa, Hilda. 2012. *iHub_ Model: Understanding the Key Factors of the iHub Model. Nairobi: iHub Research. Accessed December 28, 2016. http://research.ihub.co.ke/downloads/*iHub_Model_Report_Final.pdf.

Moraa, Hilda, and Wangechi Mwangi. 2012. The Impact of ICT Hubs on African Entrepreneurs: A Case Study of iHub (Nairobi). Nairobi: iHub Research. Accessed October 12, 2018. https://ihub.co.ke/blogs/7990/ihub-entrepreneurs-report.

Mubaiwa, Kudzai. 2015. The Case for Business Incubators and Innovation Hubs. *Herald* (Harare, Zimbabwe). June 4, 2015. http://www.herald.co.zw/the-case-for-business-incubators-and-innovation-hubs/.

Ofori, Oral. 2016. hapaSpace Is the Newest Co-Working Hub in Kumasi. *The African Dream*. February 18, 2016. http://www.theafricandream.net/habaspace-is-the-newest-coworking-hub-in-kumasi/.

Oluwagbemi, Michael. 2015. How Startup Incubators in Africa Contribute to Entrepreneur Success. Venture Capital for Africa. July 1, 2015. https://vc4africa.biz/

blog/2015/07/01/how-startup-incubators-in-africa-contribute-to-entrepreneur-success/.

Oxford, Adam, and Duncan Jeffries. 2013. Why Tech Hubs Are a Key Part of Africa's Future. *Green Futures Magazine*. July 3, 2013. https://www.forumforthefuture.org/greenfutures/articles/why-tech-hubs-are-key-part-africas-future.

Perren, Lew, and Peter L. Jennings. 2005. Government Discourses on Entrepreneurship: Issues of Legitimization, Subjugation, and Power. *Entrepreneurship Theory and Practice* 29 (2): 173–184. doi:10.1111/j.1540-6520.2005.00075.x.

Phan, Phillip H., Donald S. Siegel, and Mike Wright. 2005. Science Parks and Incubators: Observations, Synthesis and Future Research. *Journal of Business Venturing* 20 (2): 165–182. doi:10.1016/j.jbusvent.2003.12.001.

Rose, Gillian. 2012. *Visual Methodologies: An Introduction to Researching with Visual Materials*. 3rd ed. London: Sage.

Ruef, Martin. 1999. Social Ontology and the Dynamics of Organizational Forms: Creating Market Actors in the Healthcare Field, 1966–1994. *Social Forces* 77 (4): 1403–1432. doi:10.1093/sf/77.4.1403.

Sanderson, Owen. 2015. On Hubs, BRCKs, and Boxes: The Emergence of Kenya's Innovation and Technology Ecosystem. Master's thesis. Fletcher School, Tufts University, Medford, Massachusetts. http://fletcher.tufts.edu/~/media/Fletcher/Microsites/IBGC/pdf/Student%20Research/Final%20Sanderson.pdf.

Saraswati, Jyoti. 2014. Konza City and the Kenyan Software Services Strategy: The Great Leap Backward? *Review of African Political Economy* 41 (sup1): S128–S137. doi:10.1080/03056244.2014.976189.

Shapshak, Toby. 2016. Africa Will Build the Future Says Zuckerberg, Visits Kenya on First African Trip. *Forbes*. September 1, 2016. http://www.forbes.com/sites/tobyshapshak/2016/09/01/africa-will-build-the-future-says-zuckerberg-visits-kenya-on-first-african-trip/.

Sherman, Hugh, and David S. Chappell. 1998. Methodological Challenges in Evaluating Business Incubator Outcomes. *Economic Development Quarterly* 12 (4): 313–321. doi:10.1177/089124249801200403.

Steyaert, Chris, and Jerome Katz. 2004. Reclaiming the Space of Entrepreneurship in Society: Geographical, Discursive and Social Dimensions. *Entrepreneurship and Regional Development* 16 (3): 179–196. doi:10.1080/0898562042000197135.

Toivonen, Tuukka, and Nicolas Friederici. 2015. Time to Define What a "Hub" Really Is. *Stanford Social Innovation Review*. April 7, 2015. http://www.ssireview.org/blog/entry/time_to_define_what_a_hub_really_is.

Treisman, Loren. 2014. Indigo Trust, DOEN Foundation and Hivos Foundation Launch a Fund for African Tech Hubs. *Indigo Trust.* June 13, 2014. https://indigotrust.org.uk/2014/06/13/indigo-trust-foundation-doen-and-hivos-foundation-launch-a-fund-for-african-tech-hubs/.

Treisman, Loren. 2015. The Long Road to Stimulating Tech Innovation in Africa. *Huffington Post UK.* September 2, 2015. http://www.huffingtonpost.co.uk/loren-treisman/tech-innovation-in-africa_b_8071288.html.

VC4Africa. 2014. *2015 Venture Finance in Africa.* Venture Capital for Africa. December 2014. http://www.aspeninstitute.org/sites/default/files/content/docs/resources/Summary%20VC4Africa%202015%20Report%20-%20Venture%20Finance%20in%20Africa.pdf.

Vourlias, Christopher. 2015. Lowered Expectations for Ghana's Hope City? *Aljazeera America.* April 19, 2015. http://america.aljazeera.com/articles/2015/4/19/lowered-expectations-for-ghanas-hope-city.html.

Wakoba, Sam. 2014. You Are the Hope of Africa, Ban Ki-Moon Tells iHub. *TechMoran.* October 31, 2014. http://techmoran.com/hope-africa-ban-ki-moon-tells-ihub/.

Wangari, Brenda. 2015. Inside the iHub, Kenya's Widely Acclaimed "Tech Headquarters." *Techpoint.ng.* September 15, 2015. https://techpoint.ng/2015/09/15/inside-the-ihub-kenyas-widely-acclaimed-tech-headquarters/.

9 Hackathons and the Cultivation of Platform Dependence

Lilly Irani

Introduction

How do you popularize an infrastructure? The question may seem strange. Infrastructure, and especially digital infrastructure, can seem an unalloyed good—that which enables others' productive activities. Development projects often present places as in need of infrastructure—power lines, phone lines, computers, or Internet networks. Infrastructures promise circulation and mobility. They promise the transformation of ideas, the movement of bodies, and the possibility of progress (Larkin 2013). But what if infrastructure exists but people have not taken it up? This has been the case in many parts of the world with the networks of computers, protocols, and fiber optic cable we call the Internet. For the last three decades, the Internet has marked the edge of modernity for many policymakers and users (Burrell and Anderson 2008). To policymakers and economists, the Internet was a "general purpose" technology (Brynjolfsson and McAfee 2014); they cast it as more than a tool—a technology that enables a wide range of innovations. Others read it as inherently democratic (Chan 2013). This, as Friederici argues earlier in this volume, is the "hope and hype" of the Internet.

Despite this "hope and hype," policymakers have faced a largely unused Internet in many parts of the "developing" world. A 2011 report by the World Bank's Independent Evaluation Group (IEG) looked back on almost a decade of information technology investment and found contradictory results. Where people had Internet access, they still often did not use it (IEG 2011, 11; Best and Kenny 2009). This contrasted with mobile usage (Best and Kenny 2009) and mobile penetration, which grew from 10 percent in 2005 to 85 percent in 2009 (IEG 2011, xiii). The IEG recommended that the World Bank encourage the development of IT applications that would "capitalize" on the extensive investment in Internet infrastructure already made. The Internet was supposed to be a global technology. But its globality seems late in arriving.

For those heavily invested in the Internet—financially, materially, or ideologically—the problem has been how to fulfill its promise. The core question of this article becomes, then, how do powerful institutions and firms popularize an infrastructure in which they have already invested? Hackathons offer one powerful technique. The hackathon is a short event—often lasting a day or two—where organizers invite people to imagine and prototype software applications. The format provides people an opening to tinker, play, network, and create prototype technologies, often for organizations other than those that employ them. With hackathons, organizations open up technical production to more public participation. The World Bank, for example, has organized a water hackathon to attract entrepreneurial coders to create apps that make the Internet more useful than it has been (World Bank 2012b). Crowdsourcing companies hold hackathons to popularize the idea of the data-processing gig economy. The Gates Foundation, Facebook, and the US State Department hold hackathons to enlist people all over the world to rapidly build new applications using sanctioned infrastructures and, implicitly, to address conveners' problems. These powerful hackathon conveners invite programmers to solve problems in ways that valorize conveners' infrastructures, co-opting hope and labor in one hacked experiment after another (Zukin and Papadantonakis 2018).

In this chapter, I examine what happens when organizations ask people to imagine the Internet and its platforms as answers to social needs, including a look at the limits of prototyping apps in short time frames as a way of addressing those needs. My analysis focuses on three cases of hackathons, which I approach both ethnographically and historically. Two cases are drawn from an ethnographic study of development practices spanning South Asia and Silicon Valley. The third is from an examination of primary World Bank documents that report on the challenges of ICT policy and interventions at the bank—challenges to which hackathons appeared as one best practice solution. From the vantage point of these three case studies, I argue that hackathons invite participants to innovate, but on a set of platforms and infrastructures that, while enabling fast prototyping, also heavily delimit the range of technologies the event can produce. In some cases, the infrastructural investments of hackathons are the material background conditions that enable the event but that escape the notice of participants. In many cases, however, organizations convene hackathons precisely to mine participants' activities for "legitimate futures" that extend the organizations' digital reach and use.

What Are Hackathons?

Through hackathons, entrepreneurial producers experiment with how to make something useful out of existing bodies of digital code and infrastructures. These intense software production events have instantiated a cultural form that originally developed in open source production communities. Hackathons began as a way for participants in globally distributed open source projects—those already invested in an infrastructure—to work together, face-to-face, for a short time. Face-to-face, programmers who normally only connect online can quickly locate and fix bugs in project code by pointing, talking, and guiding attention and collaboration with their whole bodies. These hackathons have allowed for intense collaboration among programmers with pre-existing deep ties to the open source community (Coleman 2013, 209).

While the early, open source hackathons often focused on improving, repairing, and maintaining shared infrastructures, hackathons have grown to include speculation about technological futures. Facebook regularly hosts both internal and public hackathons to explore future projects and to inculcate in employees the ethos to "move fast and break things" (Fattal 2012, 940). Institutions as large as the United Nations or as small as a coworkspace might put on hackathons to brainstorm about organizational problems, energizing volunteers to generate large numbers of approaches to the issue at hand. Such hackathons might generate ideas for social ventures, tools for mapping water in crisis regions, or prototypes of future startup offerings. These events often fail to produce actual working technologies (Lodato and DiSalvo 2016), but they are more than just a way of exploring possible futures—hackathons can also become rehearsals for future employment, partnerships, or investments. The events often end with participants showing off their work to venture capitalists, philanthropists, or recruiters—those with the power to invest money, time, and connections in the software futures on display.

In recent years, companies, NGOs, universities, and even government agencies have taken up hackathons as a means to recruit volunteer labor, generate interest in social or technological platforms, and develop new partnerships. In 2012, Infosys partnered with the World Bank as part of a global sanitation hackathon (Infosys 2012), and, in 2013, nonprofits and government bodies across the United States participated in a National Civic Day of Hacking, an intense Saturday of coordinated digital volunteerism (Knell 2013). Independently, the Gates Foundation and Facebook organized HackEd, a hackathon to turn massive data sets (and the background specter of surveillance) into the promise of education apps (O'Dell 2013). More recently, the Government of India has offered up education data sets at an OpenEducation AI hackathon sponsored by IBM, Amazon Web Services, Google, and Indian education startups (OpenEd.ai 2017). Hackathons proliferate as a space that allows firms to explore hires, investment, and ideas that might not otherwise readily emerge within the culture of the organization. Crucially, conveners promote their own and partners' infrastructures as the bases for this exploratory labor.

Tapping Labor, Expertise, Relationships, and Political Hope

In this section, I explore the uses to which organizations put hackathons as a widely deployed organizational form. Broadly, I argue that we can examine hackathons as sites that allow conveners to access labor, affective knowledge, and relationships.

First, labor. Most simply, hackathons invite participants to provide free experimental labor. Cultural scholar Melissa Gregg argues that civic hackathons invite citizens to donate labor to governments starved by austerity measures (2015). With Gregg, I argue here that hackathons solicit donated labor, specifically research and development labor. But even where austerity does not hold—such as in profitable corporations and design firms—organizations seek donated labor to tap the resources of those outside their bounds. Hackathons are not only austerity measures.

Second, hackathons invite participants to develop their own projects, resources, and desires in response to the convener's agendas and provocations. The events become a kind of postmodern software laboratory. As shown in critiques of top-down software design and top-down development initiatives alike, development institutions and companies employ a range of techniques to characterize market preferences, symbolic meanings, and practices. Hackathons invite participants to generate varied experiments in meaningful technology. With hackathons, conveners tap participants' varied imaginations and tacit knowledge to point the hosting organization in novel directions.

Third, hackathons allow conveners to explore potential relationships with participants without commitment (Jones, Semel, and Le 2015). The events enable convening organizations to explore potential relationships through concrete joint activity that might reveal something of the viability of the partnership in the longer term. In this sense, hackathons masquerade as participation but might more accurately function as interview and evaluation spaces. This becomes particularly crucial when international development projects rely on partnerships and networks among NGO, firm, and state actors.

Fourth, hackathons tap political hope and redirect it into exploratory research and development. Several scholars of hackathons have pointed to histories of issue-based activism that make hackathons meaningful for participants (Lodato and DiSalvo 2016; Schrock 2016). Hackathons also appear as one response to critiques of development as a form of universalizing top-down expertise (Elyachar 2012, 117). In place of experts sent by foreign countries and agencies, hackathons invite citizens to act as innovators of their own lives, as well as those of their neighbors. They channel people's own frustrations with development into donated research and development labor. Elsewhere, I argue that hackathons are a mechanism to train citizens to become entrepreneurial agents of development, with potentially antidemocratic consequences (Irani 2015b).

With hackathons, then, conveners tap local labor, cultural knowledge, relationships, and hope to search for what value can be made from their existing infrastructural investments.

Searching for Value at the Margins of Platforms

The three hackathons that form the focus of this chapter took place in 2011. What connects them is the Internet, its platforms, and digital toolkits as promising infrastructures. The first case is a hackathon organized by a design studio in Delhi, which demonstrates how hackathons translate expansive political hope into more limited projects that extend the value of already existing infrastructures. In the second case, the World Bank coordinated a set of simultaneous water hackathons in cities around the world. Like the Delhi hackathon, the World Bank events drew participants with diverse hopes and relationships nearer to the bank and its partners. The bank case demonstrates how organizations can use hackathons to locate partners already amenable to pre-existing agendas. I conclude with the case of a hackathon organized by a nascent crowdsourcing startup in Silicon Valley. This case makes clear that hackathons do more than extend the value of existing infrastructures; they can also legitimize infrastructures whose validity is in question.

OpenGovernment in Delhi: Hitting the Limits of Extant Infrastructure

A Delhi "innovation and strategy" studio, DevDesign, served as translator for the first hackathon, searching the lives of the marginalized for their needs and desires. Though DevDesign staff members were usually employed as ethnographers for hire, they also spun off their own initiatives to find opportunities for projects to pitch to funders. The hackathon was one such initiative, organized to generate possible futures. At other times, the staff would casually brainstorm, develop hypothetical project proposals, or even pursue side projects at work. Studio members could later winnow these proliferating projects to those worth pursuing at any given time.

Vipin, a senior consultant at DevDesign, organized this hackathon around the theme of open governance. Vipin was a former Accenture management consultant who had long dreamt of ways to engineer improvements in society. He consulted with the Gates Foundation. He lunched with Ford Foundation officers. As a graduate of the Indian Institute of Technology (IIT) and Indian Institute of Management (IIM), both government-funded schools, Vipin was a product of long-term state investments in technical and organizational education. He carried PDF slide decks on his laptop, ready to show training programs and digital platforms to potential funders and partners as he moved through Delhi's development worlds. Vipin was the grandson of an Indian Administrative Service officer, carrying on a family occupation in its postliberalization form—the public-private partnership.

With the hackathon, he solicited applications through his professional and personal networks, as well as via email lists for development workers, like Idealist.org. Out of about thirty applicants, he chose three software engineers from Delhi and Hyderabad, an Ivy League political anthropologist based in Delhi, an American designer working between Nairobi and London, and me. I had worked at the DevDesign studio for ten months at that point. We would all work together for five days.

British Council sponsorship funded travel for the American living in London. The studio poured some of its own resources into funding accommodations for hackathon participants. The hackathon was just one of several workshops at a Delhi festival celebrating and teaching activism, entrepreneurialism, and development to students and professionals. Other workshops included designing craft programs for an ashram NGO in Ahmedabad and developing solar power initiatives in Auroville, a UNESCO-recognized experimental community in Tamil Nadu. The workshops all brought together people who did not know each other to spend a few days dreaming up development projects and then making those dreams concrete, as demos, plans, and presentations. They were sites of experimental production. The hackathon was just the digital version of these promissory sites of experiment.

The festival and the hackathon within it were funded by the DevDesign studio and by several European cultural institutions. The studio invested in the festival as a way of building a "scene" of like-minded people interested in development, nonprofit work, and cultural experiments in filmmaking, literature, and innovation. Through a scene, the studio would be able to find potential clients, potential contractors (animators, artists, and translators, for example), and potential funders, whether for-profit or nonprofit. The scene brought together a set of people, resources, and sensibilities around forms of life, entertainment, and reason. The Delhi festival as a whole also offered an audience for European cultural institutions and their soft diplomacy efforts to build up "creative economies" by finding Indian business partners for Switzerland, France, Germany, and the UK. For DevDesign members, these European institutions offered connections to "global" perspectives and links to networks with resources and potential clients. Attendees paid 1,000 rupees, or US$20, to connect to the scene, to learn of others' work, and to sustain their hope amid entrepreneurial precarity and the apparent high failure rate of development projects.

Many of these motivations were at play for those of us at the hackathon. As we ambled into the studio at nine o'clock the first morning, the cook handed us chai, and we sat with laptops open at a long table. The convener had us introduce ourselves and share our motivations for attending. Many of us spoke to the seduction of tangible action—of making and doing something that goes beyond mere description, or complaint. A young software consultant from Bangalore wanted to quit "cribbing" about governmental inefficacy to "see if we can make a difference." An IIT-trained designer, he wanted to see if design could actually "save the world" instead of just "making posters" for clients about doing so. The convener (Vipin), a startup founder, wanted to help citizens "like him" direct their energy into "good governance." I was there to see what would happen if I brought together my anthropological sensibilities, which are critical of development, and my coding skills to attempt technology as a critical practice—I came to the hackathon with genuine hope. Prem, a legal anthropologist, came because, in his words, "anthropologists sit and critique things but they never get around to doing anything." All the speech act theory in the world left him still wanting to experiment with other forms of intervention.

The hackathon was a device for translating these various desires, backgrounds, and political sensibilities into experimental labor and promising demos. Three of us were consultants in various capacities who hoped to sustain our demo through some indeterminate form of financial support: maybe a startup, a grant proposal, or a state contract.

Each of us brought different forms of expertise. Three of us had worked as professional software designers. Three were working software engineers. Prem, the anthropologist, came to the table with two years of research with land rights movements, both in remote Uttar Pradesh and among Delhi activists. His accounts of political struggle grounded our collaborative imagining (Murphy 2005) of what a social media system might do, in what form, and for whom.

We each also drew on different networks of resources and social ties. I set up a meeting with a consultant to the Government of India Planning Commission. We peppered him with questions to explore how we might locate a partner within the government. To explain the law-making process to us, Vipin brought in a friend whose NGO worked with the Indian Parliament. He suggested that his NGO might be interested in finding funders to carry forward what we prototyped in those five days. Vipin later told us he also knew program officers at the Ford Foundation who were "looking for inspiration from a good project"—he promised these social connections to diversify his portfolio of potential investors in the future of whatever software we would produce.

Hackathons and the Cultivation of Platform Dependence 231

Crucially, we also came to the table with varied visions of politics. Vipin was, in the end, a technocrat. He described the law as codes—encodings of incentive and punishment. He wanted to "open government" by allowing citizen-experts—lawyers, consultants, and other highly trained citizens—to find the loopholes, bugs, and design flaws in the law. He was interested in governing, not politics. Another software engineer, Ravi, mostly tinkered quietly on his computer. Occasionally, he raised his head to ask when we would be done deliberating so we could schedule deadlines to engineer the prototype by the end of the hackathon. Others, however, leaned toward a messier sense of politics as struggle. Prem, an avowed Marxist, studied people's struggles to win and gain rights to land from the state, despite face-offs with police, mining companies, and henchmen. Prem, and many of us with him, did not share Vipin's faith in elite experts as a substitute for the politics of the poor. Dinesh, a programmer with a penchant for painting and feminist science studies, told tales of his bicycle tours of rural Maharashtra, arguing for the technological and political savvy of the villagers he met along the way.

Our challenge was to converge on a project recognizable as open governance—the theme of the hackathon—that each volunteer would be willing to labor toward. Prem and Vipin staked out opposing positions. The process of debate, however, brought to the surface different accounts of reality, different theories of politics, and different imaginations of what *could* be possible. Out of our conflictual assembly of competing epistemologies—the technocrat and the Marxist, for example—we found a concept that most of us were excited to pursue. In Vipin's absence, the rest of us decided to work on a platform that tracked Indian parliamentary debates on bills. The platform would enable highly literate activists to track issues affecting movements they were involved in and allow organizers to document the face-to-face deliberations of poorer constituencies around central government issues.

These intense debates are central to how hackathons generate innovation. Sociologist David Stark characterizes innovation as the process of the search for opportunities amid multiple possible orders of worth (Stark 2009, xvii–xviii). The sorts of tensions we marked and managed are common features of such gatherings. The challenge was to make sure that arguments about facts did not get in the way of arguments about what could come to be. DevDesign actually benefited from the differences of opinion among us; out of those conflicts, we identified risks to the projects and ways of recognizing potential value in a complex world.

The hackathon seemed to accommodate more leftist politics, but the manufactured urgency and discipline of the demo pressed these politics in service of entrepreneurial insights. The activist support software, Prem warned us, would require "some REAL footwork" to get "on the street" and work with existing organizations that were thinking in terms of political participation. That week, we weren't on the street. We were in the studio. The hackathon afforded us little time to reach out to NGOs or activist networks. We had little time to understand their information practices or to build trust with them. We could not even promise maintenance of any demo to come out of a potential collaboration. Our work in the hackathon could only draw on the knowledge, desires, and relationships we brought into the room with us. Out of such materials and existing alliances, we were to fashion promising opportunities for philanthropists, investors, and volunteers. The time, tools, and skills in the room were geared toward prototype work, not footwork.

Even the kinds of prototype work we could undertake were limited by the political economies of Internet production in a country where few have direct access to the Internet. When we learned that only 10 percent of Indians have Internet access, we thought about alternative ICT infrastructures—phones or radios, for example. Krish, a software engineer, explained to us that in the long term, the project could get into rural areas through interactive voice response phone systems, rural kiosks, or SMS-based systems. "In Andhra Pradesh, there's a women's radio station," he told us. "The scope of what we want to envision is THAT. What we implement in five days is probably a website." The skills in the room were of the web; web tools were those most at hand for urgent hacking. He continued, "So we're going to go to a conversation where we'll chop off everything. Cut. Cut. Cut. Cut. But if there's a master document that accompanies this chopped up little thing …" he trailed off. The hackathon was an experiment in making prototypes of promising projects, constructing "opportunities" by drawing on bonds and resources already in hand. In the momentum of the hackathon, we had to build on existing infrastructural orders; there was little time to critique let alone challenge the power relations produced by the elite infrastructures on which the hackathon depended.

The World Bank "Water Hackathons": Searching for the Value of Infrastructural Investments

Like the DevDesign, the World Bank also wanted to generate new ideas, networks, and knowledge for projects. They accomplished this by organizing a global hackathon held simultaneously in ten cities in 2011.

A 2011 IEG report that looked back on almost a decade of information technology investment found contradictory results. On the one hand, the report found that mobile penetration grew globally from 10 percent in 2005 to 85 percent in 2009 (IEG 2011, xiii). On the other hand, Internet access had largely not reached the poor (11–13). Where people actually had Internet access, they often didn't use it (11).

The applications that could turn all that connective infrastructure into something useful have proven difficult to produce and manage. The bank has found it challenging to shift gears from large-scale government and corporate infrastructure projects to smaller ICT applications that need to fit into and gain buy-in from myriad users in diverse contexts to work (e.g., IEG 2011, xvi). Further, the bank lacked enough of its own IT experts who could support projects as need arose.

The World Bank's ICT group, one of the subjects of the evaluation, responded with a 2012–2015 strategy that explained how the group would address these issues. It stated that the bank needed to cultivate pools of external experts, "stimulate private sector and civil society development of applications," and focus on "service delivery"—an area of ICT investment where bank managers could offer the sorts of expertise and connections that venture capitalists, other banks, and most private investors could not. The ICT group also needed a strategy that would enable it to collaborate with other sectors of the bank. ICTs, the report noted, affect services across sectors—water, education, e-governance, employment—but IT staff, however few, were contained in one group.

The ICT group seized on the hackathon as a means to respond to these challenges and generate promising loan targets (see Weaver 2008, 735) across the private sector and civil society. In its 2012 strategy, the ICT group described hackathons as a way to "co-create services and applications with citizens and businesses" (World Bank 2012a, 7). The strategy described hackathons and app contests as ways to mobilize citizens and technologists as "a pool of creativity" to close a "service delivery gap" that many governments did not even know existed (7). These events called on citizens to translate their tacit knowledge and frustrations into investable applications. These applications, the bank hoped, would make good on the promise of all the ICT and broadband infrastructure that other World Bank projects had funded.

In October 2011, the group convened a global hackathon in ten cities, including Nairobi, Bangalore, Cairo, Tel Aviv, and Washington, DC (World Bank 2012b, 55). The organizers subsequently published a report, *Water Hackathon: Lessons Learned*, as a World Bank Research Paper, explaining how the hackathon could become a model approach to development. With the hackathon, the organizers sought to raise "awareness of water sector challenges ... among technical communities in-country and globally" (vi). In addition to awareness, the report continued, the organizers sought to create "a network of atypical partners engaged in finding solutions to water-related challenges," a "preparation of a list of challenges facing the water sector," and "adoption of new applications" in World Bank projects. The hackathon white paper describes the role of these partners as not only working on "locally identified problems" but also supporting "local community building by leveraging existing networks and recognized local champions" (5). The events, then, allowed the bank to bring existing, local, trusted, and productive relationships into its orbit to generate investable futures. The hackathon was thus a way for the bank to create a map of challenges, opportunities, coders, and relationships that could make the Internet matter locally.

These zones of experiment allowed representatives of private sector and "expert" organizations to discipline the dreams of hackathon participants. The city hackathons invited sponsors to offer problem statements as well as prizes. In Cairo, Pepsi offered cash prizes, while the agribusiness Farm Frites, Egypt's largest potato grower (World Bank 2012b, 29), posed irrigation problems that programmers might tackle. In Bangalore, Hewlett Packard, government ministries, and Pepsi were among the local partners (54). In Lagos, organizers consulted water experts and decided to focus on gray water recycling and borehole sharing. Judges from Google and Nokia guided and ultimately judged participants' projects.

The hackathons offered a way for the bank to search for futures. The futures were not just bits of software or even demonstrations of software. They were demonstrations of particular assemblages, or comings together, of people and skills, passions, and relationships.

CrowdHack: Legitimizing Crowdsourcing Infrastructures

In Silicon Valley, another hackathon attempted to make crowdsourcing infrastructures relevant to engineers and the public. The startup Cloud-Factory organized a two-day hackathon in 2011 around the question of what can be done with a programmable workforce—a way of organizing the labor process referred to as "crowdsourcing" or "human computation" in high tech industries. CloudFactory, a human computation company, staged the event as a competition held before an industry and academic conference called CrowdConf.

CrowdConf convened engineers, academics, investors, journalists, and managers in imaginative and discursive work with financial implications. The conference, and the hackathon held as part of it, was a place to both explore and hype the value of crowdsourcing. Crowdsourcing as a high tech sector was still in formation in 2010. It refers to various ways of producing value out of networked digital labor. Journalist Jeff Howe coined the term in 2006 to describe Web 2.0 companies that solicited work from people through their computers and phones. "Human computation" services allowed programmers to outsource large volumes of data-processing work on demand and pay-as-you-go. Moreover, coders could outsource the work by algorithm, incorporating human work output directly into their code. CrowdConf, convened four times between 2010 and 2013, assembled those curious about and heavily invested in the "past, present, and future of crowdsourcing" (CrowdConf 2010, n.p.). The events, the press releases, the talks, and the hackathon all generated substance and created significance for crowdsourcing as something more than just a fancy name for outsourcing. CloudFactory and its competitors, like CrowdFlower and Amazon Mechanical Turk, collaborated in staging these events to build up public legitimacy and to engineer interest and investor taste for the sector.

One thing the crowdsourcing industry has had to fight is the perception that it is just another way to outsource anxious Americans' jobs. And there certainly are continuities with outsourcing; one CrowdConf speaker had spent years at McKinsey advising corporations how to outsource their work to India. At the conference, he outlined the gaps in outsourcing that more fragmented, contractual, and unpaid crowdsourcing workforces could fill.

CloudFactory, the hackathon host, stressed the ethical dimensions of its business model. During one of the conference sessions, the founders of CloudFactory described the company's origin in their travel from the United States to Nepal, where they "discovered an amazingly talented group of people" living in villages but making very little money (Sears 2017, n.p.). The CloudFactory founders built the company around enabling programmers to build automated processes that call on those talented Nepalis to do work. Their story echoed that of Samasource, an outsourcing company that promises to create jobs, rather than aid, for women, refugees, and youth living in poverty (Lehdonvirta and Ernkvist 2011). The CEO cited her first job managing an Indian call center as her inspiration to place call centers directly in slums (Abate 2014).

Crowdsourcing advocates emphasized the new kinds of technologies crowdsourcing made possible. "A lot of people don't get it," Karl, the CEO of an ethical crowdsourcing company, griped to me. "They're just trying to do outsourcing cheaper," he explained. He went on to explain how his company paid workers decent wages in India and hoped to make new kinds of programming possible. His goal, he explained, was to "create something with real value—apps that benefit everyone." His optimism was common among those at the conference who saw their love of technology as an interest in human well-being. Yet, this vision of "everyone" elided questions of which people labored and who reaped the benefits (Irani 2015a; see also Vora 2015).

Karl had attended CloudFactory's hackathon to explore just what "human computation," as a platform for programmers, could make possible for humanity. The rewards were few. In invitations to the event, organizers promised, "All hackers get caffeine (loads of it), pizza, glory (of course), and a limited edition CloudFactory t-shirt" (Allick 2011). They offered winners an "on-stage shout out" at the single-track conference and the chance to demonstrate their app in the exhibit hall. The hackathon began with workshops to teach participants how to use the platforms of choice—CloudFactory, CrowdFlower, Twilio, and GitHub. Ten teams spent the day intensely coding, absorbing the energy of their fellow hackers, and developing prototypes of computer applications that incorporate human computation. Hackathon participants went home to sleep while crowds of workers across the world worked through the data-processing tasks that were designed into the apps.

Hackathons and the Cultivation of Platform Dependence 237

The winning projects, later described on stage and in press releases, drew on commonsense notions of good or "cool" circulating at the San Francisco Bay Area conference. One winning team built an app that rated photographs of moles for melanoma; the app employed CrowdFlower's APIs to connect to workers in Nepal, who rated each photo for signs of melanoma. Another winning project used barometers on people's Android phones to collect and aggregate weather data. A third winner developed an app called "Clean up India." The developer used CloudFactory's APIs to recruit people in India to go outside and tidy up a park or street. Workers sent before and after photographs as evidence of their labors. Press releases after the conference advertised the apps. The conference organizers also announced the winning applications immediately after a panel on how crowdsourcing generated "philanthropy" by hiring workers in poorer countries.

CloudFactory explored how programmers—from different companies, with varied cultural imaginations—might make use of the digitally mediated labor platforms evangelized by the tech industry. Like the design studio and water hackathons, this hackathon invited participants to draw on their own knowledge, networks and desires to generate the seeds of future technologies. Like those other hackathons, this one asked participants to dream in forms that made existing infrastructural investment—here, in crowdsourcing APIs—relevant and valuable. Specifically, the infrastructure here was not only the Internet but the computationally organized labor of far-flung others—people available to work at costs lower than those already in the organized sector. By spectacularly demonstrating what good could come of crowdsourcing, hackathon winners bolstered the legitimacy of an industry and an infrastructure hampered by concerns about the ethics of globalized IT and labor.

Hackathons and the Production of Inclusion

The hackathons described here offer an insight into the politics of inclusive development through processes of software production. In each of the three cases, the conveners framed hackathons as sites of participation and inclusion. Inclusion, a watchword of development since the first years of this century, can mean many things and head off many possible critiques of globalization and development.

The Legitimacy of the Inclusion

In one sense, hackathons promise inclusion by opening innovation to the desires of those beyond the walls of private firms. A hackathon can convene people to make a technology seem like a platform that empowers local actors to create, rather than being perceived as the imposition of a mediating technology in a social space.

CrowdHack promised this form of inclusion by inviting academics and Bay Area software engineers to play on CloudFactory's platform in the presence of the company's founders. As people hacked on CloudFactory's platform, they could make suggestions to the company about how to improve it. Their hacking generated knowledge with which CloudFactory engineers could valorize their platform. The company selected and publicized the top hacks to publicize, legitimize, and hype their platform and crowdsourcing sector systems more widely (Sunder Rajan 2006; Chan 2013). Inclusion, here, functioned to harvest tacit knowledge and cultural sensibilities from those beyond the firm's walls. The hackathon was an instrument to facilitate this harvesting for the valorization of the crowd platform. Here, hackathons fit with a wide range of corporate techniques for harvesting innovative uses and knowledge from beyond the firm. These techniques are often popularized and formalized as "open innovation" (Von Hippel 2005).

The World Bank hackathon generated not only knowledge, but also legitimacy for the enterprise of development. The bank had long faced criticism for the performance and politics of its top-down projects. Decisions from the top frequently mismatched the needs and social desires at the grassroots (Rao and Walton 2004). This discourse of development located the West as the source from which development knowledge and modern forms of life diffused (Escobar 1991). The shifts to participatory development (Cornwall 2000) and "community-centered approaches" (Escobar 1991) were two responses to this. Microfinance and "bottom of the pyramid" approaches that invest in the poor as entrepreneurs were another (Elyachar 2012). The World Bank hackathon also appeared to empower middle-class professionals—another answer to these critiques. The bank, in *Water Hackathon: Lessons Learned*, emphasized the importance of the "local" in making the global through the language of "authenticity" (World Bank 2012b, 11). "A local tech partner," the paper advised, could not only help with local arrangements but also "lend authenticity" to the event (11).

Furthermore, "Hackathons," the guide warned, "should not come across as a branding exercise." The organizers designed these hackathons as sociotechnical devices to harness and fabricate authentic, local, and inventive energy and vision.

Of the three hackathons described in this chapter, the Delhi hackathon appears to be the most "authentic" following the language of the bank's white paper. The event was associated only with small Delhi firms, NGOs, and the British Council—a European cultural funder, but hardly one with the clout of the bank. Even then, the relative modesty of the hackathon bound participants to pre-existing Internet infrastructures that had been developed for wealthy places and people.

Leveraging the Local

The World Bank white paper framed hackathons as a way to "leverage" the local: local knowledge, local networks, and "local champions" in the service of bank goals and policy agendas (however negotiated). The bank acted not alone but in partnership with a range of multinational technology organizations. The bank commissioned an organization called Random Hacks of Kindness (RHoK) to organize the events and report lessons in the white paper. RHoK draws together resources from Microsoft, Google, Yahoo!, NASA, and the World Bank. What does this coalition leverage when it leverages the local?

From local knowledge, hackathon participants can generate possible ideas for software. They can identify risks to the success of the software, as well as possible desires the software might speak to. Hackathons are also a way by which conveners can open themselves to people from different social worlds. This contact zone between social worlds is not just a matter of good politics. It is a matter of recognizing value. Innovation is not the making of new things alone, Stark argues. Instead, it is recognizing what might be of value among many new things. Hackathons are one kind of organization where people come together in "heterarchy" (Stark 2009), bringing their varied understandings of worth to bear on the direction a project should take. Stark analyzed New York startups in the early days of the web, as workers, investors, and CEOs scrambled to search for what the web might be worth to US customers (Stark 2009, 81–111; see also Neff 2012). Teams within the startups Stark studied heterarchically convened designers, programmers, and marketers to assess germs of products according to varied regimes of worth. As ephemeral convenings, hackathons allow an organization to draw near a wider range of perspectives than available within the firm. For the convening organization, the talk and demonstrations at a hackathon can bring previously unrecognizable forms of value into view. It is a mistake, then, to see hackathons as only generating innovation from participants. They also allow conveners to innovate by allowing them to "leverage" varied local epistemes and cultural understandings.

Hackathon participants also bring their local networks and relationships into the room, including business relationships, trusted friendships, and family members—people through whom knowledge, investment, patronage, and regard might flow. Anthropologists have drawn attention to how development enterprises need to understand and mobilize existing social relationships—social relationships that exceed the developmental and economic templates of individuals in modern society. Jamie Cross and Anita Chan, for example, show how the One Laptop Per Child and solar lantern projects have become occasions for NGOs and companies to explore and create partnerships (Chan 2013, 189; Cross 2013). Julia Elyachar draws attention to how NGOs map and mobilize social relations among the poor, whether in Cairo's neighborhoods or in self-help groups in India, as "phatic" program infrastructure (2010). Elyachar argues that we ought to recognize these social relations as the product of "phatic labor"—the labor of everyday sociality that creates potential value (2010, 457). Through hackathons, conveners hope to draw close partners that might also bring near other pre-existing social relationships that can diversify the reach of the conveners' platforms. Hackathons need not completely subsume social relationships into the production of capital. Innovation requires difference; hackathons offer one way for capitalist production to tap into difference without taking responsibility for its shape or sustenance.

Hackathons leverage the local in a third way, as they convene what the bank report calls "local champions." A champion is an individual—driven by passion—who pushes, pushes, pushes to see an innovation adopted. The language of champions comes from Peters and Waterman (1984), eminent business consultants, and earlier, from theorist of innovation Donald Schoen (1963).[1] In searching for champions, institutions attempt to locate and marshal individuals, not for their labor time alone, but for the intensity of that labor as affective drive (see Vora 2015). A champion does not simply offer affective labor. A champion is one who will navigate obstacles, scheme, and hustle to pursue a goal. From among hopeful hackathon participants, the World Bank sought those motivated local translators and "non-traditional partners" who could move the bank's institutional interests forward (World Bank 2012b, 7).

These passions offer no guarantees of progressive outcomes. The World Bank taps human capacities to care for others through technology. It draws on locality to generate novel differences that might matter to people—those relevant information products that some might adopt. But these "local" forms of knowledge and affect can equally be humanitarian passion, ethnonationalist affect, an impulse to order others, or personal aspiration. Those affects are already stirred up through histories of capitalism, neoliberalism, and postcolonial nationalism. Hackathons channel those affects toward valorizing organizations' infrastructural investments.

Who Mediates the Local?

In asking who wins and who loses with hackathons, we should also ask who can participate in hackathons at all. Who mediates the "localization" of a global form (Mazzarella 2003)? Hackathon teams rely on easy and fast social relations to proceed. The Delhi hackathon allowed no time to do the "real footwork" of developing partnerships with other organizations and activists, work that did not fit within the scope of the hackathon. Though we could build some software in a couple of days, we had little time to explain our developing goals to members of activist networks. There was no time to build coalitions, align frames (Snow et al. 1986), or build trust with activists, NGO workers, landless villagers, or frustrated city dwellers. To get to the demo in five days, the people coming together had to be sufficiently similar, sufficiently flexible, and sufficiently few. The hackathon required fast trust and fast talk. The participants all spoke English fluently.[2] Even if hackathon team members share an alternative common language, English is the dominant language in programming worlds of practice (Takhteyev 2012). Major operating systems, programming languages, and toolkits require some interaction in English.

Hackathons also pull people away from spending time at home, getting rest, and caring for those not at the hackathon. The events rarely provide alternative care arrangements to substitute for the time participants put into the events. By contrast, hackathons often celebrate the self-sacrifice of actors who are willing to hack away a weekend with only pizzas as payment.[3] The Delhi hackathon attracted young, college-educated people without family obligations. In the name of participation, hackathons often fail to account for the forms of habitus and networks of care that enable some to participate while others cannot afford the luxury. Similarly, Crowd-Hack invited programmers, not human computation workers, to imagine the future of technology. The capacity to hack for days is, in part, the capacity to deprioritize one's obligations to others and direct one's attention to a landscape of IT infrastructures that have already been shaped elsewhere. At hackathons, institutions and firms stage openness while eliding histories of privilege that enable people to participate.

Cultivating Platform Dependence
In the name of local innovation, the three hackathons described in this chapter ultimately relied on pre-existing platforms to innovate. The World Bank hackathon and CrowdHack explicitly evangelized platforms. The Delhi hackathon conveners had no intention to evangelize a platform. We discovered in the work of hacking, however, that we had to rely on existing Internet and Web 2.0 code libraries and platforms. The very premise of a hackathon is that one can build intensely and quickly by drawing on a large stock of extant platform infrastructure. When our interests shifted to more broadly accessible and maintainable radio technology, there was no time to build, extend, or sustain such an infrastructure. The time pressures common to hackathons required us to forge ahead with infrastructures that were already dominant and ready-to-hand. Limited time forced us to pursue what the World Bank white paper calls "the low hanging opportunities" (World Bank 2012b, 15).

In computer science, this mode of problem solving is referred to as the "greedy algorithm." The bias to choose the easiest path often leads to less optimal solutions. The Delhi hackathon made clear that low-power radio would never be the lowest hanging fruit. This strategy of problem solving leads entrepreneurial technology makers to reaffirm the dominance of already dominant players, extending their reach into new niches of culture, imagination, and life rather than creating alternatives to such platform dependence.

Conclusion

In "Gens: A Feminist Manifesto for the Study of Capitalism," Laura Bear, Karen Ho, Anna Tsing, and Sylvia Yanagisako (2015) argue against accounts of capitalism that homogenize the multiple temporalities, spatialities, and relationalities that constitute life. Anna Tsing shows how global projects of capital generate "friction" when they hit the ground (2005). Supply chains are one way in which capitalists and their agents organize the movements of labor, materials, and people to manage these frictions and differences (Tsing 2009). Hackathons, I argue, are another.

Projects around digital economies often claim what scholar Anita Chan (2013) calls "the myth of digital universalism." Information technologies can seem multifunctional and "general purpose" (Brynjolfsson and McAfee 2014). As these projects work to commodify knowledge as code, patents, and information objects, they too encounter differences that can reveal the universalism as myth. Chan calls for a "digital interrupt" to draw attention to the frictions, protests, and difference that refuse to be subsumed into knowledge economy projects (2013, 177–194).

Hackathons, I argue, are one technique by which those invested in the Internet attempt to make it a global technology. Elsewhere, I have argued that hackathons can be pedagogical mechanisms. The hackathon in Delhi was part of a large festival of arts, technology, and even NGO events that evangelized an entrepreneurial ethos (Irani 2015b). The hackathon unfolded in a wider context of social impact competitions, philanthrocapitalism, and the rearticulation of Indian nationalism as the success of technology capitalists. The event offered an embodied, temporally compressed education in how to collaborate in small groups to take authoritative, visionary action. These hackathons build capabilities but also tap into the capabilities and relationships people have to expand the capacities of a given infrastructure. In convening participants to hack around institutional challenges, hackathons immerse participants in the problem framings offered by the institution. The manufactured urgency of these events recasts a highly delimited call to work on an institution's terms as effervescent challenge and journey. This urgency compresses deliberation. It celebrates those who can adapt to entrenched interests and make opportunity out of austerity.

Organizations invite people to bring difference out into the open and make it available for software innovation through the hackathon. Hackathons are one way organizations make difference knowable, manageable, and even profitable (Sanyal 2007, 96–97). Difference might be varied knowledge or the diverse social relations people mobilize in their local worlds. Conveners of hackathons might glean ideas and knowledge from event participants. They might hire promising teams, drawing closer members' existing social relations and cultural knowledge, and possibly even neutralizing them as competitors. Hackathons are not threatened by difference. They are one way institutions can selectively cultivate and support certain forms of difference as a mode of governance far softer than enforcement or discipline. Difference, then, is not necessarily a "digital interrupt" (Chan 2013). By drawing difference near, hackathons help convening firms and institutions expand their influence by incorporating difference into their engines of value.

Sociologists of hackers Johan Söderberg and Alessandro Delfanti (2015) locate a hacker ethos in the desire to turn technologies toward ends other than those originally intended. But as hackers lose their definition as a social movement group, with common goals and identity, they fragment into a multiplicity of users, causes, and issues. The more they become a divided multiplicity, Söderberg and Delfanti argue, "the more reliable source of innovation for firms they become" (2015, 795). These participants become resources for organizations, but hackathons offer paper thin resources for the participants. Participants offer their hope, their energy, and their knowledge. Yes, they experience the jouissance and craft of hacking (Coleman 2013). But what should participants do when they hit the limits of fast work? What should they do when they run up against the limits set by existing infrastructural investments? Perhaps we can turn the hackathon from a site of experimental, innovative production to a site of movement building. When we run up against the limits of what we can accomplish by accepting the resources already given, perhaps we can organize beyond our teams to demand more from development than making value out of what more powerful entities have assented to provide.

Notes

1. By the early twenty-first century, the canonical *Diffusion of Innovations*, by Everett Rogers (2003), had drawn the concept into the 5th edition.

2. Despite India's global visibility as an English-language service exporter, English skills are rare. Only 4 percent of Indians between eighteen and sixty-five spoke English fluently in 2005, and those fluent speakers were primarily members of the upper castes (Azam, Chin, and Prakash 2013).

3. Gloria Lin's (2016) undergraduate thesis argues that hackathons leave little time and room for the care of self and others, preventing participation from a more diverse range of people.

References

Abate, Carolyn. 2014. Leila Janah Helps People in the Developing World Find Work—near Home. *Christian Science Monitor.* October 23, 2014. https://www.csmonitor.com/World/Making-a-difference/2014/1023/Leila-Janah-helps-people-in-the-developing-world-find-work-near-home.

Allick, Mollie. 2011. Register for CrowdHack at CrowdConf 2011. Email. September 16, 2011.

Azam, Mehtabul, Aimee Chin, and Nishith Prakash. 2013. The Returns to English-Language Skills in India. *Economic Development and Cultural Change* 61 (2): 335–367.

Bear, Laura, Karen Ho, Anna Tsing, and Sylvia Yanagisako. 2015. Gens: A Feminist Manifesto for the Study of Capitalism. *Fieldsights—Theorizing the Contemporary* (blog). Cultural Anthropology Online. March 30, 2015. https://culanth.org/fieldsights/652-gens-a-feminist-manifesto-for-the-study-of-capitalism.

Best, Michael L., and Charles Kenny. 2009. ICTs, Enterprise and Development. In *ICT4D—Information and Communication Technology for Development*, edited by Tim Unwin, 177–205. Cambridge: Cambridge University Press.

Brynjolfsson, Erik, and Andrew McAfee. 2014. *The Second Machine Age: Work, Progress, and Prosperity in a Time of Brilliant Technologies.* New York: Norton.

Burrell, Jenna, and Ken Anderson. 2008. I Have Great Desires to Look Beyond My World: Trajectories of Information and Communication Technology Use among Ghanaians Living Abroad. *New Media & Society* 10 (2): 203–224.

Chan, Anita Say. 2013. *Networking Peripheries: Technological Futures and the Myth of Digital Universalism.* Cambridge, MA: MIT Press.

Coleman, E. Gabriella. 2013. *Coding Freedom: The Ethics and Aesthetics of Hacking.* Princeton, NJ: Princeton University Press.

Cornwall, Andrea. 2000. *Beneficiary, Consumer, Citizen: Perspectives on Participation for Poverty Reduction*. Stockholm: Sida.

Cross, Jamie. 2013. The 100th Object: Solar Lighting Technology and Humanitarian Goods. *Journal of Material Culture* 18 (4): 367–387. doi:10.1177/1359183513498959.

CrowdConf. 2010. Sponsor post: CrowdConf: 1st Annual Conference on the Future of Distributed Work. GigaOM. https://web.archive.org/web/20110302131841/ https://gigaom.com/2010/09/02/crowdconf-1st-annual-conference-on-the-future-of-distributed-work/.

Elyachar, Julia. 2010. Phatic Labor, Infrastructure, and the Question of Empowerment in Cairo. *American Ethnologist* 37 (3): 452–464.

Elyachar, Julia. 2012. Next Practices: Knowledge, Infrastructure, and Public Goods at the Bottom of the Pyramid. *Public Culture* 24 (1): 109–129.

Escobar, Arturo. 1991. Anthropology and the Development Encounter: The Making and Marketing of Development Anthropology. *American Ethnologist* 18 (4): 658–682.

Fattal, Alex. 2012. Facebook: Corporate Hackers, a Billion Users, and the Geo-politics of the "Social Graph." *Anthropological Quarterly* 85 (3): 927–955.

Gregg, Melissa. 2015. FCJ-186 Hack for Good: Speculative Labour, App Development and the Burden of Austerity. *Fibreculture Journal*, no. 25. http://twentyfive.fibreculturejournal.org/fcj-186-hack-for-good-speculative-labour-app-development-and-the-burden-of-austerity/.

Independent Evaluation Group (IEG). 2011. *Capturing Technology for Development: An Evaluation of World Bank Group Activities in Information and Communication Technologies*. Washington, DC: Independent Evaluation Group, World Bank.

Infosys. 2012. Sanitation Hackathon 2012 Announces Winning Solutions for India. Infosys press release. December 3, 2012. https://www.infosys.com/newsroom/press-releases/Pages/sanitation-hackathon-2012.aspx.

Irani, Lilly. 2015a. Difference and Dependence among Digital Workers: The Case of Amazon Mechanical Turk. *South Atlantic Quarterly* 114 (1): 225–234.

Irani, Lilly. 2015b. Hackathons and the Making of Entrepreneurial Citizenship. *Science, Technology & Human Values* 40 (5): 799–824.

Jones, Graham M., Beth Semel, and Audrey Le. 2015. "There's No Rules. It's Hackathon.": Negotiating Commitment in a Context of Volatile Sociality. *Journal of Linguistic Anthropology* 25 (3): 322–345.

Knell, Noelle. 2013. National Civic Day of Hacking Event Takes Many Forms Locally. *Government Technology*. May 30, 2013. http://www.govtech.com/National-Civic-Hacking-Event-Takes-Many-Forms-Locally.html.

Larkin, Brian. 2013. The Politics and Poetics of Infrastructure. *Annual Review of Anthropology* 42:327–343.

Lehdonvirta, Vili, and Mirko Ernkvist. 2011. *Knowledge Map of the Virtual Economy*. Washington, DC: World Bank.

Lin, Gloria. 2016. Masculinity and Machinery: Analysis of Self Care Practices, Social Climate and Marginalization at College Hackathons. Master's thesis. University of California, San Diego.

Lodato, Thomas James, and Carl DiSalvo. 2016. Issue-Oriented Hackathons as Material Participation. *New Media & Society* 18 (4): 539–557. doi:10.1177/1461444816629467.

Mazzarella, William. 2003. *Shoveling Smoke: Advertising and Globalization in Contemporary India*. Durham, NC: Duke University Press.

Murphy, Keith. 2005. Collaborative Imagining: The Interactive Use of Gestures, Talk, and Graphic Representations in Architectural Practice. *Semiotica* 1 (4): 113–145.

Neff, Gina. 2012. *Venture Labor: Work and the Burden of Risk in Innovative Industries*. Cambridge, MA: MIT Press.

O'Dell, J. 2013. Facebook & Gates Foundation Expand Their Education Hackathon. *VentureBeat*. April 2, 2013. https://venturebeat.com/2013/04/02/facebook-gates-foundation/.

OpenEd.ai. 2017. AI for Education Global HackWeek: July 28-August 11, 2017. OpenEd.ai. Accessed September 21, 2017. http://opened.ai/index.html.

Peters, Thomas J., and Robert H. Waterman. 1984. *Search of Excellence: Lessons from America's Best-Run Companies*. Read by Ian Jones. Audiobook. Melbourne: Royal Victorian Institute for the Blind Tertiary Resource Service.

Rao, Vijayendra, and Michael Walton. 2004. *Culture and Public Action*. Stanford, CA: Stanford University Press.

Rogers, Everett M. 2003. *Diffusion of Innovations*. 5th ed. New York: Free Press.

Sanyal, Kalyan. 2007. *Rethinking Capitalist Development: Primitive Accumulation, Governmentality, and Post-colonial Capitalism*. New Delhi: Routledge.

Schoen, Donald A. 1963. Champions for Radical New Inventions. *Harvard Business Review* 41 (2): 77–86.

Schrock, Andrew R. 2016. Civic Hacking as Data Activism and Advocacy: A History from Publicity to Open Government Data. *New Media & Society* 18 (4): 581–599.

Sears, Mark. 2017. The Company behind Silicon Valley's Dirty Little Secret—with Mark Sears. *Mixergy*. February 27, 2017. https://mixergy.com/interviews/cloudfactory-with-mark-sears/.

Snow, David A., E. Burke Rochford, Jr., Steven K. Worden, and Robert D. Benford. 1986. Frame Alignment Processes, Micromobilization, and Movement Participation. *American Sociological Review* 51:464–481.

Söderberg, Johan, and Alessandro Delfanti. 2015. Hacking Hacked! The Life Cycles of Digital Innovation. *Science, Technology & Human Values* 40 (5): 793–798. doi:10.1177/0162243915595091.

Stark, David. 2009. *The Sense of Dissonance: Accounts of Worth in Economic Life*. Princeton, NJ: Princeton University Press.

Sunder Rajan, Kaushik. 2006. *Biocapital: The Constitution of Postgenomic Life*. Durham, NC: Duke University Press.

Takhteyev, Yuri. 2012. *Coding Places: Software Practice in a South American City*. Cambridge, MA: MIT Press.

Tsing, Anna Lowenhaupt. 2005. *Friction: An Ethnography of Global Connection*. Princeton, NJ: Princeton University Press.

Tsing, Anna. 2009. Supply Chains and the Human Condition. *Rethinking Marxism* 21 (2): 148–176. doi:10.1080/08935690902743088.

Von Hippel, Eric. 2005. *Democratizing Innovation*. Cambridge, MA: MIT Press.

Vora, Kalindi. 2015. *Life Support: Biocapital and the New History of Outsourced Labor*. Minneapolis: University of Minnesota Press.

Weaver, Catherine. 2008. *Hypocrisy Trap: The World Bank and the Poverty of Reform*. Princeton, NJ: Princeton University Press.

World Bank. 2012a. *ICT for Greater Development Impact: Sector Strategy*. Washington, DC: World Bank.

World Bank. 2012b. *Water Hackathons: Lessons Learned*. Water Papers. Washington, DC: World Bank.

Zukin, Sharon, and Max Papadantonakis. 2018. Hackathons as Co-optation Ritual: Socializing Workers and Institutionalizing Innovation in the "New" Economy. *Research in the Sociology of Work* 31:157–181. doi:10.1108/S0277-283320170000031005.

10 Meeting Social Objectives with Offshore Service Work: Evaluating Impact Sourcing in the Philippines

Jorien Oprins and Niels Beerepoot

Introduction

Impact sourcing has recently emerged as a subfield of the global ICT-ITES (information and communication technology–information technology enabled services) sector and is premised on the potential of balancing commercial interests with socioeconomic development (Malik, Nicholson, and Morgan 2013). As defined by Carmel, Lacity, and Doty (2016, 19), impact sourcing involves the practice of hiring and training marginalized individuals, who normally would have few opportunities for good employment, to provide information technology, business process, or other digitally enabled services. Its ambition is to deliver high-quality information-based services produced by marginalized groups in (predominantly) the Global South. Impact-sourcing service providers mediate between clients and employees to balance the dual objectives of providing high-value services at low cost for clients and meaningful employment to marginalized individuals by giving them access to IT-enabled service jobs (Madon and Ranjini 2016). Because ICTs connect workers to work irrespective of their location (Friedman 2005; Levy 2005), these technologies could help overcome the social, cultural, and physical barriers that might otherwise exclude marginalized groups from participating in the labor market (Monitor Group 2011; Everest Group 2014).

The Rockefeller Foundation has been the leading global institution promoting impact sourcing. It launched its Digital Jobs Africa Initiative in 2013 and commissioned key reports by the Monitor Group (2011), Avasant (2012), Accenture (Bulloch and Long 2012), and Everest Group (2014). These reports have mainly focused on the incentives for clients to purchase services from impact-sourcing service providers. In addition, a growing number of scholars have recently taken an interest in impact sourcing. Some studies have approached the model from an entrepreneurial angle and analyzed the seemingly contrasting social and commercial aspects of impact sourcing on service providers' strategic decision making (e.g., Gino and Staats 2012; Nicholson et al. 2015; Sandeep and Ravishankar 2015b) and on how they position themselves in the local community (e.g., Sandeep and Ravishankar 2015a). Other studies have examined the effects of impact sourcing on service workers (e.g., Heeks and Arun 2010; Lacity, Rottman, and Carmel 2014; Madon and Sharanappa 2013; Malik, Nicholson, and Morgan 2013) and on their local communities (see Madon and Ranjini 2016). Some scholars suggest that impact sourcing has the potential to foster socioeconomic development in the Global South by providing (direct and indirect) employment to marginalized communities and by enhancing their knowledge and skill sets (Heeks and Arun 2010; Madon and Sharanappa 2013; Malik, Nicholson, and Morgan 2013; Madon and Ranjini 2016).

Beyond the evidence provided by these pioneering case studies of individual impact-sourcing initiatives, knowledge is still limited on its success in implementation across different local contexts and its effectiveness in reaching out to marginalized individuals. In this chapter, we examine how impact sourcing operates at the intersection of a commercial logic and a social welfare logic. As a pro-poor model, it offers a good exemplar of how the information and communication technologies for development (ICT4D) debate has evolved, from an initial focus on ICT availability to help the poor become users of digital content (Heeks 2009; Avgerou 2010) by using ICTs as a tool for income generation, such as by doing work online (Heeks 2009), to a focus on their impact, by achieving social and economic development goals (Heeks 2009). As argued by Heeks (2009, 11), ICTs seem well understood as tools for delivering information and services to the world's poor. Where they have so far been little understood is as tools the poor can use to create new incomes and new jobs (Heeks 2009).

As a case study for examining whether impact-sourcing initiatives are effective in providing employment to marginalized individuals, we focus on the experiences of an impact-sourcing venture in the Philippines. Where previous studies have mainly looked at initiatives in India, we look at a venture established by Visaya Knowledge Process Outsourcing (Visaya KPO) in Tanjay, a small city located in the central Philippines. The country is one of the largest beneficiaries of international offshoring of ICT-ITES activities

(IBPAP 2012; Tholons 2014; Usui 2012). Yet, only a few impact-sourcing ventures have started so far. As the most prominent example, Visaya KPO provides a useful case for examining who benefits from this new initiative and how a balance is sought between simple contract fulfillment and creating positive societal change. This chapter is based on semistructured interviews with the management and workers involved in Visaya KPO. At the managerial level, we examine the choices and rationale behind the initiative. At the workers' level, we consider the service workers involved in the initiative and their perceptions of their employment status. An important issue in impact sourcing is whether it reaches the poorest and neediest people (e.g., Heeks and Arun 2010; Nicholson et al. 2015; Sandeep and Ravishankar 2015a).

In this chapter, we first review the current state of the literature on impact sourcing and examine how it functions at the intersection of ICT4D and mainstream ICT-ITES service delivery. We then explain the research methodology we used for this study before elaborating on the present state of impact-sourcing initiatives in the Philippines. Subsequently, we concentrate on Visaya KPO's impact-sourcing initiative and how it finds its balance between a commercial logic and a social welfare logic. The chapter then focuses on the management's rationale behind the geographic location selected and the socioeconomic profile and perceptions of the service workers it employs.

Literature Review: Impact Sourcing

In the past two decades, millions of IT-enabled services jobs have been relocated, or "offshored," from the United States and Europe to low-cost economies around the world (Lambregts, Beerepoot, and Kloosterman 2016). What started with low-end IT activities being moved from the United States to India has grown into the large-scale migration of multivarious service production activities from advanced to developing countries. Digital technologies have enabled new activities to take root in developing countries and have encouraged these nations to envision new strategies for national development (see Graham 2015). One concern with the rise of the ICT-ITES sector is how access to employment in the sector is highly uneven, and how the sector might even perpetuate inequality in developing countries (Krishna and Pieterse 2008; D'Costa 2011). Jobs are highly concentrated in urban areas and, in most cases, require a college education

(Beerepoot and Hendriks 2013; Kleibert 2015). For ICTs to provide inclusive employment requires experimenting with new delivery models for IT-enabled services, such as impact sourcing, and identifying new production locations.

Various authors who have studied impact sourcing view it as an ICT4D model because its ambition is to provide ICT-enabled employment to marginalized groups (see Malik, Nicholson, and Morgan 2013; Nicholson et al. 2015). The general objective of ICT4D is to improve socioeconomic conditions in developing countries through the use of ICTs (Avgerou 2010). In early conceptualizations of ICT4D, the focus was on providing marginalized communities access to ICTs as a source of information and knowledge (Heeks 2009; Avgerou 2010). More recently, rather than viewing marginalized communities as passive consumers, ICT4D regards them as active innovators and producers of digital content (Heeks 2009). It is here that impact sourcing, if it pursues a social welfare logic, can be positioned in fulfilling ICT4D objectives. Some authors, however, have highlighted the tension within the impact-sourcing model between pursuing a commercial logic (e.g., effectively competing with mainstream ICT-ITES service providers) and adopting a social welfare logic (e.g., providing income and training to marginalized groups; Nicholson et al. 2015; Sandeep and Ravishankar 2015b). To clarify how stakeholders in impact sourcing balance these logics, we provide an overview of existing studies on impact sourcing in this section. Herewith we focus on four groups of stakeholders involved in the impact-sourcing ecosystem: the clients of impact-sourcing services, service providers, service workers, and the communities in which they reside (Carmel, Lacity, and Doty 2016).

Consultancy reports commissioned by the Rockefeller Foundation have focused mostly on the first two stakeholders. Following Emerson (2003) and Porter and Kramer (2011), they have high expectations for impact sourcing to scale up over time because it offers a "win-win" scenario (Bulloch and Long 2012; Everest Group 2014). For clients, the Everest Group (2014) and the Monitor Group (2011) highlight the business value proposition when buying services from impact-sourcing service providers. Compared with mainstream ICT-ITES service providers, impact sourcing enables employing a low-cost untapped talent pool. At the same time, impact sourcing helps businesses meet internal corporate social responsibility (CSR) objectives by providing meaningful and relatively high-income employment to marginalized individuals (Bulloch and Long 2012; Everest Group 2014). Optimistic accounts highlight how impact-sourcing service providers aim at "doing well by doing good." They "do good" by reaching out to marginalized communities, as a result of which they "do well" by having a competitive advantage over mainstream ICT-ITES service providers (Bulloch and Long 2012; Avasant 2012; Monitor Group 2011).

Various scholars have added to these reports by examining how impact-sourcing service providers balance a social welfare and a commercial logic. For example, Gino and Staats (2012) focused on the business model of Samasource, one of the most prolific service providers in impact sourcing. They found that rather than putting social welfare at the core of the business model, the organization operates like any mainstream ICT-ITES service provider, by aiming at the delivery of high-value services at low cost. According to Samasource's mission statement, the company's goal is "to connect poor people to the digital supply chain so that they can earn a living and build valuable skills. But it accomplishes that goal by running a business that delivers high value at low cost" (Gino and Staats 2012, 96). In a similar vein, Nicholson and colleagues (2015) and Sandeep and Ravishankar (2015b) argue that impact-sourcing service providers find themselves having to operate between a social welfare logic and a commercial logic. The researchers refer to impact-sourcing service providers as "hybrid organizations," which speak a different language to clients than to service workers and the communities in which they reside. While emphasizing local social and economic development in their interaction with workers and their communities (Sandeep and Ravishankar 2015a), to clients they speak the language of competition and profit (Nicholson et al. 2015). Although their long-term existence as impact-sourcing service providers hinges on their capability to speak both languages, they consider it most important to master the latter one (Nicholson et al. 2015; Sandeep and Ravishankar 2015a). Accordingly, Nicholson and coauthors (2015) find that when choosing the location for impact-sourcing ventures and recruiting service workers, impact-sourcing service providers often let commercial motivations prevail.

Service Workers and Their Communities

Notwithstanding these compromises, in examining social and economic development, scholars have identified a range of benefits for service workers involved in impact sourcing. For example, Lacity, Rottman, and Carmel (2014) focused on US prison inmates and found that in-prison employment with an impact-sourcing initiative increases incomes, elevates inmates' social status in prison, builds their self-efficacy, and strengthens their human capital. When studying impact sourcing in the Global South, scholars have been guided by the ICT4D debate and international development literature. Malik, Nicholson, and Morgan (2013) and Madon and Sharanappa (2013) assessed the social development implications of impact-sourcing organizations in India by using Amartya Sen's capability framework (see Sen 2000), and Heeks and Arun (2010) drew from the sustainable livelihoods framework (see DFID 1999). They all found that impact sourcing brings various developmental benefits to service workers, including rising incomes, as well as strengthening social networks and human capital, that is, operational computer skills, English language skills, and knowledge of ICT (Heeks and Arun 2010; Malik, Nicholson, and Morgan 2013; Madon and Sharanappa 2013). Moreover, they found that involvement in impact sourcing helps service workers build their self-esteem and improves the social empowerment of female workers, who experience greater respect, recognition, and acceptance within their families (Heeks and Arun 2010; Madon and Sharanappa 2013; Malik, Nicholson, and Morgan 2013).

Evidently, the social and economic benefits gained by service workers spill over to their communities (Heeks and Arun 2010; Madon and Ranjini 2016; Madon and Sharanappa 2013; Malik, Nicholson, and Morgan 2013). Employment in an impact-sourcing initiative could positively influence children's futures if it increases spending by service workers on their children's education (Heeks and Arun 2010; Malik, Nicholson, and Morgan 2013). Moreover, the increase in income among service workers provides an injection into the local economy through their consumer expenditures (Heeks and Arun 2010; Madon and Ranjini 2016). As a result, small-time local vendors may flourish. To date, however, no structural impact assessments have been conducted on these spillover effects and how the recipient communities benefit.

Although various authors have examined the developmental outcomes of impact sourcing on service workers and, to a lesser extent, on the community in which they reside, more research is needed to understand more precisely who enjoys the aforementioned benefits. This issue is important to address because of the hybrid nature of the impact-sourcing business model, in which service providers find themselves having to compromise on the degree to which their employees are socially or geographically marginalized (Nicholson et al. 2015). Yet, before examining how impact sourcing balances social welfare and commercial logics, and how this affects the extent to which the model offers employment to marginalized individuals, we provide the methodological underpinnings of this chapter.

Research Methodology

The case study for this research is an impact-sourcing initiative established and run by Visaya KPO in Tanjay, Negros Oriental, a province in the central Philippines. We conducted our field research between September and November 2015, selecting Visaya KPO out of four impact-sourcing ventures that we identified in the Philippines. The other three initiatives were Data-Motivate, Digisource, and Mynd Consulting. Together they employ around 450 workers. We approached all four initiatives and formally interviewed or informally spoke with their managerial staff. For pragmatic reasons, we then selected Visaya KPO as a case study. The managers of this venture, as well as one of its clients and its service workers, were most accommodating and willing to contribute to this research.

We became aware of the existence of Visaya KPO via the Department of Information and Communications Technology (DICT), a Philippine government department responsible for supporting the country's ICT-ITES sector. With its help, we established contacts with the CEO, vice president, and operations manager of Visaya KPO as well as with two stakeholders from Accenture, who had initiated the initiative and played a key role in its establishment. With them, we held semistructured interviews to gain an understanding of the managerial choices and rationale behind the initiative. We asked about the interplay of a commercial logic and a social welfare logic as well as about the long-term prospects of employment offered by Visaya KPO. In addition, we conducted semistructured interviews with 30 of the 116 workers employed by Visaya KPO to identify their socioeconomic profiles, their professional histories, and their perceptions of their employment status. We employed purposive sampling to get a diverse group of respondents in terms of employment duration (ranging from only a few months to two years). Thirteen respondents were male, and seventeen were female. The interviews, on average, lasted one hour and were recorded, transcribed, and analyzed by grouping and cross-comparing the data under the key themes that the research focused on. Finally, informal conversations and observations in Tanjay served to cross-check the findings derived from the interviews and helped us get an initial understanding of the socioeconomic impact that this venture has on the local community.

Impact Sourcing in the Philippines

The Philippines is a lower-middle-income country with 25.2 percent of the population living on less than US$1.25 per day in 2012 (World Bank 2016). Starting at the beginning of the current century, the country followed in the footsteps of the world's number-one service offshoring destination, India, and became a major service offshoring hub (IBPAP 2012; Tholons 2014; Lambregts, Beerepoot, and Kloosterman 2016). Over the years, the Philippine ICT-ITES sector advanced from a multimillion-dollar industry ($350 million in export revenues in 2001) to a multibillion dollar industry ($18.4 billion in export revenues in 2014; Satumba 2008, 14; Remo 2015), employing around one million workers in 2014 (De Vera 2014). These workers mainly conduct services in business process outsourcing (BPO) centers at the lower end of the value chain (e.g., customer services, back-office services, and data processing; Kleibert 2015). The sector initially concentrated in a few large urban areas, with an estimated 80 percent located in Metro Manila (IBPAP 2012; Kleibert 2015), and provided employment to only a narrow labor market segment of young, urban, highly educated people (Mitra 2011).

Compared to the country's mainstream ICT-ITES sector, impact sourcing is still of negligible size in the Philippines (and elsewhere). This can be attributed to the offshoring of IT-enabled services to developing countries still being a recent phenomenon (see Lambregts, Beerepoot, and Kloosterman 2016). In this initial stage, Metro Manila in the Philippines and the six largest metropolitan areas in India emerged as the first entry points into both countries. More recently, firms have started to look beyond the common labor pool (Marasigan 2016), expanding production to class two and even class three cities (Tschang 2011).[1] Impact sourcing is part of the process to look for new production locations and to test new delivery models (see also Monitor Group 2011). As a CSR strategy, it fits the ambition of many companies for their CSR activities to stay close to their core competencies and result in a deeper engagement with local communities than traditional philanthropy could achieve (Emerson 2003; Porter and Kramer 2011).

Visaya KPO established an impact-sourcing initiative in Tanjay in July 2013. The multinational Accenture backed the rural impact-sourcing efforts of Visaya KPO by guaranteeing that it would outsource some of its service tasks to the company. Visaya KPO took on the role of for-profit impact-sourcing service provider and started with 20 impact workers. Thereafter, it attracted additional for-profit clients, which led the number of workers to increase to 116 by late 2015. For the Accenture account, workers operate as virtual assistants (e.g., making room reservations) for the offices in Metro Manila or Cebu City, a low-skilled task, which suggests that Accenture is not (yet) too ambitious in terms of delegating work to its impact-sourcing unit. Other work includes medical transcription for a US-based company and outbound sales for college education programs in the United States.

Empirical Results

This section focuses on the organizational motivations underpinning Visaya KPO's impact-sourcing initiative. We discuss the profile of service workers it reaches, as well as the socioeconomic effects on the local community.

Balancing Commercial and Social Welfare Logics

Visaya KPO's impact-sourcing initiative was initially triggered by a social welfare logic. Like the impact-sourcing ventures examined by Sandeep and Ravishankar (2015b), and following Porter and Kramer (2011), Visaya KPO emanates from Accenture's desire to create economic value in a way that also has positive outcomes for the community surrounding its operations. Accenture is one of the largest international providers of ICT-ITES services. In the Philippines, it employs approximately thirty-five thousand workers in Metro Manila and Cebu City (Hidalgo 2015). Inspired by the Rockefeller reports and based on the success of its earlier initiatives in India, Accenture presented to Visaya KPO the idea of a "rural BPO" (Accenture manager interview, October 7, 2015). Accenture's interest in impact sourcing illustrates how the model is gaining recognition among mainstream ICT-ITES service providers. For the multinational firm, this meant meeting its CSR objectives by providing high-income employment in an area where people had few such opportunities. The salaries in Visaya KPO range from

10,000 to 17,000 Philippine pesos ($200–$340) per month and are well above the average local monthly salaries of 3,000 to 4,000 PHP ($60–$80). Provincial differences in wage standards (based on variety in the local cost of living) make it attractive for companies to look beyond Metro Manila, where entry-level positions in the ICT-ITES sector have a salary of 21,000 to 24,000 PHP ($420–$480) per month. Although the rationale behind the initiative was to stimulate social and economic development, in its practical establishment, this logic was intersected with one that is "rooted in profit and competition" (Nicholson et al. 2015, n.p.), as best illustrated by the cost savings from lower local salaries in Tanjay.

Accenture demanded that the quality standards applicable to all of their offices in the Philippines be met. For Visaya KPO, rather than recruiting personnel among marginalized communities, the focus was on establishing a venture that could compete with mainstream ICT-ITES service providers in, for instance, Metro Manila. To be cost effective, Visaya KPO needed to attract other clients as well. The management of Visaya KPO feared that a rural location, where operating costs would be relatively high and the pool of qualified workers smaller, would complicate this. One manager said, "It's a lot harder sell for me to tell my client, 'I will do your process in the town of Tanjay.' Then they're going to say, 'Where is that?' ... They have to be concerned about downtime. They have to be concerned about service-level quality" (Visaya KPO manager interview, October 15, 2015).

A similar story was conveyed by a representative from DataMotivate, who argued, "One of the things that we try to balance is [to] tell the story of the social impact. Because the experience is that most clients don't care about that. So we basically compete on quality and price, just like any other BPO" (DataMotivate manager interview, November 26, 2015).

These anecdotes indicate how, contrary to Accenture, which refers to the venture in Tanjay as "impact sourcing," service providers can be anxious to sell their initiatives to clients under that label. Similar to impact-sourcing ventures in India (Sandeep and Ravishankar 2015b), they feared that to market themselves as impact-sourcing service providers would raise doubts about the quality of their service delivery.

To establish a profitable and competitive venture, Accenture and Visaya KPO let a commercial logic prevail when selecting a location. They required a place that had high-quality electricity, telecommunications, and Internet infrastructure. Moreover, they insisted on colleges and a university being nearby to serve as feeder pools for workers. This pool needed to be sufficiently large as to allow the initiative to scale up over time. Although many smaller provincial cities would have met these requirements, personal motivations eventually led to the selection of Tanjay (the hometown of one of the initiators).

Tanjay is a class-four provincial city of approximately seventy-nine thousand inhabitants (PSA 2010). Its main sources of income are sugarcane, farming, and fishing (PSA 2016). It accommodates two colleges, whose graduates, because of the lack of local opportunities, tend to leave the community to find jobs elsewhere in the Philippines or abroad. A Visaya KPO manager explained the city selection: "So why tier four? We wanted to go for the least capable [of] economic output but also … able to sustain the size of manpower that we need. So if you go to a class five or a class six. … if you go too low, you won't be able to get enough manpower" (Visaya KPO manager interview, October 21, 2015).

The choice of Tanjay aligns with location choices in impact-sourcing ventures studied by Madon and Sharanappa (2013) and Malik, Nicholson, and Morgan (2013). They examined initiatives in India, located twenty kilometers from Ranchi, capital of Jharkhand state, and forty kilometers from New Delhi. The closest larger city to Tanjay is Dumaguete, the provincial capital, located thirty kilometers away. Dumaguete is home to a range of colleges and universities, as well as BPOs run by mainstream ICT-ITES service providers. This demonstrates how impact-sourcing ventures are restricted in their geographic outreach, as digital connectivity and availability of human resources are critical for their establishment.

Profiling Service Workers and Their Communities

With Accenture and Visaya KPO purposely selecting a location with colleges and a university, all respondents had a high school degree and had proceeded to college after graduating. Their demographics resemble service workers employed by an impact-sourcing venture in India, who, on average, had 13.5 years of education (Heeks and Arun 2010). Moreover, the respondents' educational qualifications are similar to those employed in mainstream ICT-ITES contact centers in the Philippines (Bird and Ernst 2009; Mitra 2011; Beerepoot and Hendriks 2013). Among the thirty respondents in our research, two-thirds are college graduates with a background in information technology, business administration, nursing, marketing, or communication. For workers who had not finished a college education, Visaya KPO offered income opportunities unavailable to them in the mainstream Philippine job market. In line with general observations on the Philippine BPO sector (see Marasigan 2016), academic qualifications are less of a requirement for employment with Visaya KPO.

The majority of respondents classified their family as being part of the middle class, similar to the backgrounds of service workers employed in the mainstream ICT-ITES sector (Mitra 2011). Moreover, like mainstream service workers (see Bird and Ernst 2009; Mitra 2011), most respondents were thirty-five years old or younger. Against the background of impact-sourcing objectives, it was surprising to find that most respondents had formerly been employed in mainstream ICT-ITES contact centers for a period ranging from seven months to five and a half years. Before the establishment of Visaya KPO in Tanjay, they used to be employed in Metro Manila, Cebu City, Bacolod City, or Dumaguete. As such, the initiative was given a head start, as many workers already had experience in BPO work. Training on the job took just one month and focused on the specific line of business that they worked on. All respondents viewed Visaya KPO as just another mainstream ICT-ITES service provider and were not familiar with the term "impact sourcing," nor were they aware of their part in a multinational firm's CSR initiative.

Regardless of whether the initiative directly reaches marginalized workers, any sizable new business venture in a small town like Tanjay has a positive socioeconomic impact, especially when salaries are much higher than the local average. Respondents mentioned that they favored not having to commute to Dumaguete or move to larger cities like Cebu City or Metro Manila. Employment in their hometown enabled workers to spend more time with their families. Female respondents with young children or family members who needed care were particularly likely to mention that because of Visaya KPO, they are able to earn a relatively high income while staying close to their families. One woman noted that the Visaya KPO office in Tanjay was "very accessible to me, in our house. If it's our break, I can check my kids, and then go back to the office" (Visaya KPO service worker interview, October 26, 2015). Another recalled, "When I was four years old, my mother went to Hong Kong to work. I don't want my son to experience that. ... It's not easy to grow up without a mom. ... As long as I can have a living here in Tanjay" (Visaya KPO service worker interview, October 27, 2015).

Along with social development, the spread of BPO work to smaller towns adds to economic growth beyond the country's congested major cities. While we lack evidence on the number of indirect jobs generated in Tanjay, optimistic accounts elsewhere (NASSCOM 2010; Kite 2014) suggest that every job in the ICT-ITES sector generates four additional jobs in support services (e.g., security, housekeeping) and through workers' expenditures. In Tanjay, small local vendors and restaurants have fared well since the establishment of Visaya KPO. Businesses have been improved, and some gradually started selling their wares at night to target service workers on the "graveyard shift." Accordingly, respondents expressed that the outlook of Tanjay has changed, for example, "Tanjay is more industrialized right now" (Visaya KPO service worker interview, November 14, 2015), and "When people knew that there's a call center here, there are a lot of restaurants being built. Soon we will be having a Jollibee here" (Visaya KPO service worker interview, November 14, 2015).[2]

While these are only anecdotal examples of indirect local socioeconomic developments, they are illustrative of how ICT advancements provide new employment opportunities in what used to be the periphery of the global economy (see also Mann and Graham 2016). Whether this takes place through specific impact-sourcing ventures or mainstream service provision, it is part of an economic transition in which digital work enables educated young people, who live in more remote areas in developing countries, to compete on the global labor market without having to move to capital cities or abroad.

Conclusion

The ICT4D literature has recently started to address the question of how marginalized people can utilize ICTs for income generation by doing ICT-enabled work. Impact sourcing is built on this premise, and this chapter has examined how impact-sourcing initiatives in the Philippines fulfill these objectives. Our study found that only a few impact-sourcing initiatives exist in the Philippines, and, compared to mainstream service outsourcing, their total employment is negligible. Given the recent interest in impact-sourcing initiatives among philanthropic foundations and business representatives, however, common practices among the current initiatives, which are the pioneers in this field, need to be explored. Understanding their successes and setbacks provides valuable lessons for new ventures and, more generally, about the potential of impact sourcing as a development tool. Even a small portion of work in the global ICT-ITES sector being carried out in impact-sourcing ventures would positively affect the lives of many workers and their families.

In line with findings from various studies in India (see Gino and Staats 2012; Nicholson et al. 2015; Sandeep and Ravishankar 2015b), a commercial logic is prevalent in the impact-sourcing case that we examined. To effectively compete with mainstream service providers, the founders of our principal case study, Visaya KPO, required a location with high-quality digital connectivity and available qualified staff. This led to the recruitment of workers who have, in most cases, a college degree and previous work experience in the ICT-ITES sector. By recruiting these workers, Visaya KPO compromises on its ambition to employ workers with more marginalized backgrounds. Critics could argue that impact sourcing is mainly driven by motivations to tap into a lower-cost labor pool, hiring educated workers in lower-cost locations rather than integrating workers from more marginalized backgrounds into existing operations in Metro Manila or Cebu City. In the latter case, Philippine national labor law would require companies to pay marginalized workers at par with the existing workers, which is not attractive for companies given the extra training these workers require. Most workers in this research were not even aware that their workplace was regarded as an impact-sourcing venture. Rather, they viewed it as a convenient workplace close to their homes, which provided them with a better income compared to local alternatives. Even toward clients, the initiative was only selectively advertised as an impact-sourcing venture.

The case of Tanjay illustrates the commonalities that impact sourcing has with mainstream ICT-ITES service delivery, as impact-sourcing ventures effectively operate in the same market. Impact sourcing mainly adds to the already ongoing geographic spread of ICT-ITES service work. The relocation of service jobs to class-two cities in the Philippines starting about a decade ago, driven by high competition among BPOs in Metro Manila and the necessity for strategic diversification, has been well documented (see Kleibert 2015; Beerepoot and Vogelzang 2016). Via impact sourcing, the sector is now moving to more remote destinations (albeit not too peripheral) but is not reaching new groups of workers who fall outside the scope of mainstream service outsourcing. Herewith impact sourcing resembles observations made about microcredit (see Morduch 2000; Banerjee et al. 2015), where beneficiaries are not always the most financially strained or those without access to other financial institutions.

Whereas some prolific international impact-sourcing organizations (e.g., Samasource, Digital Divide Data, RuralShores) explicitly state their objective to provide employment to marginalized communities, the empirical evidence from various studies referred to in this chapter points to ambiguity in the use of the concept. Contrary to ICT4D conceptualizations that unambiguously declare the aim to use digital technologies to provide employment to marginalized people, employing college-educated workers in a rural setting is, in some key reports, enough to classify as impact sourcing (see Monitor Group 2011; Bulloch and Long 2012). This shows how the model falls short in its role as a manifestation of ICT4D. Another shortcoming of impact sourcing is that it relies too much on an existing business model (doing IT-enabled work in an office-based setting), which is cost sensitive and requires economies of scale to be viable. The high start-up cost in remote locations means that a commercial logic prevails, which makes it hard to distinguish impact-sourcing ventures from mainstream ICT-ITES service providers.

Despite the deficiencies of the impact-sourcing model, the case of Tanjay illustrates how it provides new employment opportunities in places that, until recently, were not on the radar of the global ICT-ITES sector. Beyond the evidence provided in this study, the longer-term effects and consequences of impact sourcing on the workers and their communities require further, more systematic, investigation (see also Carmel, Lacity, and Doty 2016). Since most impact-sourcing ventures that have been studied started only recently, another subject for investigation is whether, in the medium term, impact-sourcing initiatives will make more effort to recruit marginalized groups rather than (mainly) searching for new locations. At the crossroad of both logics, hybrid forms of organization could emerge that employ workers from marginalized communities. Only through such efforts can impact-sourcing ventures claim distinctiveness from mainstream ICT-ITES service delivery. Doing so would also make them more attractive to socially minded clients who are willing to pay a premium for employing marginalized workers.

A final subject for investigation is the challenges faced by lower-end ICT-ITES service providers and, hence, by impact-sourcing ventures. Impact-sourcing ventures might be confronted with the increasing automation of service work (see Brynjolfsson and McAfee 2014; Ford 2015), which requires a strategic reorientation of their activities. Dual goals of moving up the value chain (by doing higher value-added activities) and meeting social objectives could eventually be too much to combine, but those visions are illustrative of how impact sourcing is filled with ambition.

Notes

1. In the Philippines, cities are classified based on their average annual income. Classes range from one to six, with class-one cities (e.g., Cebu City) earning the highest average annual income, and class-six cities earning the lowest average annual income (Bureau of Local Government Finance 2008, order no. 23-08).

2. Jollibee is a Philippine multinational chain of fast food restaurants similar to McDonald's.

References

Avasant. 2012. *Incentives and Opportunities for Scaling the "Impact Sourcing" Sector*. September 2012. https://assets.rockefellerfoundation.org/app/uploads/20120901233822/Incentives-Opportunities-for-Scaling-the-Impact-Sourcing-Sector.pdf.

Avgerou, Chrisanti. 2010. Discourses on ICT and Development. *Information Technologies and International Development* 6 (3): 1–18.

Banerjee, Abhijit, Esther Duflo, Rachel Glennerster, and Cynthia Kinnan. 2015. The Miracle of Microfinance? Evidence from a Randomized Evaluation. *American Economic Journal: Applied Economics* 7 (1): 22–53.

Beerepoot, Niels, and Mitch Hendriks. 2013. Employability of Offshore Service Sector Workers in the Philippines: Opportunities for Upward Labour Mobility or Dead-End Jobs? *Work, Employment and Society* 27 (5): 823–841.

Beerepoot, Niels, and Emeline Vogelzang. 2016. Service Outsourcing to Smaller Cities in the Philippines: The Formation of an Emerging Local Middle Class. In *The Local Impact of Globalization in South and Southeast Asia: Offshore Business Processes in Services Industries*, edited by Bart Lambregts, Niels Beerepoot, and Robert C. Kloosterman, 196–207. New York: Routledge.

Bird, Miriam, and Christoph Ernst. 2009. Offshoring and Employment in the Developing World: Business Process Outsourcing in the Philippines. Employment Working Paper No. 41. International Labour Organization, Geneva. http://www.ilo.org/wcmsp5/groups/public/---ed_emp/---emp_elm/---analysis/documents/publication/wcms_117922.pdf.

Brynjolfsson, Erik, and Andrew McAfee. 2014. *The Second Machine Age: Work, Progress, and Prosperity in a Time of Brilliant Technologies*. New York: Norton.

Bulloch, Gib, and Jessica Long. 2012. *Exploring the Value Proposition for Impact Sourcing: The Buyer's Perspective*. Accenture. https://assets.rockefellerfoundation.org/app/uploads/20120314232314/Exploring-the-Value-Proposition-for-Impact-for-Impact-Sourcing.pdf.

Bureau of Local Government Finance (Philippines). 2008. Department order no. 23-08. July 29, 2008. http://nap.psa.gov.ph/activestats/psgc/articles/DepOrderReclass.pdf.

Carmel, Erran, Mary C. Lacity, and Andrew Doty. 2016. The Impact of Impact Sourcing: Framing a Research Agenda. In *Socially Responsible Outsourcing: Global Sourcing with Social Impact*, edited by Brian Nicholson, Ron Babin, and Marcy C. Lacity, 16–47. London: Palgrave London.

D'Costa, Anthony. 2011. Geography, Uneven Development and Distributive Justice: The Political Economy of IT Growth in India. *Cambridge Journal of Regions, Economy and Society* 4 (2): 237–251. doi:10.1093/cjres/rsr003.

De Vera, Ben O. 2014. Employment in BPO Sector Hits 1-M Mark. *Business Inquirer*. August 22, 2014. http://business.inquirer.net/177150/employment-in-bpo-sector-hits-1-m-mark.

DFID. 1999. *Sustainable Livelihoods Guidance Sheets*. London: Department for International Development.

Emerson, Jed. 2003. The Blended Value Proposition: Integrating Social and Financial Returns. *California Management Review* 45 (4): 35–51.

Everest Group. 2014. *The Case for Impact Sourcing*. September 2014. https://assets.rockefellerfoundation.org/app/uploads/20140901161823/The-Case-for-Impact-Sourcing.pdf.

Ford, Martin. 2015. *The Rise of the Robots: Technology and the Threat of a Jobless Future*. New York: Basic Books.

Friedman, Thomas L. 2005. *The World Is Flat: A Brief History of the Twenty-First Century*. New York: Farrar, Straus and Giroux.

Gino, Francesca, and Bradley R. Staats. 2012. The Microwork Solution. *Harvard Business Review* 90 (12): 92–96.

Graham, Mark. 2015. Contradictory Connectivity: Spatial Imaginaries and Techno-Mediated Positionalities in Kenya's Outsourcing Sector. *Environment & Planning A* 47:867–883.

Heeks, Richard. 2009. The ICT4D 2.0 Manifesto: Where Next for ICTs and International Development? University of Manchester. Development Informatics

Working Paper Series, No. 42. Institute for Development Policy and Management, University of Manchester. https://www.oecd.org/ict/4d/43602651.pdf.

Heeks, Richard, and Shoba Arun. 2010. Social Outsourcing as a Development Tool. The Impact of Outsourcing IT Services to Women's Social Enterprises in Kerala. *Journal of International Development* 22:441–454.

Hidalgo, Vanessa B. 2015. Accenture Seeks Ways to Keep Its Workforce Happy. *Business Inquirer*. March 30, 2015. http://business.inquirer.net/189533/accenture-seeks-ways-to-keep-its-workforce-happy.

IBPAP. 2012. *Philippines: IT-BPO Investor Primer*. DOST-ICT Office, IT and Business Process Association, Philippines. http://www.ibpap.org/publications-and-press-statements/research-initiatives/investorprimer2012.

Kite, Grace. 2014. Linked In? Software and Information Technology Services in India's Economic Development. *Journal of South Asian Development* 9 (2): 99–119.

Kleibert, Jana. 2015. Expanding Global Production Networks: The Emergence, Evolution and the Developmental Impact of the Offshore Service Sector in the Philippines. PhD diss., University of Amsterdam.

Krishna, Anirudh, and Jan N. Pieterse. 2008. Hierarchical Integration: The Dollar Economy and the Rupee Economy. *Development and Change* 39 (2): 219–237. doi:10.1111/j.1467-7660.2007.00477.x.

Lacity, Mary C., Joseph W. Rottman, and Erran Carmel. 2014. Impact Sourcing: Employing Prison Inmates to Perform Digitally-Enabled Business Services. *Communications of the Association for Information Systems* 34 (1): 913–932.

Lambregts, Bart, Niels Beerepoot, and Robert C. Kloosterman, eds. 2016. *The Local Impact of Globalization in South and Southeast Asia: Offshore Business Processes in Services Industries*. New York: Routledge.

Levy, David L. 2005. Offshoring in the New Global Political Economy. *Journal of Management Studies* 42 (3): 685–693.

Madon, Shirin, and C. Ranjini. 2016. The Rural BPO Sector in India: Encouraging Inclusive Growth through Entrepreneurship. In *Socially Responsible Outsourcing: Global Sourcing with Social Impact*, edited by Brian Nicholson, Ron Babin, and Marcy C. Lacity, 65–80. London: Palgrave London.

Madon, Shirin, and Sandesh Sharanappa. 2013. Social IT Outsourcing and Development: Theorising the Linkage. *Information Systems Journal* 23 (5): 381–399.

Malik, Fareesa, Brian Nicholson, and Sharon Morgan. 2013. Assessing the Social Development Potential of Impact Sourcing. Paper presented at 6th Annual SIG GlobDev Workshop ICT in Global Development, December 14, 2013, Milano, Italy. Paper 4.

Mann, Laura, and Mark Graham. 2016. The Domestic Turn: Business Process Outsourcing and the Growing Automation of Kenyan Organisations. *Journal of Development Studies* 52 (4): 530–548.

Marasigan, Mary Leian C. 2016. How Work in the BPO Sector Affects Employability: Perceptions of ex-BPO Workers in Metro Manila. In *The Local Impact of Globalization in South and Southeast Asia: Offshore Business Processes in Services Industries*, edited by Bart Lambregts, Niels Beerepoot, and Robert C. Kloosterman, 138–152. New York: Routledge.

Mitra, Raja M. 2011. *BPO Sector Growth and Inclusive Development in the Philippines*. World Bank Group. January 26, 2011. http://www-wds.worldbank.org/external/default/WDSContentServer/WDSP/IB/2011/12/22/000333037_20111222002222/Rendered/PDF/660930WP0P122100B0BPO0Sector0Growth.pdf.

Monitor Group. 2011. Job Creation Through Building the Field of Impact Sourcing. Working Paper. June 2011. Rockefeller Foundation. https://assets.rockefellerfoundation.org/app/uploads/20110314231120/Job-Creation-Through-Building-the-Field-of-Impact-Sourcing.pdf.

Morduch, Jonathan. 2000. The Microfinance Schism. *World Development* 28 (4): 617–629.

NASSCOM. 2010. *Impact of IT-BPO Industry in India: A Decade in Review*. New Delhi: NASSCOM.

Nicholson, Brian, Fareesa Malik, Sharon Morgan, and Richard Heeks. 2015. Exploring Hybrids of Commercial and Welfare Logics in Impact Sourcing. Paper presented at the 13th International Conference on Social Implications of Computers in Developing Countries, Negombo, Sri Lanka, May 2015.

Porter, Michael E., and Mark R. Kramer. 2011. Creating Shared Value. *Harvard Business Review* 89 (1/2): 62–77.

PSA. 2010. City of Tanjay, Negros Oriental. Philippine Statistics Authority. Last updated December 31, 2017. http://nap.psa.gov.ph/activestats/psgc/municipality.asp?muncode=074621000&regcode=07&provcode=46.

PSA. 2016. List of Cities. Philippine Statistics Authority. Last modified December 31, 2017. http://nap.psa.gov.ph/activestats/psgc/listcity.asp.

Remo, Amy R. 2015. IT-BPO Sector Posted 18.7% Revenue Growth in 2014. *Business Inquirer*. March 19, 2015. http://business.inquirer.net/188861/it-bpo-sector-posted-18-7-revenue-growth-in-2014.

Sandeep, M. S., and M. N. Ravishankar. 2015a. Impact Sourcing Ventures and Local Communities: A Frame Alignment Perspective. *Information Systems Journal* 26 (2): 127–155.

Sandeep, M. S., and M. N. Ravishankar. 2015b. Social Innovations in Outsourcing: An Empirical Investigation of Impact Sourcing Companies in India. *Journal of Strategic Information Systems* 24:270–288.

Satumba, Ahmma Charisma L. 2008. *Business Process Outsourcing in Financial and Banking Services in the Philippines*. ILS Discussion Paper Series. http://ilsdole.gov.ph/business-process-outsourcing-in-financial-and-banking-services-in-the-philippines/.

Sen, Amartya. 2000. *Development as Freedom*. Oxford: Oxford University Press.

Tholons. 2014. 2015 Top 100 Outsourcing Destinations: Rankings. December 2014. http://www.tholons.com/nl_pdf/Tholons_Whitepaper_December_2014.pdf.

Tschang, F. Ted. 2011. A Comparison of the Industrialization Paths for Asian Service Outsourcing Industries, and Implications for Poverty Alleviation. ADBI Working Paper Series, No. 313. October 2011. http://www.adb.org/sites/default/files/publication/156168/adbi-wp313.pdf.

Usui, Norio. 2012. *Taking the Right Road to Inclusive Growth: Industrial Upgrading and Diversification in the Philippines*. Manila: Asian Development Bank. https://think-asia.org/bitstream/handle/11540/908/taking-right-road-to-inclusive-growth.pdf?sequence=1.

World Bank. 2016. Philippines. World Bank: Data. Accessed October 29, 2018. http://data.worldbank.org/country/philippines.

11 Digital Labor and Development: Impacts of Global Digital Labor Platforms and the Gig Economy on Worker Livelihoods

Mark Graham, Isis Hjorth, and Vili Lehdonvirta

The Rise of Digital Labor

Work has historically been geographically bounded. Workers and the work they perform have always been inexorably linked, with labor being the most place bound of all factors of production (Hudson 2001). As David Harvey (1989, 19) famously noted, workers are unavoidably place based because "labor-power has to go home every night."

But the widespread use of the Internet has changed much of that. Clients, bosses, workers, and users of the end products of work can all be located in different corners of the planet. This chapter is about what the spatial unfixing of work means for workers in some of the world's economic margins.[1] It provides examples illustrating who performs much of the digital work that is carried out today and reflects on some of the key benefits and costs associated with these new digital regimes of work.

The rise of digital labor has come about at a confluence of two trends. First, in much of the world, unemployment (and underemployment) is a major social and economic concern for policymakers, for people with jobs, and for those looking for jobs (ILO 2015). The International Labour Organization (ILO 2014) estimates that between 2014 and 2019, there will be 213 million new labor market entrants.

Second, much of the world is increasingly characterized by rapidly changing connectivity. We have gone from a world, only ten years ago, where less than 15 percent of humanity was connected to the Internet, to one today where around 50 percent of the world's population is connected. There are now over three billion connected people on the planet. Furthermore, ten years ago, less than 8 percent of people in low-income countries were connected. Today, the figure is over one-third (ITU 2016).

In response to this confluence of a need for more jobs in places where they do not currently exist and the spreading of digital connectivity among billions of the world's population, millions of people have turned to outsourced digitally mediated work as a way to transcend some of the constraints of their local labor markets. Many governments, third sector organizations, and private sector actors see significant developmental potential in digital labor: jobs can be created for some of the world's poorest by taking advantage of connectivity and the willingness of an increasing number of firms to outsource business processes. Underpinning some of these hopes is an idea that, in a global market for labor, the actual locations of workers are irrelevant. Anyone can, in theory, do any work from anywhere—an idea that, if true, could bring significant economic benefits to workers in parts of the world where good jobs are hard to come by.

This chapter challenges that notion by highlighting four key concerns addressed (alongside other themes) in a multiyear program of research into digital labor at the world's economic margins. Building on those concerns, the chapter concludes with a reflection on four broad strategies—certification schemes, digital labor organizing, regulatory strategies, and democratic control of online labor platforms—that could be employed to improve conditions and livelihoods for digital workers.

Empirical Foundation

This chapter draws on preliminary findings from ongoing research that is presented with more detailed methods and further context in a longer open-access article (see Graham, Hjorth, and Lehdonvirta 2017). The data sources we refer to in this chapter consist of transaction log data from one of the world's largest digital labor platforms and interviews conducted with workers, managers, and policymakers in Southeast Asia and sub-Saharan Africa.

The transaction data consist of transaction records of all 61,447 projects completed during the month of March 2013. These records were provided to us by the platform in an anonymized, privacy-protected form.

The qualitative data consist of semistructured interviews conducted in person with 125 digital workers and 27 digital work stakeholders (policymakers, platform owners, and third sector organizations) carried out by the authors during fieldwork in Manila, Kuala Lumpur, Vietnam, Johannesburg,

Digital Labor and Development

Cape Town, Nairobi, and Lagos between September 2014 and October 2015. The main sampling goal was to ensure varied representations of (primarily) low-skilled labor experiences in the countries of interest. In this chapter, we present selected cases from the data rather than a representative view. In the sections below, we first outline the theories and hopes pertaining to each area of concern, and then interrogate them with supporting and contrary examples from the research articles and data.

Four Concerns for Digital Labor

Bargaining Power

A key feature of digital work platforms is that they attempt to minimize outside regulation of the relationship between employer and employee (Lehdonvirta 2016). Workers are, for instance, generally classified as independent contractors (even though their work sometimes more closely resembles that of an employee, a finding discussed in more detail in Wood et al. 2016), and national labor laws are rarely applied to digital workers. These issues are particularly acute when transactions cross national borders, as it becomes unclear which jurisdictions' regulations apply to the work being transacted.

If we have a world in which work is a commodity that can be bought and sold (as a result of standardizing and disembedding tasks as well as a lack of regulations and protections for workers), much of this work can, in theory, be done from anywhere. Concomitantly, if work can be done from anywhere, competitive dynamics (in which there is more demand for work than supply of it) could lead to a situation in which low-cost, low-capability suppliers of work (for instance digital workers) could be disadvantaged and become clear price takers with little bargaining power (Kaplinsky 2004; Manning 2003). We summarize some of the findings and provide further examples and discussion below.

Drawing on anonymized transactional data of tasks carried out by members of a pool of more than 4.5 million registered workers over the course of one month, we have identified distinct geographies emerging in the context of global trade in digital labor. These structural characteristics provide insights into the competitive production relations that digital workers in the Global South must navigate when seeking to move beyond their local labor markets to engage in digital work.

A key pattern emerging relates to imbalances in the relationship between supply and demand of digital work. Figure 11.1 maps the geographic distribution of buyers involved in the more than 60,000 transactions performed in March 2013. Most buyers of work are located in high-income countries (with the darkest shade on the map). Among the top-twenty list of countries with the highest number of purchases, the only countries not considered to be high-income nations are Malaysia (ranked fifteenth) and India (ranked nineteenth).

The geography of sales (see figure 11.2) reveals a very different pattern. Even though most demand comes from the Global North, most work is carried out in low-income countries. India and the Philippines, in particular, perform much of the work on the platform. Yet, a significant amount of work continues to be carried out in wealthy countries such as the United States, Canada, and the United Kingdom. Figure 11.2 also illustrates the supply of work's broad geography. The fuller context, therefore, is one whereby demand is relatively concentrated geographically, but supply is relatively diffuse, with workers from low- and high-income countries ending up competing in the same contexts—a situation that is likely to have influence the relative degrees of bargaining power exerted by individual digital workers. (Note that at the time of writing, the platform hosted nine

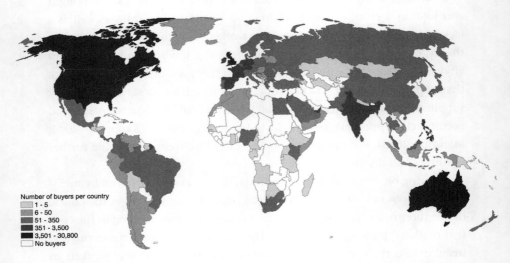

Figure 11.1
Number of buyers of digital work per country. *Source*: Authors.

Digital Labor and Development

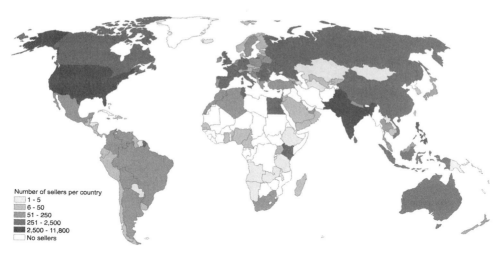

Figure 11.2
Number of sellers of digital work per country. *Source*: Authors.

million registered workers and only four million registered clients.) Not all of these registered workers and clients are active users. But we can also draw from evidence in Kuek et al. (2015) that demand for work far outstrips supply.

Finally, we can explore the spatial variance of hourly pay rates requested by digital workers. The cartogram in figure 11.3 depicts each country as a circle sized according to the dollar inflow over the course of a month (March 2013). The shading of the inner circle indicates the median hourly rate requested by digital workers (i.e., published on their individual online profiles on the platform) in that country; the rates published are not necessarily identical with actual hourly rates or pay received, as evidenced through our fieldwork. Nonetheless, the graphic broadly reveals that median wages are, perhaps unsurprisingly, low in low-income countries and significantly higher in medium- and high-income countries. The cartogram also reinforces that the market for work is highly international, with the United States being the only country in which a majority of work is commissioned by domestic clients.

Despite visions of global labor platforms rendering the locations of workers irrelevant, the differences between places seem to be precisely what encourage particular networks of digital work to be brought into being. As such, we wanted to explore how workers themselves, at some of the

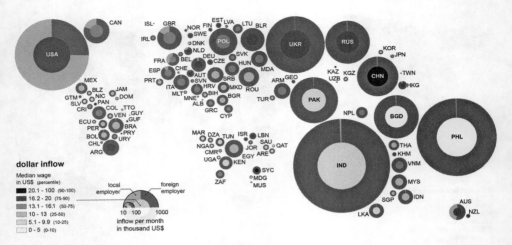

Figure 11.3
Dollar inflow and median requested hourly pay by country. *Source*: Authors.

world's economic margins, experience digital practices in a platform with such global, but uneven, geographies.

Feelings of disempowerment in the context of rate setting are a key theme in many of our interviews with digital workers. The fierce competition between digital workers seeking earning opportunities through digital labor platforms has directly influenced many workers' strategies for securing work, often resulting in underbidding practices. These patterns relate to the skewed distributions of supply and demand for digital work, when digital labor platforms can significantly expand the pool of potential workers available to employers (Beerepoot and Lambregts 2014). No longer limited to the local market, or to physically moving to a lower-cost labor market, many employers can easily practice "labor arbitrage," that is, buy labor from where it is cheapest. This can reduce the market power of workers relative to employers and put downward pressure on labor prices.

Numerous workers framed their digital work experiences within these dynamics of downward spiraling wages. For example, Nu, a Vietnamese software tester, explained that she would look at bids offered by other freelancers from different geographic locations to ensure her rate was lower.[2] Asked how she secures digital work, she explained: "Actually it's very simple, and I think that if I set the minimum [hourly rate] so I will have more job to do. ... There are many freelancers from around the world. I see a country like Philippines—they have very low rate so I need to compare to them."

Narratives of a race to the bottom emerge from interview data even when participants have explicitly reflected on what they consider to be a fair rate for the services they offer. We heard stories in Southeast Asia from workers who have been willing to lower their rates beyond what they considered fair given their qualifications and experiences. For instance, Vi, a Vietnamese translator, was often willing to lower his fee, despite having set his hourly rate based on digital research and careful consideration: "I think five cents [US$0.05 per word] is the right one. I don't want to work for less. Sometimes I will, if I really need, if I really want the job, I will ask for less. Maybe, three cents." Similarly, the twenty-six-year-old Filipina virtual assistant Tala explained: "I first set it [hourly rate] at US$8 because that's what my previous client was paying me. But I found it quite difficult to find jobs. So I set it at US$4. And I think I even set it at US$3.50 currently. So I mean, if you don't get a lot of invitations, you don't have any other choice but to lower down your expectations, I guess." These and other examples point to the potential for a pronounced lack of bargaining power for digital workers. When explaining factors that go into decisions to lower rates, interviewees often mention the visibility of the global pool of supply within digital labor platforms (i.e., competitors).

Our qualitative data thus reveal stories of disempowerment, an inability of workers to exert any significant bargaining power, and a "race to the bottom" in wage rates. These factors sometimes have a negative impact on the lives of workers. Jocelyn, a Filipina transcriber, noted, "Sometimes, I feel really worried where I can get work. What only consoles me is the thought that it's not me who is to blame why I don't get work. ... It's only that there's no client available—no project available. ... it's really unpredictable."

Despite these challenges, the interviews offer limited evidence of digital workers seeking to strengthen their position through collective action or acts of solidarity. Interviewees often described the global supply pool of digital workers in terms of "competition" rather than "colleagues," being fearful that other workers will take clients away from them. The imbalance between the supply and demand of work thus seems to disempower many digital workers. Concomitantly, the dispersed geography of digital work reveals examples of employment being disembedded from local norms and local moral economies that would traditionally regulate an employment relationship, and that lean toward what might be seen as a more internationally operating entrepreneurial moral economy based more singularly on competition.

Economic Exclusion

In geographically circumscribed labor markets, certain groups of people may be excluded entirely from the market or from one of its segments as a result of discrimination or occupational segregation. For example, workers may be discriminated against based on religion, ethnicity, or disability (Reskin 2000), or segregated into certain segments of the market based on their gender or ethnicity (Maume 1999). Digital labor platforms can potentially change some of these dynamics in two ways. First, they can allow workers to access geographically distant markets where there is less discrimination or segregation. Second, they can allow workers to access their local market through a veil of anonymity provided by the digital medium, masking the characteristics likely to provoke discrimination. Indeed, the marketing literature produced by digital labor platforms describes cases when this has reportedly happened (e.g., Elance 2013). Our research revealed evidence of economic inclusion of this sort, as well as examples of exclusion and discrimination (detailed in Wood et al. 2016).

Some interviewees relayed stories of how they were unable to obtain employment or earning opportunities through their local labor markets. Temporary or permanent migrants (who moved for study or other reasons) were particularly likely to speak about how they could now do work from places where they could not legally work before because of a lack of appropriate visas or permits.

Equally, digital labor platforms may, to some extent, offer economic inclusion for individuals who do not hold the educational qualifications necessary to secure traditional employment in local labor markets. For example, Jean, a Filipina transcriber in her late twenties, became pregnant while studying for a university degree in mass communication. As a consequence, she had to give up her studies and she found herself having to rely on financial support from her extended family. She explained how she had applied for jobs in the business process outsourcing (BPO) sector in Manila over the course of seven years without ever being shortlisted for an interview. Jean explicitly stated that her inability to secure a job in the industry was because she did not meet the requirement of having a university degree. When she received her first transcription contract through a digital labor platform in 2012, it thus marked her first experience of professional work since her late teens.

Some of the women we spoke to lived at home with their parents or extended families. They were able to combine wage labor with caring labor (see McDowell 2015), though it was difficult to ascertain whether they were being paid a full reproductive wage. Digital labor platforms can thus improve economic inclusion by allowing people to combine paid work with other commitments, though this can also indirectly support the continuation of gendered divisions of labor.

Besides examples of increased economic inclusion in the contexts of digital work, we also found evidence of different types of economic exclusion and discrimination. Some of the discrimination was explicit (for instance, the blatant request for South Asians not to apply for the vacancy shown in figure 11.4). Other instances were somewhat less explicit. For example, Martin, a thirty-one-year-old content writer from Lagos, believed that workers from the United States or UK were far more likely to get job offers and thus spoke about ways to mask his Nigerian profile location in digital work platforms. Similarly, William, a twenty-six-year-old SEO writer from the Nairobi slums often changed the geographic location listed on his profile. He explained, "It's very discriminatory. … It forces you sometimes to realign your profile to fit that job description." Many of William's clients continue to believe he is based in Australia. This is a necessity, he feels: "You have to create an identity that is not you. If you want to survive

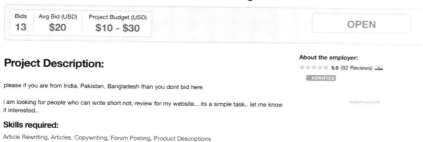

Figure 11.4
Screenshot from a major digital labor platform. *Source*: Authors.

online, you have to do that. If you don't do that, I'm telling you, nothing will come."

Other types of discrimination are even less obvious. In South Africa, Kenya, and Nigeria, some digital workers highlighted clients who had a poorly informed understanding of the African context. Specifically, workers mentioned clients who were unable to distinguish one African country from another; who assumed that African workers did not speak international languages fluently, like English or French; who assumed that African digital workers were uneducated; and who assumed that African workers would be willing to work for whatever pay was offered. When asked what she would change about digital work, Janette, a South African administrative assistant in her early thirties, responded: "People's perception of Africa. ... I have come up against people whose perception of this continent as a whole is just, it's downright ignorant. ... You'll talk to people and they think Nigeria is next door to South Africa, or we're all neighborly, or the whole continent has got Ebola." Tatiana, a Cameroonian virtual assistant living in Johannesburg with her husband and four children, encountered similar misperceptions: "People think that when you're from Africa ... whenever they hear Africa, Africa is somewhere where people are poor, people can't even afford [an] Internet connection. ... That is why when I applied for a job, I never send my resume."

Digital labor platforms clearly do allow many people who are disadvantaged in their local labor markets to obtain earning opportunities. Given their limited opportunities for more conventional forms of employment, however, these workers may have little choice but to accept unfavorable positions in their digital work. As illustrated above, discrimination and economic exclusion can also play out in digital labor markets and can be experienced by a range of workers supplying their labor to global clients.

Intermediation

In development studies, significant attention has been paid to how value chain structures influence outcomes from international trade. A consistent finding has been that value capture is the most important imperative for actors in production chains (Coe and Yeung 2015), and that a significant part of the value of trade in terms of earnings is captured not by producers themselves, but by intermediaries who use geographic location, networks, and other positional advantages to mediate between buyers and sellers, potentially contributing to (and reinforcing) global inequalities (Pietrobelli and Saliola 2008). Although ICTs have contributed to the reintermediation of some commodity chains of physical products (Graham 2011; Murphy and Carmody 2015), because of the direct worker-client interactions that digital labor platforms facilitate, they are often expected to allow workers to circumvent some intermediaries and obtain more direct access to foreign demand (Beerepoot and Lambregts 2014; Lehdonvirta and Ernkvist 2011). This could allow workers to capture a larger share of the revenues created. In our research, we identified evidence of disintermediation but also more surprising network patterns, such as reintermediation. These findings are presented in Lehdonvirta et al. (2015); what follows is a summary of selected findings.

Some of the digital workers we interviewed have been able to take advantage of disintermediation, for example, by evading the unfair practices of locally based companies. A more surprising finding to emerge from the interviews, however, were numerous stories about reintermediation, often leading to exploitation of digital workers with limited visible experience and feedback on digital platforms. Interviewees suggested that, in many cases, the relatively direct connection between the client and the worker is only temporary. Some successful digital workers become intermediaries themselves, taking on more work than a single person can handle and hiring other workers on the platform to carry out the work for them.

Dalale, a twenty-six-year-old Mauritian woman, studying for a master's degree in English literature at a university in Kuala Lumpur, offered an example of this. She had done digital work for the past two years, writing articles and blog posts to improve the search engine optimization of various businesses. Yet, Dalale rarely worked directly with the end clients. Her clients were, for the most part, other digital freelancers who had developed strong digital profiles, characterized by high numbers of positive feedback ratings, making them able to attract a much larger number of tasks at much higher rates than Dalale was able to.[3] Dalale knew this because she often noticed that jobs she had unsuccessfully applied for were reintroduced to the market by another digital freelancer. For example, once she applied for an SEO writing task, suggesting a price of US$15 rather than the listed suggestion of US$50. She later discovered that the job went to another digital freelancer who had requested a price of US$23. This contractor subsequently offered the job to Dalale for just US$3.50. While Dalale accepted tasks from these reintermediaries (although not without explicitly pointing out that they were unfair), she found that the lack of direct interaction and communication with end clients made it very difficult to understand the full task requirements, making the writing process more challenging.

There are two ways to interpret these types of reintermediation of work. They can be viewed as rent-seeking behaviors, where contractors who have a competitive advantage in attracting clients use that advantage to position themselves between the end client and the digital worker who delivers the actual work. Many digital workers suggested that the greatest source of such competitive advantage on digital labor platforms is the official track record automatically displayed in each contractor's profile, namely their reputation score and list of previous projects completed. Given the limited means to evaluate candidates over the Internet, clients are very likely to pick a candidate with the most impressive track record. That candidate can then forward the task to a competitor with less reputational capital, adding no value to the process but gaining yet another entry into his or her own track record. This creates a positive reinforcement loop that greatly favors the first mover.

Sometimes the new intermediaries do add value to the process. For instance, they can perform quality control over the subcontractors' deliverables to retain a strong reputation rating. They also have to break larger tasks into smaller pieces, find subcontractors for each piece, and manage work schedules.

Broadly, reintermediation appears to be important for the completion of high volumes of tasks that require a high level of trust. And while online labor markets attempt to treat labor as a commodity, often they are mediating labor power. In other words, buyers need a way of ensuring that labor power can actually be translated into labor. The reintermediations that we have observed thus appear to be part of a process of capital transforming labor markets, which have been designed with ideological views of how markets should operate (something that Marx [1867] 1990 and many after him have observed). Those transformations provide clients with a trusted intermediary, while intermediaries take on the role of project manager, supervising the tasks of lower-level workers. These examples of functional upgrading (a supplier taking on new roles in the chain at higher added value) are desirable inasmuch as they allow low-income workers to capture more value. But if only a small number of functionally upgraded suppliers are able to establish themselves as chokepoints in the chain, the developmental effects of this sort of work can be highly uneven.

Skill and Capability Development

Disintermediation is conceptually linked to functional upgrading, or increasing the scope of functions performed by the producer in the value chain (Kaplinsky and Morris 2001). In other words, disintermediation provides producers with the opportunity to attempt to perform higher-value-added services. Simply being positioned closer to customers can give producers opportunities to learn more about customer needs and to develop corresponding skills and capabilities (Dicken 2015). Since work carried out through digital labor platforms is usually associated with disintermediation and the potential to link up with customers more directly, it is often expected to result in functional upgrading and movement toward higher-value-added work in service chains (Graham and Mann 2013; Lehdonvirta and Ernkvist 2011).

One example of functional upgrading in our data is the case of twenty-seven-year-old Joseph living in Nairobi. After completing his university degree, Joseph was never able to get a job relating to his expertise area, procurement. The only job he could get was as a cashier in a supermarket, where he worked for two years (twelve to fourteen hours per day, seven days a week) earning a monthly salary of US$300. Since 2012, he has been doing "lead generation" and currency (Forex) trading. Asked if he had learned anything from his digital work, he responded, "Yeah, like the forex knowledge. I have learnt new skills on computers. ... I am considering trading with my own account since I consider myself now qualified. ... For the Forex client, I have seen him make a lot of money."

ICT-enabled outsourcing however, can also make it easier for workers to be kept at arm's length from core business processes, hindering knowledge flow from the core to the periphery and thus perpetuating rather than erasing skill and capability disparities (Pietrobelli and Rabellotti 2011). A key theme to emerge in interviews with digital workers was the common practice of clients withholding contextual information about their business or the tasks they outsource through digital labor platforms. Many digital workers explained that they know very little about the clients they are working for. Mindo, a Filipino data entry worker in his midtwenties, for example, said about his client, "I really don't have any idea on what kind of e-commerce site that he has. ... We only talked about how it will be done and the output that he needs."

Some workers also expressed a pronounced reluctance to probe clients for information relating to their core businesses, suggesting that they are only entitled to the knowledge volunteered by clients in task descriptions. Moreover, in numerous cases when digital workers have asked their clients for further clarification, they report on being met with silence.

Digital workers are thus, in many cases, kept at arm's length, unable to access information about the wider chain their labor forms part of. Those digital workers are unsure of what function their tasks serve, what the tasks mean, or how their work is used by end clients. Furthermore, only some digital workers were able to articulate or make qualified guesses as to how their clients derived value from the labor they performed. Despite the theoretical potentials for digital markets to afford disintermediated connections between workers and end clients, many workers remain unaware about not just the purpose of the work they do, but also who exactly ultimately requests it.

These information asymmetries afford little in terms of providing digital workers with opportunities to upgrade their skills so that they can take on new functions or positions in the value chains in which they are embedded. Rather, information asymmetries enforced by clients inhibit workers' ability to increase their skill sets, something they are able to do only if knowledge is available about the end uses to which their labor is being put.

Possible Implications for Policy and Practice: Four Strategies

This chapter has shown that a global, but uneven, market for digital labor exists, with a significant imbalance between the supply and demand of work. Frictions of distance have not been eliminated (Graham, Andersen, and Mann 2015); they have rather been warped to enable new spatial fixes for digital work. As Harvey (2008, n.p.) notes, "The perpetual need to find profitable terrains for capital-surplus production and absorption shapes the politics of capitalism." In the contexts of scarce labor and high wages, capital needs to find ways of disciplining labor power. Digital labor is effective in this regard because it encourages both "technologically induced unemployment" in high-wage economies (through the offshoring of work) and, in some cases, the "proletarianization of hitherto independent elements of the population" (Harvey 2008, n.p.).

Furthermore, the interviews we undertook have demonstrated that not only do market mechanisms seem to serve clients more effectively than workers, but the market itself is skewed in ways that can further exacerbate inequality among those seeking jobs. Those who make it through the barriers of reputation and ranking systems experience tangible benefits. Those who do not suffer harm. There is tedium, loneliness, alienation, but also empowerment. There are new jobs for many who crave and need them but inherent precariousness and nothing at all resembling job security. And structurally, there is a transfer of risk and responsibility. As Coe and Yeung (2015, 110) have argued, "Global production networks are fundamentally an organizational platform for economic actors to mitigate the different forms of risk." Virtual production networks are no different; but in this case, the presence of a labor reserve of millions of potentially replaceable workers attempting to underbid each other through markets makes it easier to place the burden of risk on workers themselves. Clients absolve themselves of most of it, platforms absolve themselves of most of it, and workers are left as the ones most exposed to it.

The ultimate goal of the research on which this chapter is based is to allow for insights into who ultimately benefits from contemporary practices of digital labor. There is no simple story of exploitation, and many of the workers we spoke to were indeed happy to have a job and happy with the wages they received. But nor is digital work a straightforward pathway to economic development for a broad base of workers. If we accept that practices of work in the capitalist world system have always been characterized by exploitation and power imbalances between labor and capital, then it seems odd to even suggest that digital mediations of work would do anything other than amplify those processes.

Digital tools and digital connectivity have certainly allowed new digital divisions of labor to be brought into being. But this does not mean that a flat marketplace, in which all participants have access to a perfect amount of information, has been created. Rather, geography has been bridged in some key ways, allowing work and money to be almost seamlessly transferred and transacted between anywhere in the world. But, in other ways, this dispersed geography is used against workers: opaque production networks conceal exploitative work practices from end clients, and an international labor pool of digital workers alongside a lack of copresence makes it hard to organize place-based struggles for worker rights (e.g., picket lines) or to enact solidarity with fellow workers on the other side of the planet (see Graham 2016; Lehdonvirta 2016). Furthermore, the ownership and control of labor platforms in just a few unaccountable hands means that work tends to be performed outside the purview of national governments: minimum wages, worker protections, and even taxes (which very few workers we interviewed admitted to paying) seem to be optional rather than required for both the platforms and the clients who source work through them.

But this is just one vision for the Internet and digital labor. Building on Peck's (2002) call to avoid painting the "global" as an unruly domain that is effectively beyond regulation, our mappings have shown digital work to be trans- and multiscalar but characterized by distinct networks and geographies: transnational, but never geographically disembedded. And that starting point is what is needed to rethink what alternate futures for digital labor might look like. To achieve better wages and ways of attaining more stable contracts, we outline below four broad possibilities (that combine class politics, occupational politics, identity politics, and reproduction politics) that might alter what Piven and Cloward (2000) refer to as the existing "power repertoires" between capital and labor.

Market-Based Strategies
Because transnational flows of commodities and work frequently involve long, complex, mediated, and opaque production networks, a range of infomediaries have emerged to analyze critically working and production conditions in upstream nodes on supply chains. For instance, consumer watchdog magazines like *Which?*, *Consumer Reports*, and *Stiftung Warentest* seek to reveal information that sellers of end products often wish to conceal. Organizations involved in certification schemes (such as Fair Trade and the Rainforest Alliance) attempt to ensure that minimum standards are adhered to, and activist organizations like Sourcemap and Wikichains aim to increase informational transparency in supply chains (Cook 2004; Kleine 2015).

The idea underpinning all this work has been that ICTs could be used to facilitate not just the easy geographic movement of products and services, but also a more transparent geographic flow of information about those products and services. If consumers or buyers have more information about products and production practices, then firms are less likely to engage in ethically dubious practices (Hartwick 2000; Graham and Haarstad 2011).

Strategies of consumer watchdogs, certification schemes, and activist organizations could thus be emulated and applied to the contexts of digital work. The International Association of Outsourcing Professionals (IAOP) could, for instance, update its ethical standards to be more in line with the worker protections needed in a digital economy. An organization could also be established to certify that core ILO labor standards are obtained (see Burchell et al. 2014), but also that workers are paid living wages, have appropriate social and economic protections, and are not saddled with an undue amount of risk. It could be argued that a lot of exploitative digital labor occurs because end users and even private buyers of work are unaware of the nature of practices upstream in the production network. Organizations committed to transparency and identifying best practices could do much to improve working conditions (see Graham and Shaw 2017; Graham and Woodcock 2018; and some of the lead author of the chapter's initial efforts at http://fair.work/).

Labor Rights Strategies

In the history of labor struggles, workers have been able to withdraw labor in order to secure improved working standards. Yet, the very nature of digital labor means that workers can find it hard to do so. Digital workers have been unable to build any large-scale or effective digital labor movements. This is not only because many of them simply don't know each other, but also because there is an understanding that if they withdraw their labor, then workers in other parts of the world are quickly able to replace them. Digital work platforms are always designed to remind workers that they are a market—and one in which workers from all over the world are supposed to compete with one another to offer the most favorable terms possible to clients.

What, then, can be done to counter systems that make it so challenging for workers to mount any sort of place-based activism? First, current conditions for workers mean that this could be a fruitful time for efforts that attempt to foster common class consciousness among digital workers (Huws 2009; Graham and Wood 2016), and perhaps even the creation of a transnational digital workers union. Indeed, the uproar among workers in 2016 after Upwork unilaterally raised the commission that it takes for contracts reveals some of the channels that workers from very different parts of the world used to coordinate with one another. (They were, however, unable to mount any effective response.)

Current attempts at unionizing digital workers still often take geography as a starting point for organizing strategies (see, for instance, the app-based driver's association http://www.teamstertnc.org/) and have thus far been unable to empower a transnational group of workers to bargain collectively. But as Wright and Brown (2013) and Burawoy (2011) alternatively point out, the internationalizing of product markets has undermined the possibility for multiowner collective bargaining. Such strategies could alternatively follow Moody's (1997) vision of transnational "social movement unionism," which calls for loose, but inclusive, alliances among various social movements to campaign on single issues or causes. Large organizations that contract out digital work could be encouraged or pressured to work only with union workers. But this strategy will only get so far, as there will always be employers and clients threatened by such a tactic.

A second strategy could be built on what Hyman (1999, 94) refers to as "imagined solidarities" to enact digital "spaces" of resistance, or what Harvey (1995) terms "militant particularisms." Although a lack of physical copresence inhibits workers' ability to identify one another, the same networks that are mediating their work can be harnessed to create digital picket lines. In the same way that copresent picket lines aim to disrupt the ability to conduct business as usual, digital ones could be formed to disrupt the digital presence of employers. Digital workers already make extensive use of affordances like Facebook groups, subreddits, Zello (an Internet-based walkie-talkie app), and the innovative Turkopticon (a browser plugin that allows workers to rate employers using Amazon Mechanical Turk) to coordinate, share complaints, pass on work opportunities, and give feedback to one another. Those same networks could then potentially be used for practices like "Google-bombing" the web presence of irresponsible employers (that is, use search engine optimization strategies that are designed to manipulate the algorithmic filters to make certain topics more or less visible and findable); mass action to encourage other workers to avoid a particular employer temporarily, and the mass messaging of both workers and business partners explaining the key reasons for the action—in short, a full and targeted attempt at digital disruption.

The question remains, however, whether proximity and physical copresence may indeed be needed for mass and effective forms of worker solidarity. The digital contexts in which digital disruption would need to take place are highly controlled, regulated, and algorithmically opaque—factors that make it challenging to disrupt or to picket any employer. Furthermore, because of the nontransparent nature of digital production networks, the strategies mentioned above are unlikely to work for larger employers. As such, what may be needed is a reconsideration of how the digital means of production are governed and regulated.

Regulatory Strategies

The dispersed and global nature of digital work platforms has made it extremely challenging for digital workers not only to organize effectively, but also to lobby politicians to represent their interests. Unlike global networks of digital work, policymakers are confined by political boundaries and can therefore only regulate a piece of a much larger network.

As figure 11.2 demonstrates, however, only a handful of countries are home to the majority of demand for digital work. In those strategic points (because of their network centrality), both labor and consumers potentially have more agency (Coe and Jordhus-Lier 2011; Selwyn 2012), thus opening up space in those places for regulations to be enacted that govern how clients should treat their workers irrespective of location. Regulations could cover minimum hourly rates based on a living wage in the worker's country of residence (see Galbraith 1995) and rights to additional protections and severance packages after workers have been employed for a predefined period. In short, regulations could be built on top of a more inclusive definition of employment and a vision that digital labor platforms should be re-embedded into the norms and moral economies of material labor markets. There is currently very little political will to achieve these objectives in core buyer countries, but that does not mean any of them are impossible.

Political Economy Strategies

Finally, it is worth remembering that the existence of the global-scale many-to-many trading of labor is only possible because of the existence of digital platforms. These platforms extract rents from every transaction and set key rules that govern how workers and clients interact with each other.[4] (It is worth noting that the platform's power can usually trump that of not just workers, but also clients.) Platforms also design their digital contexts to provide some kinds of affordances and not others, encouraging competitive production relations through reverse auctions instead of cooperative production relations, but also potentially reducing inequalities, such as when prejudice based on nationality is overcome by the provision of verifiable information on workers' skills (Agrawal, Lacetera, and Lyons 2013).

More broadly, it is worth remembering that the existence of platforms themselves are not creating demand for digital work. Clients in some parts of the world need to complete certain tasks, and workers in other parts of the world need an income. Platforms play a key role in organizing relationships between the two parties. But other types of organization are possible.

It would therefore be feasible to reconsider who owns the digital means of production. Just as there have previously been both consumer- and worker-led pressures to transact with cooperative building societies and cooperative supermarkets instead of privately held banks and shops, there could similarly be movements to work with cooperatively managed platforms (see, for instance, Scholtz 2016). The desire to connect geographically disparate clients and workers is not one that will go away, and digital platforms are central nodes of control and extraction. We therefore can ask what greater democratic control over the production and use of surplus would look like.

Concluding Remarks

In this chapter, we have demonstrated that although digital work is now a global phenomenon, it is characterized by distinct geographies. Some workers are able to thrive in platforms that reward entrepreneurialism by skillfully building their ranking scores, aligning their self-presentation with the needs of clients, and re-outsourcing tasks to be performed for even lower wages. These positive effects on the lives of digital workers, which are often touted by promoters and supporters of digital labor in the contexts of international development, are grounded within discourses of individualization (Murphy and Carmody 2015) and are often framed in contrast to the alternative: mass unemployment. Yet, we argue that a focus on structural issues is also needed. By highlighting four key concerns in a global, but uneven, marketplace for digital labor, we can begin to address some of the ways in which digital labor might not best serve economic development goals.

Some of the frictions we have identified here (for example, imperfect information and alienation, discrimination, and the liability of foreignness), discriminate against or otherwise harm workers who are unable to navigate the complexities of a digital work marketplace. The bargaining power of workers is undermined by the size and scope of the global market for labor. The anonymity that the digital medium affords is a double-edged sword, facilitating some types of economic inclusion, but also allowing employers to discriminate at will. Disintermediation is occurring in some instances, but the large pool of people willing to work for extremely low wages as well as the importance of rating and ranking systems are also encouraging enterprising individuals to create highly mediated chains. Those mediated and opaque chains are, in turn, restricting the abilities of workers to upgrade their skills within them.

These findings have important implications, as digital labor has been presented as a tool for economic development. Governments like those of Nigeria, Malaysia, and the Philippines, as well as large organizations like the World Bank, are increasingly coming to view digital labor as a mechanism for helping some of the world's poorest escape the limited opportunities for economic growth in their local contexts. As Coe and Yeung (2015, 193) note, however, uneven power relations existed long before global production networks were brought into being, and they are necessarily "enmeshed in relations of inequality." It is therefore worth asking why we might expect digital labor and the platforms mediating it to level the field. At this nascent stage, it is important to reflect not just on what we already know about the uneven geographies of digital labor and the frictions faced by digital workers, but also to envision alternatives and strategies that might bring into being a fairer world of work.

Acknowledgments

This work was made possible by research grants from IDRC (107384-001), and the European Research Council under the European Union's Seventh

Framework Programme for Research and Technological Development (FP/2007–2013) / ERC Grant Agreement n. 335716. The authors are also grateful for the help of all the freelancers and workers who graciously contributed their time and patiently answered interview questions.

Notes

1. This chapter is an abridged version of the following article (reprinted with permission): Mark Graham, Isis Hjorth, and Vili Lehdonvirta, "Digital Labor and Development: Impacts of Global Digital Labor Platforms and the Gig Economy on Worker Livelihoods," *Transfer: European Review of Labour and Research* 23, no. 2 (2017): 135–162.

2. All interviewee names have been changed.

3. It is hard to overstate just how important feedback scores are to the process of finding work. Some workers revealed that it took them years of constant effort to find their first job because most clients do not trust workers with no feedback.

4. We do not mean just financial rents. As Sipp (2015) notes, unlike almost any other type of work, digital work platforms do not allow workers to even own their own reputational capital.

References

Agrawal, Ajay, Nico Lacetera, and Elizabeth Lyons. 2013. Does Information Help or Hinder Job Applicants from Less Developed Countries in Online Markets? NBER Working Paper Series, No. 18720. National Bureau of Economic Research, Cambridge, Massachusetts.

Beerepoot, Niels, and Bart Lambregts. 2014. Competition in Online Job Marketplaces: Towards a Global Labour Market for Outsourcing Services? *Global Networks* 15:236–255.

Burawoy, Michael. 2011. On Uncompromising Pessimism: Response to My Critics. *Global Labour Journal* 2:73–77.

Burchell, Brendan, Kirsten Sehnbruch, Agnieszka Piasna, and Nurjk Agloni. 2014. The Quality of Employment and Decent Work: Definitions, Methodologies, and Ongoing Debates. *Cambridge Journal of Economics* 38:459–477.

Coe, Neil M., and David Cristoffer Jordhus-Lier. 2011. Constrained Agency? Re-evaluating the Geographies of Labour. *Progress in Human Geography* 35:211–233.

Coe, Neil, and Henry W.C. Yeung. 2015. *Global Production Networks: Theorizing Economic Development in an Interconnected World*. Oxford: Oxford University Press.

Cook, Ian. 2004. Follow the Thing: Papaya. *Antipode* 36:642–664.

Dicken, Peter. 2015. *Global Shift*. 7th ed. London: Sage.

Elance. 2013. *Elance Annual Impact Report: Work Differently*. June 2013. https://www.elance.com/q/sites/default/files/docs/AIR/AnnualImpactReport-small.pdf.

Galbraith, John Kenneth. 1995. A Global Living Wage. In *Reinventing Collective Action*, edited by Colin Crouch and David Marquand, 54–60. Oxford: Blackwell.

Graham, Mark. 2011. Disintermediation, Altered Chains and Altered Geographies: The Internet in the Thai Silk Industry. *Electronic Journal of Information Systems in Developing Countries* 45 (5): 1–25.

Graham, Mark. 2016. Digital Work Marketplaces Impose a New Balance of Power. *New Internationalist*. May 25, 2016. https://newint.org/blog/2016/05/25/digital-work-marketplaces-impose-a-new-balance-of-power/.

Graham, Mark, Casper Andersen, and Laura Mann. 2015. Geographical Imagination and Technological Connectivity in East Africa. *Transactions of the Institute of British Geographers* 40 (3): 334–349.

Graham, Mark, and Håvard Haarstad. 2011. Transparency and Development: Ethical Consumption through Web 2.0 and the Internet of Things. *Information Technologies and International Development* 7:1–18.

Graham, Mark, Isis Hjorth, and Vili Lehdonvirta. 2017. Digital Labour and Development: Impacts of Global Digital Labour Platforms and the Gig Economy on Worker Livelihoods. *Transfer: European Review of Labour and Research* 23 (2): 135–162.

Graham, Mark, and Laura Mann. 2013. Imagining a Silicon Savannah? Technological and Conceptual Connectivity in Kenya's BPO and Software Development Sectors. *Electronic Journal of Information Systems in Developing Countries* 56 (2): 1–19.

Graham, M., and J. Shaw, eds. 2017. *Towards a Fairer Gig Economy*. London: Meatspace Press.

Graham, Mark, and Alex J. Wood. 2016. Why the Digital Gig Economy Needs Co-ops and Unions. *openDemocracy*. September 15, 2016. https://www.opendemocracy.net/alex-wood/why-digital-gig-economy-needs-co-ops-and-unions.

Graham, Mark, and Jamie Woodcock. 2018. Towards a Fairer Platform Economy: Introducing the Fairwork Foundation. *Alternate Routes* 29:242–253.

Hartwick, Elaine. 2000. Towards a Geographical Politics of Consumption. *Environment & Planning A* 32:1177–1192.

Harvey, David. 1989. *The Urban Experience*. Oxford: Blackwell.

Harvey, David. 1995. Militant Particularism and Global Ambition: The Conceptual Politics of Place, Space, and Environment in the Work of Raymond Williams. *Social Text* 42:69–98.

Harvey, David. 2008. The Right to the City. *New Left Review* 53 (September/October). http://newleftreview.org/II/53/david-harvey-the-right-to-the-city.

Hudson, Ray. 2001. *Producing Places*. New York: Guilford Press.

Huws, Ursula. 2009. The Making of a Cybertariat? Virtual Work in a Real World. *Socialist Register* 37 (37): 1–13.

Hyman, Richard. 1999. Imagined Solidarities. In *Globalization and Labour Relations*, edited by Peter Leisink, 94–115. Cheltenham, UK: Edward Elgar.

International Labour Organization (ILO). 2014. *World of Work Report 2014*. Geneva: International Labour Organization.

International Labour Organization (ILO). 2015. *World Employment Social Outlook*. Geneva: International Labour Organization.

ITU. 2016. *ICT Facts and Figures 2016*. Geneva: International Telecommunications Union. http://www.itu.int/en/ITU-D/Statistics/Documents/facts/ICTFactsFigures2016.pdf.

Kaplinsky, Raphael. 2004. Spreading the Gains from Globalization: What Can Be Learned from Value Chain Analysis? *Problems of Economic Transition* 47:74–115.

Kaplinsky, Raphael, and Mike Morris. 2001. *A Handbook for Value Chain Analysis*. Ottawa: International Development Research Centre.

Kleine, Dorothea. 2015. Putting Ethical Consumption in Its Place: Geographical Perspectives. In *Ethics and Morality in Consumption: Interdisciplinary Perspectives*, edited by Deirdre Shaw, Michal Carrington, and Andreas Chatzidakis, 116–137. London: Routledge.

Kuek, Siou Chew, Cecilia Paradi-Guilford, Toks Fayomi, Saori Imaizumi, Panos Ipeirotis, Patricia Pina, and Manpreet Singh. 2015. *The Global Opportunity in Online Outsourcing*. Washington, DC: World Bank.

Lehdonvirta, Vili. 2016. Algorithms that Divide and Unite: Delocalization, Identity, and Collective Action in "Microwork.". In *Space, Place and Global Digital Work*, edited by Jorg Flecker, 53–80. London: Palgrave Macmillan.

Lehdonvirta, Vili, and M. Ernkvist. 2011. *Knowledge Map of the Virtual Economy*. Washington, DC: World Bank.

Lehdonvirta, Vili, Isis Hjorth, Mark Graham, and Helena Barnard. 2015. Online Labour Markets and the Persistence of Personal Networks: Evidence from Workers in

Southeast Asia. Paper presented at American Sociological Association Annual Meeting, Chicago, August 22–25, 2015.

Manning, Alan. 2003. *Monopsony in Motion: Imperfect Competition in Labor Markets*. Princeton, NJ: Princeton University Press.

Marx, Karl. (1867) 1990. *Capital: A Critique of Political Economy*. Translated from the German by Ben Fowkes. vol. 1. London: Penguin Classics.

Maume, David J., Jr. 1999. Glass Ceilings and Glass Escalators: Occupational Segregation and Race and Sex Differences in Managerial Promotions. *Work and Occupations* 26:483–509.

McDowell, Linda. 2015. Roepke Lecture in Economic Geography—the Lives of Others: Body Work, the Production of Difference, and Labor Geographies. *Economic Geography* 91:1–23.

Moody, Kim. 1997. *Workers in a Lean World*. London: Verso.

Murphy, James T., and Padraig Carmody. 2015. *Africa's Information Revolution: Technical Regimes and Production Networks in South Africa and Tanzania*. London: Wiley.

Peck, Jamie. 2002. Political Economies of Scale: Fast Policy, Interscalar Relations, and Neoliberal Workfare. *Economic Geography* 78:331–360.

Pietrobelli, Carlo, and Roberta Rabellotti. 2011. Global Value Chains Meet Innovation Systems: Are There Learning Opportunities for Developing Countries? *World Development* 39:1261–1269.

Pietrobelli, Carlo, and Federica Saliola. 2008. Power Relationships along the Value Chain: Multinational Firms, Global Buyers and Performance of Local Suppliers. *Cambridge Journal of Economics* 32:947–962.

Piven, Frances Fox, and Richard Cloward. 2000. Power Repertoires and Globalization. *Politics & Society* 28:413–430.

Reskin, Barbara F. 2000. The Proximate Causes of Employment Discrimination. *Contemporary Sociology* 29:319–328.

Scholtz, Trebor. 2016. *Platform Cooperativism: Challenging the Corporate Sharing Economy*. New York: Rosa Luxemburg Stiftung.

Selwyn, Ben. 2012. Beyond Firm-Centrism: Re-integrating Labour and Capitalism into Global Commodity Chain Analysis. *Journal of Economic Geography* 12:205–226.

Sipp, Kati. 2015. Because It Is My Name. *Hack the Union: The Future of Worker Organizing* (blog). July 7, 2015. http://www.hacktheunion.org/2015/07/07/because-it-is-my-name/.

Wood, Alex J., Mark Graham, Vili Lehdonvirta, Helena Barnard, and Isis Hjorth. 2016. Virtual Production Networks: Fixing Commodification and Disembeddedness. The Internet, Policy and Politics Conference, Oxford Internet Institute, University of Oxford, September 22, 2016.

Wright, Chris F., and William Brown. 2013. The Effectiveness of Socially Sustainable Sourcing Mechanisms: Assessing the Prospects of a New Form of 44. *Industrial Relations Journal* 44:20–37.

12 Geographic Discrimination in the Gig Economy

Hernan Galperin and Catrihel Greppi

Introduction

Scholars have long noted that labor markets are shaped by geography, creating special relationships between employers and workers in proximity. In recent years, however, a new narrative has emerged that challenges the role of geography in explaining hiring and wages. This narrative asserts that the global diffusion of high-capacity communication networks and the digitization of work is making location increasingly irrelevant, creating a globally contested market for labor (Blinder 2006). This is the narrative of the "flat world" popularized by, among other sources, Thomas Friedman's bestseller *The World Is Flat: A Brief History of the Twenty-First Century* (2005).

According to this narrative, digital work has the potential to drive wage and employment growth in emerging economies through several mechanisms (Raja et al. 2013; Rossotto, Kuek, and Paradi-Guilford 2012). Driven by salary differentials, employers in advanced economies are likely to gravitate toward countries with large pools of lower-cost white-collar workers. Further, online labor platforms significantly reduce problems of spatial mismatch between workers' abilities and opportunities in local job markets. This increases the returns to skills acquisition, thus unleashing a virtuous cycle of human capital development and job growth in developing regions.

Several stylized facts support this narrative. First, most employers in online labor platforms are based in high-income countries, while the majority of workers are based in middle- and low-income countries (Agrawal, Lacetera, and Lyons 2013; Graham, Hjorth, and Lehdonvirta 2017). This simple fact suggests that workers in developing regions may be able to earn higher (hourly) wages than in local labor markets. Second, digital work dramatically expands the number and range of labor opportunities, providing access to employers in higher-wage countries and increasing the likelihood that individual skills will be matched with available jobs. Third, online labor platforms allow employers to break down large processes into so-called microtasks, enabling individuals or small labor cooperatives to compete directly with offshoring firms that intermediate between employers and workers.

In this chapter, we seek to empirically test the narrative of a flattened global market for digital labor. In particular, we test the hypothesis that cost differentials will induce employers to offshore contracts to workers in low-wage countries. Our empirical strategy is based on examination of internal data from Nubelo, one of the largest Spanish-language online labor platforms.[1] We obtained records for all transactions in Nubelo over a forty-four-month period between March 2012 and December 2015. The data set includes basic demographic characteristics for employers and workers, as well as extensive platform-specific information about contracted jobs.

Our results suggest that information-related frictions long observed in traditional labor markets remain pervasive (and possibly exacerbated) in online labor platforms, resulting in a significant penalty for job seekers from developing countries. This penalty works in two ways. First, after controlling for observable individual characteristics and job bids, we show that foreign (i.e., non-Spanish) workers are 42 percent less likely to win contracts from employers in Spain, the highest-wage country in our sample and where most employers in our study are based. Second, we show that Spanish workers are able to command a significant wage premium, of about 16 percent, over similarly qualified foreign workers. The combination of a hiring and a wage penalty helps explain why the attrition rate is much higher for non-Spanish workers, who are significantly less likely to remain active in Nubelo within twelve months of joining the platform.

We offer two complementary hypotheses for these results. The first relates to the nature of the contracts outsourced through online labor platforms. Most of the job opportunities available in Nubelo (such as web design or article writing) require a degree of coproduction between buyer and seller. This differs from other platforms, particularly Mechanical Turk, where demand is typically for very small, low-skill tasks that require minimal communication between employer and worker (Irani 2013). Because employers in Nubelo anticipate higher communication costs when working with foreign contractors, the balance tilts in favor of domestic workers.

Yet, the countries served by Nubelo (which mainly encompass Spain and Latin America) are relatively homogeneous in culture, language, and (to a lesser extent) time zone. Therefore, our second and most relevant hypothesis relates to information asymmetries and uncertainty about worker quality. In online labor platforms, employers evaluate candidates based on very limited verifiable information. Further, employers are unable to observe latent worker characteristics, which may be inferred in traditional hiring contexts (e.g., through personal job interviews).

Under these circumstances, stereotypes that relate country of origin to worker quality provide cognitive shortcuts that orient hiring choices. This is similar to the spatial-signaling mechanism whereby employers discriminate against job seekers from US inner cities or those living in public housing (Neckerman and Kirschenman 1991). We provide empirical evidence for this mechanism by showing that Spanish employers adjust preferences based on the amount of platform-verified information available about workers. In particular, the hiring advantage for domestic (i.e., Spanish) workers falls as more information about individual workers is available, and as employers acquire experience contracting foreign workers. This suggests that employers discriminate because of information uncertainty rather than distaste for hiring workers from other countries.

This study contributes to the emerging literature on the dynamics and socioeconomic impact of digital labor. Our contribution to this literature is threefold. First, we corroborate previous findings about the continued salience of geographic location in labor markets in a setting where language and other cultural factors are by and large irrelevant. Second, we provide a novel measure to quantify wage differentials between foreign and domestic workers, and we explore how wage penalties change when information uncertainty is reduced. Third, we propose a path-dependent mechanism that suggests why, despite very low entry costs, workers from developing countries are less likely to remain active in digital labor markets in the long run.

Discrimination in Labor Markets and the Emergence of Digital Work

Online platforms that facilitate matching for contingent work (often called the "gig economy") have been the subject of significant scholarly attention in recent years. From a macro perspective, several scholars have examined how market design features embedded in these platforms are exacerbating power imbalances between employers and workers (e.g., Kingsley, Gray, and Suri 2015). A related literature has examined how contingent work, despite being framed in a discourse of entrepreneurship and family-work balance, may be eroding workers' rights, trapping job seekers from disadvantaged groups in precarious work arrangements (Huws 2015; Mann and Graham 2016).

The starting point for this body of work is that the Internet is rapidly changing how labor markets operate. Following Autor (2001), we identify three dimensions of such change. First, search costs are significantly reduced, potentially improving matching between employers and job seekers. Second, the digitization of tasks results in more workers able to perform their tasks remotely. Third, online platforms make geographic proximity between employers and workers less relevant, potentially flattening the labor market.

In this study, we examine the hypothesis of a flattening labor market in which geographic proximity between employers and job seekers is increasingly irrelevant. Before the diffusion of the Internet and the emergence of online platforms, spatial distance benefited workers who lived near the most productive firms. This resulted in better employment opportunities and higher wages for workers in advanced economies, since workers in other countries were largely prevented from competing for these jobs by barriers to migration as well as high search and communication costs. Online labor platforms are hypothesized to create a more level playing field in which workers compete for contracts regardless of place of residence, nationality, race, gender, or other characteristics unrelated to individual productivity (Agrawal et al. 2015).

The digitization of work has the potential to significantly reduce spatial mismatch between employers and job seekers. The spatial mismatch hypothesis posits that frictions in the housing market and underinvestment in transportation results in inferior labor market outcomes for racial minorities and other disadvantaged groups because of their greater geographic distance from high-wage jobs (Kain 1992). The key insight is that, under spatial mismatch, employers need not discriminate against racial minorities or other groups, since differential outcomes result naturally from the pool of job applicants. As geographic proximity becomes irrelevant, nondiscriminating employers will hire exclusively based on productivity-related characteristics, leading to better outcomes for previously disadvantaged groups.

The above argument rests on two key assumptions that deserve further examination: First, that tasks are easily codifiable and written into binding job contracts that can be monitored and enforced remotely. Second, that online employers can accurately observe the productivity-related characteristics of job seekers. Decades of scholarship on information asymmetry and transaction costs have shown that these assumptions largely do not hold true in offline labor markets (Ashenfelter, Layard, and Card 1999; Leamer 2007). Building on these findings, scholars have turned attention to the dynamics of online labor platforms.

In general, the findings suggest that information asymmetries and communication costs are far from irrelevant for explaining outcomes in online labor markets. For example, Gefen and Carmel (2008) find that most contracts in an online programming marketplace are awarded to domestic contractors. When jobs are offshored, employers prefer workers from countries with minimal cultural distance (rather than lower costs), such as US employers hiring programmers in Canada and Australia. Hong and Pavlou (2014) find that differences in language, time zone, cultural values, and levels of economic development negatively affect hiring probabilities in a global platform for IT contracts. Similar results are reported by Lehdonvirta and colleagues (2014), who also find that the hiring penalty for foreign job applicants increases when tasks require knowledge of formal institutions (e.g., legal work) or regular interaction with employers.

Other studies attempt to identify remedial mechanisms that mitigate these observed frictions. For example, Stanton and Thomas (2015) show that being affiliated with an outsourcing firm increases hiring and wages among inexperienced workers, helping them overcome the first-job barrier. The advantage dissipates over time and jobs as more information is available about the quality of individual workers. Mill (2011) finds that feedback from previous contracts significantly reduces the effect of geographic location on hiring. Similarly, Agrawal, Lacetera, and Lyons (2013) demonstrate that the benefit of platform-verified information is disproportionately large for job applicants from less-developed countries, which suggests that employers have more difficulty evaluating quality among foreign workers. Along these lines, Ghani, Kerr, and Stanton (2014) find that ethnic Indian employers based outside India are more likely to hire Indian workers. The researchers attribute the advantage to the familiarity of these employers with information regarding workers' qualifications rather than ethnicity-based preferences.

In general, these results are consistent with theories of statistical discrimination whereby employers, faced with uncertainty about worker productivity, attribute values to individual job seekers based on perceived group averages (Aigner and Cain 1977; Phelps 1972). At its core, statistical discrimination is a theory of social stereotyping. When hiring workers, employers seek information that helps predict future productivity. If this information is too noisy or simply unavailable, stereotypes provide cognitive shortcuts that help orient hiring choices. Geographic stereotyping has long been observed in the context of traditional labor markets (Fernandez and Su 2004). For example, Neckerman and Kirschenman (1991) describe employers in the Chicago area using the home addresses of job candidates as a primary screening mechanism. In a study of hiring behavior in the New York area, Newman (1999) finds a similar pattern of employer discrimination driven by job applicants' residences (more specifically associated with public housing) rather than their socioeconomic background.

Geographic stereotyping may play an even larger role in online labor platforms for several reasons. Online employers are largely unable to screen job applicants in person. Rather, they rely on two types of information generally available on a job applicant's online profile. First, platform-generated information, such as the number of previously obtained jobs and a reputation score. Second, nonverifiable information voluntarily provided by applicants, such as career experience and technical skills not validated by platform-administered tests.

Under some circumstances, the inability to screen candidates in person may reduce hiring biases—most famously when symphony orchestras started implementing "blind" auditions, as shown by Goldin and Rouse (2000). Conversely, less information may also trigger stereotyping when other credible signals of worker quality are unavailable. The amount of platform-validated information about job applicants on Nubelo (as in most online labor platforms) is very limited. In this context, as Pallais (2014) shows, even very small differences in the amount of information available can have a significant effect on future hiring and earnings.

In addition, given the relatively small value and short-term nature of a typical digital labor contract, employers are unlikely to devote many resources to screening job applicants (Horton and Chilton 2010). Faced with several dozen applicants for each contract, limited verifiable information, and a short-term contract window, employers are likely to activate cognitive shortcuts in hiring decisions. Prior beliefs about the average productivity of workers based on available signals (such as country of origin) are likely to become highly salient in such contexts.

Data and Descriptive Results

Nubelo matches employers who post contracts for short-term jobs with workers who bid for those jobs. Job postings typically describe the task required, the job category, the expected date of delivery, and the location of the employer. Employers select workers based on the proposed bid as well as other characteristics that are visible on job applicants' online profiles. These include name, country of residence, previous work experience in the platform, and a summary feedback score from previous jobs completed. In addition, job applicants can voluntarily include other information such as a CV, a brief description of skills and work experience outside Nubelo, portfolio samples, and a personal picture.

Our data set includes records for all transactions in Nubelo for a forty-four-month period between March 2012 and December 2015. They include information on all jobs posted by employers and on all bids placed by workers, both winning and unsuccessful. Unlike other platforms, Nubelo actively discourages employer-worker interaction prior to hiring. Therefore, all the information visible to employers is available in our data set, reducing concerns about omitted variables in our estimation models. Our units of observation are the bids made by job seekers. As a result, the data set is restricted to active contractors, by which we refer to those who have submitted at least one bid during the forty-four-month study period. The full data set includes 81,497 bids made by 18,356 job seekers for a total of 5,262 jobs posted by 2,517 employers. We indicate appropriately when partial data subsets are used.

Nubelo targets Spanish-speaking employers and freelance workers. While sixty-three countries are represented in our data set, Spain and a few large countries in Latin America account for the majority of job seekers

Table 12.1
Freelancers, employers, and jobs posted, by country

Country	Freelancers	%	Employers	%	Jobs	%
Spain	6,820	37.16	1,639	65.12	3,528	67.00
Argentina	4,045	22.04	390	15.49	689	13.10
Colombia	2,144	11.68	139	5.52	222	4.22
Mexico	1,326	7.22	175	6.95	419	7.96
Venezuela	1,268	6.91	8	0.32	13	0.25
Chile	648	3.53	49	1.95	147	2.79
Peru	399	2.17	9	0.36	17	0.32
Uruguay	239	1.30	10	0.40	16	0.30
Ecuador	175	0.95	9	0.36	50	0.95
Dominican Rep.	144	0.78	6	0.24	6	0.11
Others	1,148	6.25	83	3.30	155	2.95
Total	18,356	99.99†	2,517	100.01†	5,262	99.95†

Source: Author calculations based on Nubelo data.
†Percentages do not total exactly 100 because of rounding.

(table 12.1). Yet, labor demand is largely concentrated in Spain, which accounts for about two-thirds of all employers.

Descriptive results suggest that employers tend to favor job applicants based in Spain. As shown in figure 12.1, Spanish workers win a larger-than-expected share of all jobs posted. This difference is magnified when the sample is restricted to Spanish employers: job applicants based in Spain compose 37 percent of all workers but obtain 65 percent of the contracts originating in Spain. On the other hand, when not hiring domestically, Spanish employers hire equally from all other countries in our sample. In other words, the share of contracts awarded to workers across Latin America is proportional to their share of workers in the sample.

Nubelo supports outsourcing in a broad range of job categories. Four categories account for the majority of transactions: (1) software development, (2) graphic design and multimedia, (3) writing and translation, and (4) IT services. Demand is thus concentrated in relatively high-skill jobs, particularly when compared with microtask platforms like Mechanical Turk, where lower-skill tasks (such as image identification and data entry) are most common. As expected, the market is tighter in the job categories that require more technical skills, such as software development and IT

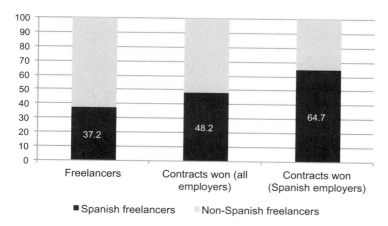

Figure 12.1
Percentage of freelancers and contracts won, by Spanish and non-Spanish. *Source*: Author calculations based on Nubelo data.

services. By contrast, competition (as per bids to projects ratio) is particularly intense for contracts in multimedia and graphic design. The intensity of competition and contract prices vary widely across but also within job categories.

Employers find two types of information in a job applicant online profile: first, information about previous jobs obtained through the platform, along with the average feedback score received by the worker in these contracts. This information is generated automatically by the platform and therefore cannot be manipulated by either party. Figure 12.2 shows the distribution of feedback scores, which is highly skewed toward the maximum of 5 ($\bar{x}$ = 4.73, SD = 0.57). This distribution is consistent with previous studies that find feedback scores in online marketplaces to be highly inflated.[2]

Recall that feedback scores are conditional on having obtained at least one job contract in Nubelo. Most active workers (i.e., those who have submitted at least one bid during the study period) have never won a contract. At the same time, a small number of successful workers concentrate much of the job volume. This results in a superstar-type distribution, which is self-reinforcing, given that, as shown below, both work experience and feedback scores are significant predictors of hiring.

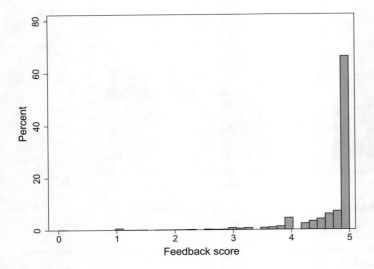

Figure 12.2
Distribution of freelancers' feedback scores (5-point scale). *Source*: Author calculations based on Nubelo data.

The second type of information available to employers is information voluntarily disclosed by workers. Nubelo encourages job seekers to complete an online profile with details about previous work experience, skills and training, a sample portfolio, and a personal picture. The platform computes the degree to which workers have completed their online profiles by assigning a certain percentage value to different data categories.[3] As shown in figure 12.3, the average worker profile is 80 percent complete. We use this threshold to examine the effect of voluntarily disclosing more (or less) information in our probability models in the next section.

Figure 12.4 shows the distribution of worker activity in Nubelo, measured by the number of bids submitted per job seeker during our study period. While the sample average is seventy-two bids, the distribution is highly skewed to the left, with a median value of just fifteen bids per worker.

Ultimately, the evidence reveals high attrition rates, with most job seekers dropping out (or remaining inactive) within the first three months of joining the platform. In fact, about 40 percent drop out within the first month. This is partly to be expected given the small probability workers will obtain a contract without validated work experience and feedback.

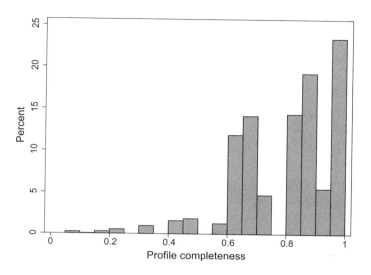

Figure 12.3
Distribution of online profile completeness. *Source*: Author calculations based on Nubelo data.

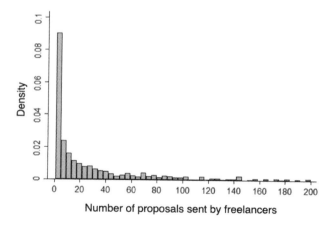

Figure 12.4
Distribution of bid activity (truncated at 200 bids). *Source*: Author calculations based on Nubelo data.

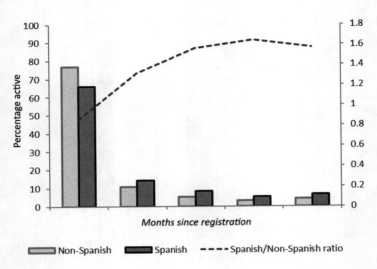

Figure 12.5
Active workers (%) by number of months since registration. *Source*: Author calculations based on Nubelo data.

Nonetheless, there are significant differences in attrition rates for Spanish and non-Spanish workers.

Workers are most active during their first six months after joining Nubelo, as shown in figure 12.5. After six months, activity levels drop sharply for both groups. Spanish workers are more likely to remain active beyond the first semester, however, with the domestic to foreign ratio stabilizing at about 1.5:1. We attribute this difference in attrition rates to the cumulative effect of hiring and wage penalties against foreign workers, as described in the next section.

Method and Results

Descriptive results suggest a hiring bias in favor of Spanish workers, particularly among Spanish employers. To formally test this proposition, we built a linear model that estimates the probability of a worker being hired, conditional on nationality and covariates that capture bid amount, bid timing (in hours after the job is posted), other worker characteristics, and country reputation. The vector of worker characteristics includes the number of previous jobs in the platform, a dummy variable for having completed the online profile at or above the sample average of 80 percent, a dummy variable for having positive feedback from previous jobs (i.e., 4 points or more on a 5-point scale), and a dummy variable that indicates whether the job seeker has previously worked with the employer.

Country reputation is measured by the number of times the employer has previously contracted a worker from the same country as the job applicant. Given high feedback scores, we hypothesize that the more previous hires from a certain country, the more likely the employer will be to hire a worker from that country. Finally, given the variance in the intensity of competition and the value of contracts across jobs, the model includes a jobs fixed-effects term, which captures both observed and unobservable differences across jobs.[4]

We restricted the sample to job postings from Spanish employers, for a number of reasons. First, as noted, contracts originating in Spain represent the bulk of job opportunities in Nubelo. Second, our main interest lies in labor offshoring from high-income to lower-income countries. With a gross national income (GNI) per capita of $28,520 in 2015 (in current US dollars), Spain's average income is about twice that of Argentina, the second largest employer. Further, during our study period, Argentine employers were prevented from hiring outside Argentina (because of government-imposed limits to international payments), which eliminates variance in our main variable of interest. Mexico, the third largest employer, with about 8 percent of jobs posted, has a GNI per capita of less than a third of Spain's.

In addition, given our interest in comparing outcomes for workers depending on country of residence, we further restrict the sample to job postings that received at least one bid from a Spanish job applicant and one from a non-Spanish (i.e., foreign) applicant. Finally, filtering for job postings that did not result in a positive match (i.e., where the employer did not hire), our restricted sample comprises 46,799 bids for 2,500 jobs.

Hiring Penalty

Table 12.2 corroborates that, after controlling for bid amount and delay, country reputation, previous contracts between applicant and employer, and observable worker quality characteristics, non-Spanish job applicants are less likely to be hired by Spanish employers. Following the full model in column 7, foreignness reduces the winning odds by 2.2 percentage points. Relative to the average winning odds of 5.3 percent in the full sample, this represents a hiring penalty of about 42 percent.

Table 12.2
Hiring probabilities (OLS with fixed effects)

Dependent variable: hiring probability	1	2	3	4	5	6	7
Foreign worker	−0.030***	−0.030***	−0.029***	−0.029***	−0.028***	−0.022***	−0.022***
	[0.002]	[0.002]	[0.002]	[0.002]	[0.002]	[0.002]	[0.002]
Controls:							
Bid amount		✓	✓	✓	✓	✓	✓
Bid delay			✓	✓	✓	✓	✓
Country reputation				✓	✓	✓	✓
Work experience					✓	✓	✓
Profile						✓	✓
Feedback						✓	✓
Worked w/employer							✓
Constant	0.152***	0.168***	0.167***	0.148***	0.128***	0.0789***	0.0743***
	[0.005]	[0.005]	[0.005]	[0.005]	[0.006]	[0.006]	[0.005]
N	46,799	46,799	46,799	46,799	46,799	46,799	46,799
R^2	0.011	0.013	0.013	0.022	0.025	0.050	0.161
Jobs	2,500	2,500	2,500	2,500	2,500	2,500	2,500
Mean DV	0.053	0.053	0.053	0.053	0.053	0.053	0.053

Source: Author calculations based on Nubelo data.
Note: Standard errors in brackets.
* $p < .1$; ** $p < .05$; *** $p < .01$

It is interesting to note how the hiring penalty for non-Spanish applicants changes as different covariates are introduced. Model 1 represents the base estimate, which includes only bid amount and nationality. In this model, the magnitude of the effect for Foreign is about 3 percentage points, which represents a 58 percent penalty relative to the sample mean. In models 2 to 7, the control variables are sequentially introduced. The penalty remains essentially unchanged until model 6, when individual reputation (i.e., feedback from previous jobs) is introduced. This strongly suggests that Spanish employers attribute quality to workers based on nationality in the absence of credible signals for individual workers. Once this information is available, the magnitude of the foreign worker penalty drops by about a third.

Wage Premium

Descriptive statistics also suggest that Spanish employers are willing to pay a wage premium for hiring domestically. To quantify this wage premium, we built a linear model that estimates bid amount (in log) conditional on nationality and a vector of worker characteristics.[5] We then restrict the sample to projects that resulted in a Spanish applicant being hired. Hence, our coefficient of interest (γ) indicates the marginal change in the number of bids submitted by foreign workers to the price of contracts ultimately obtained by Spanish workers. In other words, it quantifies the premium that the employer was willing to pay to hire domestically, relative to alternative bids by similarly qualified (per observable characteristics) foreign workers.

The results in table 12.3 indicate that, when hiring locally, Spanish employers rejected alternative bids by non-Spanish job seekers that were, on average, 14 percent lower (model 6). This translates into a wage premium for Spanish workers of about 16 percent when calculated as a premium over alternative bids.

We next examine if the wage premium varies when more information about job applicants is available (table 12.4). Our hypothesis is that the wage premium will be higher among job applicants without previous contracts, with lower-than-average feedback, and with less information on their online profiles. To test these hypotheses, we replicate the full model (model 6 in table 12.3) for different subsamples of job postings.

Table 12.3
Wage premium (OLS with fixed effects)

Dependent variable: bid amount (log)	1	2	3	4	5	6
Foreign worker	−0.120***	−0.122***	−0.124***	−0.125***	−0.140***	−0.141***
	[0.012]	[0.012]	[0.012]	[0.012]	[0.012]	[0.012]
Controls:						
Bid delay		✓	✓	✓	✓	✓
Work experience			✓	✓	✓	✓
Profile				✓	✓	✓
Feedback					✓	✓
Worked w/ employer						✓
Constant	4.738***	4.615***	4.644***	4.660***	4.746***	4.747***
	[0.008]	[0.014]	[0.015]	[0.018]	[0.019]	[0.019]
N	31,516	31,516	31,516	31,516	31,516	31,516
R^2	0.003	0.007	0.008	0.008	0.014	0.014
Number of jobs	1,626	1,626	1,626	1,626	1,626	1,626
Mean DV	281	281	281	281	281	281

Source: Author calculations based on Nubelo data.
Note: Standard errors in brackets.
* $p < .1$; ** $p < .05$; *** $p < .01$

Table 12.4
Wage premium by applicant characteristics (OLS with fixed effects)
Dependent variable: log of bid amount

	Previous contracts		Feedback score				Profile completeness			
	No	Yes	<4	3	4+	4	<80%	5	80%+	6
	1	2								
Foreign penalty	-0.119***	-0.079***	-0.164**		-0.052***		-0.122**		-0.135***	
	[0.029]	[0.020]	[0.066]		[0.017]		[0.048]		[0.016]	
Controls:										
Bid delay		✓		✓	✓			✓	✓	
Profile	✓	✓	✓	✓	✓					
Feedback	✓	✓		✓				✓	✓	
Work experience			✓	✓	✓			✓	✓	
Worked w/ employer	✓			✓	✓			✓	✓	
Constant	4.916***	4.421***	4.856***		4.482***		4.768***		4.737***	
	[0.041]	[0.049]	[0.093]		[0.026]		[0.064]		[0.022]	
Observations	7,463	8,831	1,790		13,089		2,877		17,124	
R^2	0.007	0.008	0.006		0.004		0.010		0.017	
Number of jobs	518	1,108	144		1,482		321		1,306	
Mean of DV	377.5	214.0	527.3		224.6		392.1		236.6	

Source: Author calculations based on Nubelo data.
Note: Standard errors in brackets.
* $p < .1$; ** $p < .05$; *** $p < .01$

In model 1, the sample is restricted to jobs that only received bids from workers without previous job experience. These results are compared to model 2, in which the sample is restricted to jobs that received bids only from workers with previous contracts in Nubelo. As in table 12.3, we restrict our sample to jobs for which a Spanish worker was hired, but which also received (unsuccessful) bids from non-Spanish applicants. As expected, the wage penalty is larger among job seekers without previous contracts (model 1) with respect to job seekers with validated work experience (model 2). In other words, the less validated information about individual workers is available, the higher the premium employers are willing to pay to hire domestically.

Models 3 and 4 are based on a similar exercise. In this case, the comparison is between workers with below-average feedback scores (model 3) against workers with above-average scores (model 4). The results indicate that having poor feedback from previous contracts disproportionately affects foreign job applicants. As shown, the wage penalty in model 3 is about three times larger than the penalty in model 4. Spanish employers likely interpret poor feedback as confirmation of their prior beliefs about lower average quality among foreign workers.

Finally, we replicate this exercise for jobs that received bids only from workers with below-average information in their online profile (model 5) against jobs where all bidders had average or above-average information in their profiles (model 6). As shown, the foreign wage penalty is almost identical in both cases. This suggests that employers discount this type of information regardless of worker nationality.

Conclusions

Online platforms offer a valuable laboratory to examine how discrimination operates in various social realms. In some cases, the research interest lies in understanding how the algorithms that determine what information is presented to which platform participants affects individual behavior and beliefs (Sandvig et al. 2014). This study is situated in a related line of work, which examines how biases emerge from choices by participants in multisided platforms, and how alternative market design choices promote or mitigate discriminatory behavior. Examples include scholarship about bias in short-term property rentals (Edelman, Luca, and Svirsky 2017), in peer-to-peer transportation (Ge et al. 2016), and in crowdfunding platforms (Pope and Sydnor 2011).

The focus of this study is geographic discrimination in a contract labor platform that matches employers with job seekers. We find that discrimination stems from information frictions that trigger cognitive shortcuts among employers. These cognitive shortcuts are, by definition, stereotypes that associate country of origin with expected worker productivity. We corroborate this mechanism by showing that increasing the availability of validated information about individual workers tends to deactivate geographic stereotypes, reducing the hiring and wage penalties faced by foreign job applicants.

The findings have multiple theoretical implications. The first is that the reduction in search and communication costs brought about by online platforms does not result in perfectly competitive labor markets. Rather, competition continues to be highly imperfect, with outcomes shaped by the geographic proximity between employers and job applicants. In the online environment, however, geography operates less in terms of physical distance (as it does in the spatial mismatch hypothesis) and more as a signaling mechanism that orients hiring choices. This is why the "flat world" metaphor is inadequate to describe the dynamics of online labor. The metaphor correctly describes recent changes in the underlying information infrastructure of labor markets but mistakenly identifies the source of frictions that determine employment and wage outcomes.

A flattened labor market assumes impersonal exchanges in which race, gender, nationality, and other personal characteristics become irrelevant. The results of this study indicate that this is not how digital labor platforms operate. In fact, they suggest that stereotypes may play an even larger role in determining hiring and wage outcomes, particularly when verifiable information about individual job applicants is limited and employers lack other mechanisms for screening workers.

Moreover, research has shown that most tasks cannot be easily routinized or codified, and the ones that can (such as image recognition and data entry) are increasingly being automated (Kokkodis, Papadimitriou, and Ipeirotis 2015). The limits to the commodification of work suggest that exchanges in online labor markets will continue to depend on human relationships and, as a result, be affected by communication costs and information frictions. In other words, the location and identities of the transacting parties will continue to matter.

Several platform design and policy implications emerge from our findings. First, platform operators may discourage displaying information unrelated to productivity on workers' profiles, while implementing mechanisms to validate skills and previous work experience. This would not only favor workers in developing countries (who, as shown, are penalized disproportionately when they lack verified experience) but also improve employer-worker matching. Horton (2017) estimates that about half of the job contracts posted in digital labor platforms are never filled.

Second, platforms can further develop mechanisms that help employers find and screen job applicants. As mentioned, Nubelo discourages employer-applicant interaction before hiring, but other platforms in fact promote personal interviews and direct contract negotiation between parties. Following our results, these mechanisms are likely to deactivate stereotypes in hiring choices, thus favoring foreign job seekers and more generally promoting diversity in job categories that are currently associated with specific demographic groups.

Third, employers can be nudged to hire more diversely by altering the order in which alternative candidates are presented. For example, Nubelo attempted to lower the first-job barrier (and thus reduce worker attrition) by favoring job seekers without previous contracts in the algorithm that determined the display of potential matches to employers. A similar nudge could be applied to favor foreign job seekers, or candidates from specific countries that are under-represented in certain job categories. In turn, individual hiring would help build country reputation, which, as this and other studies show, is an important predictor of hiring (see Leung 2012).

Finally, the question of governance for online labor contracts must be addressed to protect workers from the vulnerabilities associated with digital work. These vulnerabilities are multifaceted, ranging from lack of enforcement of minimum-wage legislation to the boundaries of employment relationships (De Stefano 2016). This and other questions related to the protection of workers' rights only become more pressing when hiring takes place across borders. In particular, clear jurisdictional lines must be established to enable the enforcement of existing antidiscrimination laws in hiring and compensation in the context of online labor contracting.

Support

This work was supported by the International Development Research Centre (IDRC-Canada), Project c107601-001.

Notes

1. In December 2016, Nubelo was acquired by Freelancer.com.

2. For example, Pallais (2014) finds that 83 percent of data entry workers in oDesk received a rating of at least 4, while 64 percent received a maximum rating of 5. Similarly, Stanton and Thomas (2015) find that about 60 percent of workers in oDesk received a feedback score of 5 in their first job.

3. For example, a personal picture adds 10 percent to the completeness of the profile, a description of previous work experience adds 5 percent, a description of skills adds 10 percent, and so forth.

4. More formally, the estimated model is

$$\text{Hiring}_{ij} = \alpha_{ij} + \gamma \text{Foreign}_{ij} + \delta \log \text{Price}_{ij} + \lambda \log \text{Delay}_{ij} + \eta \text{CountryRep}_{ij} + \beta Z + \sigma_j + \varepsilon_{ij}$$

where Hiring is the probability of worker's bid i being selected for job posting j, Foreign is a dummy (yes = 1) that identifies non-Spanish workers, Price denotes bid amount (in log), Delay is the difference (in hours) between the job posting and the bid submission (in log), CountryRep denotes whether the employer has previously hired from the same country of the worker submitting bid i at the time of job posting j, Z is a vector of worker characteristics that vary over time, σ controls for job fixed effects, and ε is an error term.

5. More formally, the estimated model is

$$\log(\text{Price})_{ij} = \alpha_{ij} + \gamma \text{Foreign}_{ij} + \lambda \log \text{Delay}_{ij} + \beta Z + \sigma_j + \varepsilon_{ij}$$

where Price is the bid by worker i for job posting j, Foreign is a dummy (yes = 1) that identifies non-Spanish workers, Delay is the difference (in hours) between the job posting and the bid submission (in log), Z is a vector of worker characteristics that vary over time, σ controls for job fixe effects, and ε is an error term.

References

Agrawal, Ajay, John Horton, Nicola Lacetera, and Elizabeth Lyons. 2015. Digitization and the Contract Labor Market: A Research Agenda. In *Economic Analysis of the Digital Economy*, edited by Avi Goldfarb, Shane M. Greenstein, and Catherine Tucker, 29–256. Chicago: University of Chicago Press.

Agrawal, Ajay, Nicola Lacetera, and Elizabeth Lyons. 2013. Does Information Help or Hinder Job Applicants from Less Developed Countries in Online Markets? NBER Working Paper Series, No. 18720. National Bureau of Economic Research, Cambridge, Massachusetts.

Aigner, Dennis J., and Glen G. Cain. 1977. Statistical Theories of Discrimination in Labor Markets. *ILR Review* 30 (2): 175–187.

Ashenfelter, Orley, Richard Layard, and David Card. 1999. *Handbook of Labor Economics*. North Holland: Elsevier.

Autor, David H. 2001. Wiring the Labor Market. *Journal of Economic Perspectives* 15 (1): 25–40.

Blinder, Alan S. 2006. Offshoring: The Next Industrial Revolution? *Foreign Affairs* 85 (2): 113–128.

De Stefano, Valerio. 2016. Introduction: Crowdsourcing, the Gig-Economy and the Law. *Comparative Labor Law & Policy Journal* 37 (3): 471–503.

Edelman, Benjamin, Michael Luca, and Dan Svirsky. 2017. Racial Discrimination in the Sharing Economy: Evidence from a Field Experiment. *American Economic Journal: Applied Economics* 9 (2): 1–22.

Fernandez, Roberto M., and Celina Su. 2004. Space in the Study of Labor Markets. *Annual Review of Sociology* 30:545–569.

Friedman, Thomas L. 2005. *The World Is Flat: A Brief History of the Twenty-First Century*. New York: Farrar, Straus and Giroux.

Ge, Yanbo, and Christopher Knittel. Don Mackenzie, and Stephen Zoepf. 2016. Racial and Gender Discrimination in Transportation Network Companies. NBER Working Paper Series, No. 22776. National Bureau of Economic Research, Cambridge, Massachusetts.

Gefen, David, and Erran Carmel. 2008. Is the World Really Flat? A Look at Offshoring in an Online Programming Marketplace. *Management Information Systems Quarterly* 32 (2): 367–384.

Ghani, Ejaz, William Kerr, and Christopher Stanton. 2014. Diasporas and Outsourcing: Evidence from oDesk and India. *Management Science* 60 (7): 1677–1697.

Goldin, Claudia, and Cecilia Rouse. 2000. Orchestrating Impartiality: The Impact of "Blind" Auditions on Female Musicians. *American Economic Review* 90 (4): 715–741.

Graham, Mark, Isis Hjorth, and Vili Lehdonvirta. 2017. Digital Labour and Development: Impacts of Global Digital Labour Platforms and the Gig Economy on Worker Livelihoods. *Transfer: European Review of Labour and Research* 23 (2): 135–162.

Hong, Yili, and Paul A. Pavlou. 2014. *Is the World Truly "Flat"? Empirical Evidence from Online Labor Markets*. Fox School of Business Research Paper No. 15-045. October 1, 2014.

Horton, John J. 2017. The Effects of Algorithmic Labor Market Recommendations: Evidence from a Field Experiment. *Journal of Labor Economics* 35 (2): 345–385.

Horton, John Joseph, and Lydia B. Chilton. 2010. The Labor Economics of Paid Crowdsourcing. *Proceedings of the 11th ACM Conference on Electronic Commerce*, 209–218. New York: ACM.

Huws, Ursula. 2015. When Adam Blogs: Cultural Work and the Gender Division of Labour in Utopia. *Sociological Review* 63 (1): 158–173.

Irani, Lilly. 2013. The Cultural Work of Microwork. *New Media & Society* 17 (5): 720–739.

Kain, John F. 1992. The Spatial Mismatch Hypothesis: Three Decades Later. *Housing Policy Debate* 3 (2): 371–460.

Kingsley, Sara Constance, Mary L. Gray, and Siddharth Suri. 2015. Accounting for Market Frictions and Power Asymmetries in Online Labor Markets. *Policy and Internet* 7 (4): 383–400.

Kokkodis, Marios, Panagiotis Papadimitriou, and Panagiotis G. Ipeirotis. 2015. Hiring Behavior Models for Online Labor Markets. *Proceedings of the Eighth ACM International Conference on Web Search and Data Mining*, 223–232. New York: ACM.

Leamer, Edward E. 2007. A Flat World, a Level Playing Field, a Small World After All, or None of the Above? A Review of Thomas L. Friedman's "The World is Flat." *Journal of Economic Literature* 45 (1): 83–126.

Lehdonvirta, Vili, Helena Barnard, Mark Graham, and Isis Hjorth. 2014. Online Labour Markets—Leveling the Playing Field for International Service Markets? Paper presented at IPP2014: Crowdsourcing for Politics and Policy, September 25–26, 2014, Oxford, UK.

Leung, Ming. 2012. Job Categories and Geographic Identity: A Category Stereotype Explanation for Occupational Agglomeration. Institute for Research on Labor and Employment Working Paper Series. June, 2012.

Mann, Laura, and Mark Graham. 2016. The Domestic Turn: Business Process Outsourcing and the Growing Automation of Kenyan Organisations. *Journal of Development Studies* 52 (4): 530–548.

Mill, Roy. 2011. Hiring and Learning in Online Global Labor Markets. NET Institute Working Paper No. 11-17. NET Institute, New York.

Neckerman, Kathryn M., and Joleen Kirschenman. 1991. Hiring Strategies, Racial Bias, and Inner-City Workers. *Social Problems* 38 (4): 433–447.

Newman, Katherine S. 1999. *No Same in My Game: The Working Poor in the Inner City.* New York: Knopf.

Pallais, Amanda. 2014. Inefficient Hiring in Entry-Level Labor Markets. *American Economic Review* 104 (11): 3565–3599.

Phelps, Edmund. 1972. The Statistical Theory of Racism and Sexism. *American Economic Review* 62 (4): 659–661.

Pope, Devin G., and Justin R. Sydnor. 2011. What's in a Picture? *Journal of Human Resources* 46 (1): 53–92.

Raja, Siddhartha, Saori Imaizumi, Tim Kelly, Junko Narimatsu, and Cecilia Paradi-Guilford. 2013. *Connecting to Work: How Information and Communication Technologies Could Help Expand Employment Opportunities.* Washington, DC: World Bank.

Rossotto, Carlo M., Siou C. Kuek, and Cecilia Paradi-Guilford. 2012. *New Frontiers and Opportunities in Work.* Washington, DC: World Bank.

Sandvig, Christian, Kevin Hamilton, Karrie Karahalios, and Cedric Langbort. 2014. Auditing Algorithms: Research Methods for Detecting Discrimination on Internet Platforms. Data and Discrimination preconference of the 64th annual meeting of the International Communication Association, Seattle, Washington, May 2014.

Stanton, Christopher T., and Catherine Thomas. 2015. Landing the First Job: The Value of Intermediaries in Online Hiring. *Review of Economic Studies* 83 (2): 810–854.

13 Margins at the Center: Alternative Digital Economies in Shenzhen, China

Jack Linchuan Qiu and Julie Yujie Chen

Introduction

Every world has its order, complete with its center and margins, that is imposed by nativist or imperialist authorities and internalized in one's mind. The world of digital technologies is no exception, when conventional wisdom regards the West—or more specifically, Silicon Valley—as the pivotal site of action. This, in essence, reflects the classic center-periphery framework in Wallerstein's conception of the capitalist world system (1974). But modern industries, especially the hardware, software, and services that underpin digital economies, are also very different because they morph rapidly in a "space of flows," with little respect for traditional boundaries (Castells 1996). These industries are transforming with great unpredictability as China emerges as a new global epicenter of technological growth (Schiller 2005), a trend nicely captured in the book title *From Silicon Valley to Shenzhen* (Lüthje et al. 2013).

From a Eurocentric perspective, China—until a little more than two decades ago—was often seen as belonging to the global margins. Before the rise of modern Europe, however, China boasted the world's most advanced science and technology (Needham 1979). Of the four greatest inventions in ancient China (papermaking, printing, the compass, gunpowder), the first three are actually information and communication technologies (ICTs). This historical context is crucial in suggesting that Eurocentric assumptions deserve scrutiny in order to do justice to technological developments in longstanding civilizations such as China, that we need to conceive multiple worlds with alternative and overlapping orders to grasp the reality of digital economies at the semiperipheries and peripheries—where margins turn out to be central to technological innovation. What, then, is marginality? Why does China show a tendency toward technological innovation in the digital economy in ways that differ from, sometimes even defy, innovation models in the West? What are the implications of the Chinese experiences? How do they illuminate alternative possibilities at the global level?

In this chapter, we first conceptualize marginality as phenomena occurring at the edges of geography, fringes of history, and verges of civilization. In so doing, we consider in broad strokes the connections between China and the West, going back to the Opium Wars, a fateful clash of civilizations that remains relevant when we consider the centrality of margins to contemporary digital economies. The bulk of this chapter focuses on Shenzhen's special economic zone (SEZ) in the Pearl River Delta of South China, arguably the most striking example of "margins at the center" in the Chinese context: a place full of alternative digital innovations that entail a reconsideration of what constitutes the core of global digital economy.

We discuss the regional and national contexts for Shenzhen's SEZ transformation from a technology backwater in the immediate aftermath of the Maoist era (early 1980s) to the most vibrant digital economy cluster in China at the beginning of the twenty-first century. Our main focus is on (a) economies of low-end *shanzhai* electronics in Shenzhen and (b) the app-based ride-hailing economy, energized and agitated by Uber and Didi (the market leader in China). We examine the two cases through the relationship between the center and the margins of innovation, not only geographically in the global digital economy but also from the different vantage points of labor, capital, and the state.

Using fieldwork data, interviews, and primary documents, we test our observations against the old ideas that innovations always travel from the center to the periphery, and that marginality means inevitable disadvantages. The two cases are, in this sense, selected because they are important exceptions to conventional thinking, yet because of their scale and popularity, they are central for understanding China's digital economy. Shanzhai electronics, once synonymous with low-quality copycat products, now has worldwide reach, especially for regions of the Global South. Didi China, unlike Uber, integrates both taxi and private ride services on the same platform. Uber's rhetoric of flexible hours for drivers seems to exclude incumbent taxi drivers from the emerging digital economy facilitated by online platforms. Didi's inclusion of both taxi and private drivers forces us to rethink the division between the two in the original Uber model. Furthermore, Didi taxi and private car drivers show considerable dexterity in resisting exploitation by digital platforms. They have also rallied to turn their marginality in the Chinese economy (and under Didi's rule) into a core resource for alternative development.

Together, the shanzhai business, Didi's market dominance, and the recalcitrance of its drivers all shed light on how marginality comes about in spatial, historical, and discursive terms. They also show how marginality can be transformed into centrality and into new forms of resistance and alternative development. The chapter concludes by discussing why certain margins turn out to play a central role in contemporary digital transformations, under which conditions they do so, and what the global implications are for Shenzhen and China when seen through the lens of marginal centrality, a lens through which we can see a different world, especially in non-Western contexts, whose orders are multiple and relative, rather than singular and static.

Conceptualizing Marginality

Does the center always control the periphery? Before considering the world of contemporary ICTs, it is instructive to revisit a debate among historians regarding the nature of nineteenth-century European imperialism. While the conventional view holds that imperialist expansion during this period was the result of technological prowess, politico-economic motivation, and the ideological positions of decision makers in European capitals, historians such as D. K. Fieldhouse contended, on the contrary, that "imperialism may be seen as a classic case of the metropolitan dog being wagged by its colonial tail"; that "Europe was pulled into imperialism by the magnetic force of the periphery" (1973, 81, 463).

The First Opium War of 1839–1842 offers a good example for Fieldhouse's argument of the margins pulling at the center. The key technological object used in this war was the steam-powered iron gunboat, which proved to be the critical factor in defeating Chinese troops. A surprising finding from historical research, however, shows that the British Navy, being proud of its traditional sailing warships and suspicious of the new steamer technology, was reluctant to adopt the innovation. The navy did not purchase a single gunboat until 1845, three years after the war ended (Headrick 1981, 36). Who then supplied the ten gunboats during the First Opium War? It was the Calcutta-based East India Company (EIC), the world's "first major purchaser of gunboats" (54), whose "secret committee" first used clandestine language to deceive London (especially the navy) into importing the new technological inventions from the UK before sending them to support the British fleet. Hence, "Before the age of military research and development, technological innovation (such as the gunboat) often had to sneak through the back door" (37).

The EIC's secret purchase and deployment of the gunboats during the First Opium War was one of the most prominent cases of the "colonial tail," in Calcutta, wagging the "metropolitan dog," in London. The development of gunboats was also as much a major turning point for the relationship between China and the West as it was for the invasion of the Global South in general. As Headrick (1981) documented, only with gunboats could European colonial forces move upstream and conquer the heartlands of Asia and Africa beyond strategic positions along the coast (which older sailships could conquer). Innovation from the periphery was not an isolated phenomenon, however, as we see, over and again, faster technological application at the frontiers or margins of empire. For instance, the world's first attempt to lay underwater telegraph lines took place in Calcutta in 1839, rather than in the metropolitan centers of the British Isles (158). The historical development of the telegraph into a crucial information and communication technology for the empire cannot be fully understood without recognizing the deployment of technologies in the global margins.

To inflate the role played by the margins in technological advancement is, of course, incorrect. But as we argue in this chapter, it is not uncommon for places, people, and organizations located in the semiperipheral or peripheral regions to take the lead in innovative undertakings.

It is thus necessary to reconsider margins in pluralistic terms, beyond the simple measurement of spatial distance from, or economic ties to, the center. As Doreen Massey (2005) points out, a sense of space is fundamental to our thinking about periphery and marginality, while the "power geometries" in spatial thinking are often intertwined with other positionalities of cultural, discursive, and gender relations. This is particularly the case when the innovations are so groundbreaking (e.g., iron gunboat) that they are incongruent with, or even jeopardize, conventional ways of practice, old thinking, and established business models (e.g., the sailship).

We suggest three intertwined dimensions to conceptualize marginality, all of which hinge on a dynamic and dialectic relationship and interplay between the margins and the center. Neither marginality nor centrality exists alone. They always connect to, counter with, and are constitutive of the other. Like Mezzadra and Neilson (2013), we do not think there is a clear fault line between the two. On the contrary, we contend that the shifting relationship between the center and the margins is precisely why marginality offers an important angle and site to interrogate contemporary digital economy.

The first dimension of marginality speaks to the edges of geography—that is, any system must contain certain forms of geographic discontinuities marking the inside from the outside. Edges are concentrated areas of discontinuity. One way to understand such geographic edges is the basic logic of the network society, which according to Castells (1996) means the binary operation of inclusion and exclusion. Margins are where geographic inclusion of any social, economic, and technological units approaches its limit. At the margins, forces from the core become feeble, ambiguous, easy to subvert, sometimes irrelevant, even paradoxical. As Anna Tsing remarks, "Can one be simultaneously inside and outside the state? This is the dilemma of marginality. ... Marginals stand outside the state by tying themselves to it; they constitute the state locally by fleeing from it. As culturally 'different' subjects they can never be citizens; as culturally different 'subjects,' they can never escape citizenship" (1993, 26).

Marginality is anything but static. It shifts over time and emerges as the fringes of history, as "temporal borders" (Mezzadra and Neilson 2013, 131). Human systems, including modern systems of digital economy, have to emerge from a historical context. No system can remain unchanging throughout time. On the one hand, this suggests that historical marginality is a necessary condition for the beginning and the end of any temporal period or any "world order" in a finite temporality. On the other hand, it should be unsurprising to see the multiple overlapping temporalities where it is only normal to observe the coexistence of what Raymond Williams (1978) refers to as the residual, the dominant, and the emergent—historical boundaries between cultural elements of different periods are almost always present. The emergent means great unpredictability and potentials for the history unfolding. Wallerstein, for instance, envisions that a global postcapitalist future is "intrinsically uncertain, and therefore, precisely open to human intervention and creativity" (2000, 265).

A third dimension of marginality is what we call the vector of the renounced, when disadvantaged groups turn out to enrich and redefine the mainstream. Gary Okihiro (1994) analyzes how historically marginalized groups in the United States (e.g., African Americans) struggled for freedom and justice more forcefully than whites throughout the country's two hundred-year history. The struggles of nonwhites enriched the meanings of freedom and justice, which were embraced as distinctive features of American society. Yet these two terms were often absent from the lives of nonwhites. James Scott's (2009) study of mountain people in the Zomia highlands of Southeast Asia also illustrates that marginality can sometimes reverse the logic of the center. He argues for "a deconstruction of Chinese and other civilizational discourses about 'barbarian,' the 'raw,' the 'primitive,'" because "[civilizational] discourses never entertain the possibility of people voluntarily going over to the barbarians, hence such statuses are stigmatized and ethnicized" (xi). In the following analysis, we show that the digital economy flourishing in Shenzhen is emblematic of both fighting from within and the deliberate choice to stay outside and form contingent connections to the dominant values and practices upheld by those who are at the center of digital innovations.

Regional and National Contexts

The history of Shenzhen in modern China began in 1979, when China opened up to the world after the decade-long Cultural Revolution (1966–1976). Before becoming a special economic zone (SEZ), Shenzhen was a coastal community of fewer than fifty thousand people. It was among the least developed regions in the 1970s, because during the Maoist period (1949–1979), the Chinese authorities made active preparations for another world war, which prompted Mao to move strategic industrial capacities to the hinterlands, far away from the coast. Consequently, places like Shenzhen became economic and technological backwaters within China.

Relative marginality in the Chinese state system has been intrinsic to Shenzhen for an extended period. Yet since 1978, the city has emerged as China's gateway to the capitalistic world system because of Deng Xiaoping's economic reforms. The population grew exponentially to 14.5 million by 2009 (Shenzhen Ten-year Development Report Research Group 2013), which made Shenzhen the youngest Chinese megalopolis in terms of the city's history as well as the average age of its residents.

Shenzhen soon ascended as a key hub of electronics manufacturing and technological innovation. The total industrial output of electronic manufacturing increased from RMB 6.3 billion (US$1.32 billion) in 1990 to 88.9 billion ($10.19 billion) in 1999, then to 1.2 trillion ($183.35 billion) in 2012, accounting for 23.9 percent of China's total output of electronics (Shenzhen Bureau of Statistics 1991; 2000; Guangdong Provincial Bureau of Statistics 2013). According to official statistics from the SEZ, the total revenue for Shenzhen's software industry jumped from RMB 29.1 billion (US$3.52 billion) in 2003 to 119.1 billion ($17.44 billion) in 2009, then to 378.3 billion ($61.59 billion) in 2014 (representing 9.6 percent of China's software industry). Shenzhen's software industry has grown more rapidly than hardware manufacturing in recent years. There were 72,120 patents granted to applicants from Shenzhen in 2015, representing a 34.3 percent increase over the previous year (Shenzhen Bureau of Statistics 2016).

The digital economy in Shenzhen has evolved into a complex economic system. It is home to Foxconn—the world's largest electronics manufacturer, notorious for the serial suicides committed by its workers, poor working conditions, and high-pressure management (Chan and Pun 2010; Pun, Chan, and Selden 2015). In 2010, thirteen of the fifteen Foxconn worker suicides were in Shenzhen. Foxconn represented a more aggressive capitalism, unthinkable in the developed capitalist world, a renewed model of slavery in the twenty-first century (Qiu 2016). Shenzhen is also home to Huawei, the world's largest telecom equipment maker (Economist 2012); Tencent, the parent company of QQ and WeChat (China's most popular social media platforms), and Asia's most valuable company; and ZTE, another of China's telecom equipment giants. The city is also home to groundbreaking bioinformatics companies, for instance, BGI, which plays a leading role in global bioinformatics with its unparalleled capacity in genomics sequencing. BGI's unorthodox way of collaborating with scientists was initially criticized by Western observers but has now become an industry standard worldwide (Wong 2017), yet another instance of the "tail" at the margins wagging the "dog" at the center.

Shanzhai in Shenzhen

No other Chinese city has grown its ICT industry as rapidly as Shenzhen—neither Beijing, the political center, nor Shanghai, the traditional center of commerce. Why did Shenzhen, an SEZ interfacing with mainland China and the West, succeed in fostering so many world-leading IT companies and, in so doing, become such an important center of the global digital economy, to such an extent that it is known as "the Silicon Valley of hardware" (Lindtner, Greenspan, and Li 2015)?

Shenzhen's success resides in its being an uncharted territory, with ambiguous borders and rules. Here, an enormous number of small operations coexist to work on developing, tweaking, manufacturing, packaging, marketing, selling, and recycling digital hardware and software, content and services, through a variety of formal and informal businesses. They are spatially clustered and can be networked to scale up when the need arises. Some of these interlinked businesses build trust with each other based on traditional connections, such as kinship and shared township origins. Others use online forums and newly built friendship networks to share ideas and resources. As a result, they constantly trespass and ignore the boundaries between the domestic and the foreign, the licit and the illicit, the public and the private. The most important border crossing, argue Lindtner, Greenspan, and Li (2015), is between design and manufacture, the two distinct stages in the making of digital technologies in the West. Yet, here in Shenzhen, design and manufacture happen simultaneously, with so much interplay that the two separated processes have become one (Lindtner, Greenspan, and Li 2015).

If there is a Shenzhen model for alternative digital economies, then its best manifestation must be the shanzhai electronics sector, which combines a Chinese version of the copyleft (i.e., open source technology such as Linux and the Pirate Party) with Chinese capitalism into a peculiar form of assertive marginality in the domain of open hardware. Shanzhai, meaning literally "bandit fortress in the mountains," used to be a derogatory term referring to cheap counterfeit production of branded goods, thus breaking the legal boundaries of the intellectual property rights (IPR) regime (Ho 2010; Wallis and Qiu 2012). The market value of shanzhai products derived from famous brand names such as Nokia and iPhone. The line of authenticity between the formal and the informal modes of manufacture can be blurry—our research in 2009 found that the same assembly line churning out branded batteries for Nokia phones was also used to produce shanzhai batteries for knock-off Nokia handsets of the same model. The reason was that China's official system of quality assurance operated inefficiently. As a result, by moving into the shanzhai market, the factory could generate cash revenue faster by bypassing the law (Wallis and Qiu 2012). According to Josephine Ho (2010), such informal grassroots production practices have deeper historical origins elsewhere in China, even before the designation of the SEZ, although no other copyleft practices have approached the scale and influence of shanzhai in Shenzhen.

The inefficient Chinese state is not the only reason for shanzhai. Western multinational corporations themselves are more important factors. A former Nokia engineer we interviewed in 2009 revealed that he quit his well-paid job to start his own shanzhai business because he hated the top-down approach Nokia took at the time to micromanage production processes without respecting market demands on the ground (e.g., for phones with multiple SIM-card slots). By converting to shanzhai, he and his colleagues can have more freedom to innovate. The spirit of shanzhai is thus not really about making counterfeits; it is, instead, about making better gadgets than the original, about the margins surpassing the center.

Similar dynamics were at play in a busy IT mall, where we bought screwdrivers (figure 13.1) to take iPhones apart in January 2017. The main business in this mall is to fix and repackage iPhones in all sorts of ways that are unauthorized by Apple but that meet market demands in Shenzhen. This set of tools cost RMB 30 ($4.37), and one of them even has an Apple logo on it. For those who abhor Apple's lockdown of iPhones because it excludes users, merchants, and makers from opening the device, then these tools, sold so cheaply in Shenzhen, subvert the power inequality between Apple and the users of its products. In 2016, one of us was asked to pay $100 to replace a battery on her iPhone 5S in the United States. Yet, here in Shenzhen, the asking price for the same iPhone 5S battery was 13 yuan ($1.89)! The corporate system of closed business models and the ability of local firms to circumvent such high profit margins are among the reasons those at the margins want to subvert and grow their own alternative shanzhai digital economies.

Over time, Shenzhen's shanzhai businesses have come of age, stabilized, and generated their own system at the frontiers between the authoritarian

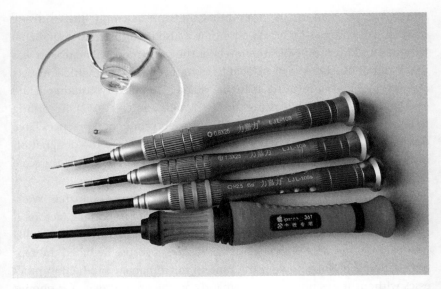

Figure 13.1
For the price of a Starbucks coffee, we bought in Shenzhen this set of tools, enabling us to open any iPhone (January 2017). *Source*: Authors.

state and global capital. This alternative system—consisting of thousands of small and large firms—is centered on chip R&D teams that work with "upstream" parts and solution providers, as well as "downstream" distributors (channel companies), while collaborating with manufacturers. Rather than a linear flow from ideation to prototyping to manufacture and then to marketing and selling, here the entire process is flexible and open, operating with much higher efficiency even compared to the most aggressively innovative brands, such as Samsung (Wallis and Qiu 2012). The result is that shanzhai devices have become not just more cheaply produced but also more innovative and effective in meeting market demands, from rural China to sub-Saharan Africa.

The latest example is Tecno, a Shenzhen-based company originating from the shanzhai scene, which targets Africa as its main market. By the end of 2016, it had beaten not only Apple, Samsung, and Nokia, but also Huawei, ZTE, and Lenovo to acquire a 40 percent market share in the African continent (Ko 2017). Why did this happen? One reason was that Tecno, in true shanzhai spirit, subverted engineering assumptions that optimize mobile phone cameras for light-skinned faces but not for Africans. As Ko

(2017, n.p.) explains, "The picture system of most mobile phones is based on white or yellow skin tones. When African users take selfies, the pictures are often either too dark or blurred. To solve this problem, Tecno collected a large number of pictures taken by African customers and tweaked the picture function of its handsets based on the data. The superior selfie quality soon became a major selling point."

While some may regard shanzhai as an outdated phenomenon in Shenzhen, the success of Tecno and the sales of iPhone-dismantling tools suggest that the alternative shanzhai system remains vibrant and has become durable. Borrowing from Raymond Williams (1978) again, we would say that shanzhai is no longer simply "emergent"; nor is it lingering as something "residual." Instead, it has become "dominant" on its own, in this region of spatial and historical marginality, and in much of the Global South, which shares similar peripheral and semiperipheral status. The marginality Shenzhen enjoys has enabled the region to become the center of an alternative world, the center of shanzhai design and production, with considerable global reach. In this sense, what we see in Shenzhen goes beyond geographic and historical margins. It belongs to the vector of the renounced in civilizational terms, too—for shanzhai has created an alternative value system, whose centerpiece is, surprisingly, about open sharing.

For the world's leading maker-entrepreneurs, Shenzhen has become the "Silicon Valley for hardware"—and not just conventional proprietary hardware, but, more important, open hardware. As Lindtner, Greenspan, and Li (2015, 5) explain, "Shanzhai is neither straightforward counterculture nor pro-system. As a multi-billion USD industry, it is deeply embedded in contemporary modes of capitalist production. At the same time, with its roots in and ongoing practices of piracy and open sharing, shanzhai challenges any inherent link made between technological innovation and the tools, instruments, and value systems of proprietary, corporate research and development." This culture of open sharing, defying conventional IPR boundaries, is best shown in the production of "public boards," or *gongban*—circuit boards designed to be given out for free to manufacturers of various types, who then need to purchase parts that would go into the "public boards" (5). Another embodiment of shanzhai culture is online information-sharing platforms such as 52RD.com, a main hub of open hardware discussion in 2009 that remains vibrant at the time of writing. The name 52RD means, in Chinese, "I love R&D." The spirit of shanzhai, as reflected here, indeed shares the sense of hobbyist dedication that characterizes copyleft activities and the "hacker ethic" (Himanen 2001). While individual designers and entrepreneurs taking part in this online platform and in the shanzhai scene indeed pursue capitalist dreams of self-enrichment, the entire social, cultural, and economic system operates as a paradise of open hardware. This paradox of conflicting ideas—capitalism versus postcapitalism, proprietary versus open sharing—has become intrinsic to Shenzhen's rise as a center of digital economy on the margins.

Didi and the Platformization of Ride Services in Shenzhen

In 2016, the most surprising technology news in China was probably Didi Chuxing defeating Uber China. The news left Western media commentators feeling both discouraged and amazed: discouraged because Uber had become yet another Silicon Valley company failing in its attempt to transplant its business to China (Mozur and Isaac 2016); amazed because of the rapid rise of the four-year-old ride-hailing company Didi Chuxing. Didi holds a greater than 90 percent market share of the ride-hailing market in China, which is projected to reach $7.6 billion by 2018 (He 2016). It manages more than 11 million rides daily across its platform. In November 2016, China became the first national government in the world to legalize ride-hailing platforms (including Didi and other apps in the Chinese context, like Uber, Yidao, and Shenzhou).

Although Didi is often seen as a competitor of Uber China, the former distinguishes itself in its strategy to include and take advantage of traditional taxi drivers. Didi, when founded in 2012, was a taxi-hailing app for passengers to book and call taxi services. The app accumulated its user base for more than two years until August 2014, when it launched its private car-booking service, followed by a carpooling service in 2015. Along these lines, Chen (2018) argued that taxi drivers have provided the essential infrastructural labor to transform taxi services in China into an app-based digital industry—that is, the platformization of ride services.

Although Chinese taxi drivers and private hires joined their respective counterparts across the globe in protesting against ride-hailing apps, a closer look at the reasons behind their collective action illustrates how marginality has been produced, acted on, and deployed by (taxi) drivers in the platform economy. Two incidents that took place in Shenzhen in 2016 represent watersheds in the platformization of ride services in China. The events also had a major influence on the relationship between drivers and ride-hailing apps. On January 5, tens of thousands of Shenzhen taxi drivers (about 90 percent of all drivers) decided to park their taxis for the day in protest against their companies' refusal to grant drivers the liberty to terminate their employment contracts or to reduce monthly rental fees for drivers' vehicles in a declining market. The strike was reported to be one of the most united in decades (Wen 2016).

The second incident is also about remonstration, but it was initiated by Didi's partner-drivers. In April 2016, Didi implemented a drive-to-own program, whereby the ride-hailing platform recruited drivers directly by offering them the chance to own their vehicles after meeting the requirements set by Didi in the next two to three years. While the hired partner-drivers each had to put down a deposit of RMB 15,000–20,000 (US$2,200–$2,900), Didi claimed to be providing partner-drivers with new vehicles without additional charges and giving them priority in allocating ride requests. After two months of the experiment, however, on June 17, thousands of partner-drivers in Shenzhen (and in the nearby city of Guangzhou) blocked traffic at the main intersections in downtown areas and on highways, posting signs that read, "VAMPIRE DIDI, RETURN MY HARD-EARNED MONEY!" (huaduzc 2016). Partner-drivers in Shenzhen also occupied Didi's branch headquarters in Shenzhen and went to local government offices that deal with complaints from the grassroots. The partner-drivers were irritated by Didi's opaque ride-allocation algorithms and the constant changes in Didi's minimum requirements on daily ride services, as well as ratings that were out of drivers' control (i100ec 2016).

Struggles by different groups of drivers for various causes in the platformization of ride services in Shenzhen, we argue, suggest Didi's strategy to create contingent marginality on the ride-hailing platform. The contributions of taxi drivers in building and sustaining the early development of the platform were wiped from the narratives about Didi's success (and the digital ride-hailing economy in China generally). In addition to being invisible in the cultural discourse, traditional taxi drivers were also in effect marginalized by the influx onto the platform of private car owners, who in turn were pushed to the margin when Didi decided to develop its own partner-driver program. The partner-driver program also failed to help Didi create a fleet of vehicles or drivers of its own. Didi soon made a U-turn in its attitude toward the traditional taxi industry by signing a deal with fifty such companies in ten cities in China, including Shenzhen, to introduce an intelligent request-dispatching system to help drivers boost their incomes and reduce the vacancy rate, and even allow drivers to pick up private car rides with the passenger's permission.

It is not coincidental that Didi kept changing its policies toward taxi drivers and potential private car drivers, nor that it turned the process into a spiral of one group of drivers competing against another. Indeed, by shifting the line between the marginal and the core workforce in the compartmentalized ride-service market, Didi has turned them all into contingent drivers by deliberately constructing marginality. As Wang (2016) pointed out, ride-hailing platforms divide labor struggles along the lines of algorithms, meaning that different groups of drivers would selectively participate in labor protest when certain algorithms worked against them. The distinction between taxi drivers and partner-drivers in the aforementioned incidents, therefore, is a convenient construction of contingent marginality in the virtual space of the ride-hailing platform.

But this does not mean drivers are powerless. In addition to protest and strikes, taxi drivers demonstrated how to improve their lives despite their marginality. Taxi drivers in Shenzhen display multiple dimensions of marginality, as we delineate at the beginning of the chapter. Shenzhen is foremost a city of migrant workers—those who travel thousands of miles from their rural hometowns to urban areas searching for jobs. Nearly 70 percent of Shenzhen's migrants-turned-taxi-drivers are from the town of Youxian in Hunan Province, about a thousand miles north of Shenzhen. Ding's (2014) study on community communications among Youxian people showed that they had built a second homeland in Shenzhen while developing their occupational identity as taxi drivers. To be more precise, the second homeland is located in a couple of "urban villages" (*chengzhongcun*) in Shenzhen. Urban villages in Chinese cities are residential enclaves for migrant workers, who are often seen as urban outsiders. Often scattered across the downtown or on a city's outskirts, urban villages are unique landscapes that present spatial ruptures and sites of tension between historical and cultural forces in China's urbanization process (Lan 2005).

Shixia is a renowned "taxi-driver village" in Shenzhen. According to Ding, urban villages like Shixia not only serve as a buffer zone for migrant taxi drivers, but provide the concrete anchor for Youxian people to develop a series of informal business practices associated with taxi driving. Their informal status as migrants and their social networks built primarily on extended family and village connections are not necessarily disadvantages for Youxian people. Youxian taxi drivers have capitalized on their trust in fellow villagers and invented what is now a popular trade practice, in which two drivers take shifts, the major driver working during the daytime and the other at night (Ding 2014, 22). This major-minor model was initially devised to maximize income and provide necessary training to new drivers. Other business practices border the legal/illegal area, such as fabricating driver history. These methods among taxi drivers have allowed them to build a strong foothold in the city.

Taxi drivers fought for the right to terminate their contracts with taxi companies in January 2016 because taxi companies often treat them unfairly through unilateral contracts that deprive drivers of collective bargaining power and sometimes employment benefits. Drivers were unhappy also because they saw a business opportunity to empower themselves using ride-hailing platforms like Didi. They wanted to exploit the tidal wave of ride-hailing apps and transpose their position of marginalized migrant workers into mainstream drivers preferred by Didi (at that time). Note that when contingent marginality is enacted on the platform, the core driver labor force is also made temporary. This challenges the dominant power relations in the traditional taxi industry working against migrant-turned-taxi-drivers in Shenzhen. The direct outcome of drivers being willing to exploit contingent marginality for their own good is the emergence of a "Didi driver village," called Daxia, which houses car fleet companies founded by former taxi drivers and thousands of private Didi car drivers (Didi Express 2017).

No one knows how long the Didi driver village will exist. The transformation, nonetheless, demonstrates that although Didi's divide-and-conquer strategy exploits both taxi drivers and private car drivers, drivers are not powerless. Their historically and economically marginal status can be deployed to their own advantage.

Conclusion

Every world has its order, as well as its limits, where alternative systems are incubated, contested, and materialized on the edges of geography, at the fringes of history, and under pressures from the vectors of the renounced. We began this chapter by developing conceptually the idea of "margins at the center," in which the three types of marginality intertwine, then applied it to China, zooming in on Shenzhen and the SEZ, both at the very frontier and arguably the centerpiece of the Chinese digital economy. In so doing, we contend that marginality has multiple dimensions and is not always a curse; under certain conditions, marginality can become a resource for technological and business innovation and for alternative movements, such as on-the-ground resistance and horizontal network formation, as can be seen in the cases of both shanzhai electronics and taxi/ride-hailing services in Shenzhen.

The rise of shanzhai and the struggle over ride-hailing apps of course represent different models of marginality, occupying the periphery of different centers or core regions. Shanzhai, as an alternative practice and open, flexible business model, is more deeply rooted in the anti-imperialist legacies of China, spanning from the Maoist era of guerrilla warfare and revolutionary zeal to today's Chinese copyleft practices. Compared to the case of ride-hailing, shanzhai is more successful commercially; has grown more mature as a large network of R&D personnel, manufacturers, and distributors; and has more global reach, especially in the Global South. Calling Shenzhen the Silicon Valley of hardware is probably an understatement because even the Silicon Valley of California so far cannot capture the sub-Sahara African mobile phone market as Tecno has, a Shenzhen-based company with its roots in the shanzhai community. Shanzhai, as such, illustrates what can happen when geographic and historical marginalities overlap with the vectors of the renounced in successfully creating and sustaining new centrality in the digital economy.

On the contrary, the struggle against Didi, Uber, and conventional taxi companies implies a tactic and a proactive movement initiated by marginalized drivers, responding to the latest expansion of digital capitalism threatening their livelihoods. There is horizontal communication among drivers, either in their physical community of urban villages or online and through mobile phones, but the network formation is confined to regions in and around Shenzhen, with little more than loose connections with similar struggles in other parts of China, cut off almost completely from the global movement against corporate platforms (Scholz 2016). Drivers did their best to resist the dominance of the platforms with a remarkable endeavor, but their attempts to reverse marginality into centrality were far less successful than the shanzhai community in achieving that goal.

The most crucial difference is ownership—whereas shanzhai businesses are all owned by grassroots entrepreneurs who share and sell their products, ideas, and data under conditions of their choosing, even conditions of their creation, Didi and taxi drivers have nothing more than token ownership, if any ownership at all, of their cars, services, and data. This is why we use the term "contingent marginality" in discussing the drivers' struggle because, innovative and proactive as many individual drivers are, as a group, they are almost completely disposable under the structural constraints of this unequal political economy. In contrast, what we see in the shanzhai model increasingly contains elements of "assertive marginality," when the people at the margins take control.

Despite the differences, in both shanzhai and the drivers' struggle we see the power of China's domestic migrant population, a most important demographic, economic, and sociopolitical category of the "informational have-less" (Qiu 2009). These are people spun off from the agrarian economy, state-owned enterprises, and the failure of the New Economy elsewhere in China. Unlike populations of the haves and have-mores, the main driving force behind the have-less population in the digital economy is their existential needs, which, as Qiu has argued, provides a more sustainable material basis for network formation and grassroots innovation (2009).

Here, it is necessary to revisit the question with which we began this chapter: why Shenzhen? Why not other SEZs? China designated five SEZs in the early 1980s, and Shenzhen is just one of them (the others are Zhuhai, Shantou, Xiamen, and Hainan). All the SEZs are along China's southeast coastline. All are recipients of sizable in-migration, including large numbers of have-less people seeking to fulfill their existential needs. Yet, only Shenzhen has made doing so possible in such a phenomenal way. Why Shenzhen?[1]

Simply being on the margins is not enough to successfully create a new center in the digital economy. What really matters in the state of marginality is the first and most basic condition: weak control from the center. If the center can enforce strong control (as described in dependency theory), then peripheral regions will end up being underdeveloped in the long run (Arrighi 2002). So a precondition has to be weak control from the center, which was an integral part of the SEZ design from the very beginning. A second necessary condition is a discrepancy between people's expectations and realistic conditions on the ground, leading to the will to struggle, or what could be called "frontier mentality." Next, there need to be structural opportunities where the will can materialize in the forms of alternative business models (e.g., shanzhai) or social movements (e.g., drivers occupying Didi office). Such materialized formations would harbor the next and arguably most crucial action: transforming individual members of the information have-less into collective actors with agency, who are networked and bonded together by not just their marginality and "frontier mentality" but, more important, their shared modes of practice and community.

As the Kenyan author Ngũgĩ wa Thiong'o put it, "The modern world is a product of both European imperialism and of the resistance waged against it by the African, Asian, and South American peoples" (1993, 4). Shenzhen, China, is but one case in the making of such a "modern world," with its order, disorder, and alternative orders. Albeit complex and probably not replicable, the case of Shenzhen is a prism for us to see possibilities of digital economies in the Global South. The acts of resistance and struggle in Shenzhen depict complex reactions to new types of Western imperialism now taking form in and through the digital economy.

Note

1. Note that the following argument is speculative and requires further research and critique to substantiate.

References

Arrighi, Giovanni. 2002. Global Inequalities and the Legacy of Dependency Theory. *Radical Philosophy Review* 5 (1–2): 75–85.

Castells, Castells. 1996. *The Rise of the Network Society*. Cambridge, MA: Wiley-Blackwell.

Chan, Janny, and Ngai Pun. 2010. Suicide as Protest for the New Generation of Chinese Migrant Workers: Foxconn, Global Capital, and the State. *Asia-Pacific Journal: Japan Focus* 8 (37): 1–33. http://apjjf.org/-Jenny-Chan/3408/article.html.

Chen, Julie Yujie. 2018. Technologies of Control, Communication, and Calculation: Taxi Driver's Labor in the Platform Economy. In *Humans and Machines at Work: Monitoring, Surveillance and Automation in Contemporary Capitalism*, edited by Pheobe

Moore, Martin Upchurch, and Xanthe Whittaker, 231–252. London: Palgrave Macmillan.

Didi Express. 2017. The Secrets of "Didi Driver Village" in Shenzhen. [In Chinese.] Didi Express. Accessed February 14, 2017. http://www.didiabc.com/news/255.html.

Ding, Wei. 2014. *Mobile Homelands*. [In Chinese.] Beijing: Social Sciences Academic Press.

The Economist. 2012. Who's Afraid of Huawei? *Economist*, August 4, 2012. http://www.economist.com/node/21559922.

Fieldhouse, David K. 1973. *Economics and Empire, 1830–1914*. Ithaca, NY: Cornell University Press.

Guangdong Provincial Bureau of Statistics. 2013. *Guangdong Statistics Yearbook*. [In Chinese.] Beijing: China Statistics Press.

He, Huifeng. 2016. Didi Partners with China's Taxi Companies to Upgrade Services. *South China Morning Post*, September 1, 2016. http://www.scmp.com/business/china-business/article/2012291/didi-partners-chinas-taxi-companies-upgrade-services.

Headrick, Daniel R. 1981. *The Tools of Empire: Technology and European Imperialism in the Nineteenth Century*. Oxford: Oxford University Press.

Himanen, Pekka. 2001. *The Hacker Ethic, and the Spirit of the Information Age*. New York: Vintage.

Ho, Josephine. 2010. Shanzha (山寨): Economic/Cultural Production through the Cracks of Globalization. Plenary Speech in Crossroads: 2010 Cultural Studies Conference, Hong Kong, June 17–21, 2010.

huaduzc. 2016. Return My Hardearned Money! Didi Drivers in Guangzhou and Shenzhen on Strike. [In Chinese.] June 18, 2016. http://wx.huadu.gd.cn/article/show-11987.html.

i100ec. 2016. SPOTLIGHT: Didi Drivers on Strike in Guangzhou and Shenzhen, Headquarters Occupied. [In Chinese.] August 8, 2016. http://www.v4.cc/News-1950703.html.

Ko, Tin-yau. 2017. How a Little-Known Brand Conquered African Mobile Phone Market. *EJinsight*, January 18, 2017. http://www.ejinsight.com/20170118-how-a-little-known-brand-conquered-african-mobile-phone-market/.

Lan, Yuyun. 2005. *Villages within the City: An Ethnographic Study on "New Village Community."* [In Chinese.] Beijing: SDX Press.

Lindtner, Silvia, Anna Greenspan, and David Li. 2015. Designed in Shenzhen: Shanzhai Manufacturing and Maker Entrepreneurs. *Aarhus Series on Human Centered Computing* 1 (1): 5. doi:10.7146/aahcc.v1i1.21265.

Lüthje, Boy, Stefanie Hürtgen, Peter Pawlicki, and Martina Sproll. 2013. *From Silicon Valley to Shenzhen: Global Production and Work in the IT Industry*. Lanham, MD: Rowman and Littlefield.

Massey, Doreen B. 2005. *For Space*. London: Sage.

Mezzadra, Sandro, and Brett Neilson. 2013. *Border as Method, or the Multiplication of Labor*. Durham, NC: Duke University Press.

Mozur, Paul, and Mike Isaac. 2016. Uber to Sell to Rival Didi Chuxing and Create New Business in China. *New York Times*, August 1, 2016. https://www.nytimes.com/2016/08/02/business/dealbook/china-uber-didi-chuxing.html.

Needham, Joseph. 1979. *Science in Traditional China*. Hong Kong: Chinese University Press.

Ngũgĩ wa Thiong'o. 1993. *Moving the Centre: The Struggle for Cultural Freedoms*. London: James Currey.

Okihiro, Gary Y. 1994. *Margins and Mainstreams*. Seattle: University of Washington Press.

Pun, Ngai, Jenny Chan, and Mark Selden. 2015. *Life and Death Behind Apple: Foxconn Workers on the Production Line*. [In Chinese.] Hong Kong: Zhonghua Press.

Qiu, Jack L. 2009. *Working-Class Network Society: Communication Technology and the Information Have-Less in Urban China*. Cambridge, MA: MIT Press.

Qiu, Jack L. 2016. *Goodbye iSlave: A Manifesto for Digital Abolition*. Urbana: University of Illinois Press.

Schiller, Dan. 2005. Poles of Market Growth? Open Questions about China, Information and the World Economy. *Global Media and Communication* 1 (1): 79–103. https://doi.org/10.1177/1742766505050174.

Scholz, Trebor. 2016. *Platform Cooperativism: Challenging the Corporate Sharing Economy*. New York: Rosa Luxemburg Stiftung.

Scott, James C. 2009. *The Art of Not Being Governed: An Anarchist History of Upland Southeast Asia*. Yale Agrarian Studies Series. New Haven, CT: Yale University Press.

Shenzhen Bureau of Statistics. 1991, 2000, 2016. *Shenzhen Statistics and Information Yearbook*. [In Chinese.] Beijing: China Statistics Press.

Shenzhen Ten-year Development Report Research Group. 2013. *Ten-Year Development Report on Fiscal Economy in Shenzhen: 2001–2010*. [In Chinese.] Beijing: China Fiscal Economy Press.

Tsing, Anna L. 1993. *In the Realm of the Diamond Queen: Marginality in an Out-of-the-Way Place*. Princeton, NJ: Princeton University Press.

Wallerstein, Immanuel M. 1974. *The Modern World-System: Capitalist Agriculture and the Origins of the European World-Economy in the Sixteenth Century.* New York: Academic Press.

Wallerstein, Immanuel M. 2000. Globalization or the Age of Transition?: A Long-Term View of the Trajectory of the World-System. *International Sociology* 15 (2): 249–265. https://doi.org/10.1177/0268580900015002007.

Wallis, Cara, and Jack L. Qiu. 2012. Shanzhaiji and the Transformation of the Local Mediascape of Shenzhen. In *Mapping Media in China: Region, Province, Locality*, edited by Wanning Sun and Jenny Chio, 109–125. London: Routledge.

Wang, H. 2016. ICTs, Sharing Economy and the Transformation of Labor Politics in China: A Case of Didi Dache. Paper presented at the International Workshop on ICT Development in East Asia, August 29–30, 2016, Hallym University, South Korea.

Wen, Yuqing. 2016. Tens of Thousands of Taxi Drivers on Strike in Shenzhen. [In Chinese.] *Epoch Times*, January 5, 2016. http://www.epochtimes.com/gb/16/1/5/n4609274.htm.

Williams, Raymond. 1978. *Marxism and Literature.* Oxford: Oxford University Press.

Wong, Winnie W. Y. 2017. Speculative Authorship in the City of Fakes. *Current Anthropology* 58 (S15): S103–S112. doi:10.1086/688867.

14 African Economies: Simply Connect? Problematizing the Discourse on Connectivity in Logistics and Communication

Stefan Ouma, Julian Stenmanns, and Julia Verne

Introduction

Discourses on Africa's (lack of) development nowadays firmly revolve around the idea of global connectivity. While historically, social (and natural) scientists, politicians, philanthropists, and development practitioners have diagnosed the various countries of the African continent as suffering from many "lacks," the lack of (global) connectivity and the implied distance to world markets has become the major developmental problematization of the twenty-first century.[1] This line of thinking about connectivity has become so powerful that the international consultancy firm McKinsey and Company recently launched a global connectedness index (Manyika et al. 2014). Here, together with many Latin American and Middle Eastern countries, much of Africa is relegated to a zone of insufficient connectivity (66). But remedies are under way. Africa's connectivity is being addressed in various domains, at various scales, inside and outside of the continent. On the one hand, a more connected Africa lies at the core of many contemporary development programs. On the other hand, Africans seem to be developing connectivity from within, and some parts of Africa are being hailed as new frontiers of locally grown technological innovations, particularly in the domain of mobile phones.

While we can observe a novel moment in development discourse, and its proponents are often eager to underline that connectivity is the prime imperative of the digital age, the underlying aspiration is not new and has indeed a longer history. In this chapter, we therefore situate the contemporary discourse of connectivity in its historical context, excavating the "living links" (Farmer 2004, 309) that connect the policy past to the policy present. We then engage with the similarities and differences between colonial and early postcolonial discourses of connectivity, as well as the contemporary one, by considering the fields of logistics and communication as two examples that are emblematic of current development efforts under the connectivity paradigm.[2]

While acknowledging the progressive and cosmopolitan potential of connectivity, we argue that contemporary discourses of connectivity in the realm of communication and logistics are problematic for their uncritical continuation of the modernist gaze, which manifests itself in an affirmative embracement of "technoliberal boosterism" (Carmody 2012, 12). Against this background, we therefore wish to propose an alternative reading of contemporary connectivity and its underlying materialities, socialities, and spatialities by bringing to the fore three key arguments that signal the problematic nature of connectivity as a blueprint for transforming economies "at the margins."

The Specter of Modernization: Situating Connectivity Historically

Celebratory accounts of global connectivity are often so focused on the narrative of a "flat world" (Friedman 2005) associated with the past thirty years that the lively (and often violent) history of the connective imperative is mostly overlooked.[3] Yet, the quest for connectivity, and the logistical and communicative technologies that arose from it, had already formed the basis of many imperial and colonial projects (Headrick 1981). Graham, Andersen, and Mann (2015), for example, illustrate the similarities between the connectivist discourse associated with the building of the Uganda Railway (1896–1901) in British East Africa and contemporary attempts to connect the region to the digital world. Albeit framed in different terms, a desire to foster connectivity also lay at the core of the post–World War II development project (Escobar 1995), which from the very beginning was heavily influenced by modernization theory in its various disciplinary disguises (Lerner 1958; Soja 1968; Rogers 1971). Technology was seen not only as a means to "amplify material progress" (Escobar 1995, 36) in order to reach higher stages of development (Rostow 1960); it was also "theorized as a sort of moral force that would operate by creating an ethics of innovation, yield, and result," contributing to "the planetary extension of modernist ideals" (Escobar 1995, 36). Geographers played a prominent role in this, as they were interested in the contact zones between "backward" traditional structures and those of the "modern" world (Peet and Hartwick 2009, 130). A major reference work of the time illustrates this well: "Unlike former days ... people ... act today in response to the new foci of change, the towns and the cities. Modern transport systems extend the length and breadth of the country [Sierra Leone], bringing new ideas, new methods, new people even to the most remote corners. ... These changes which affect all spheres of life—political, social, economic, and psychological—constitute the modernization process" (Riddell 1970, 43–44)

These accounts often celebrated the wonders of technology and infrastructure brought by the "civilizing force" of empire and colonialism, and were silent on the violence and acts of dispossession inherent in empire's modernization project.[4] The uncritical stance of those works was furthermore often paired with a positivist geographic analysis that measured and benchmarked the spatially varied effects of modernization (e.g., the use of indices on the development of transport networks, the expansion of communication and information media, the growth of integrated urban systems or physical mobility; Peet and Hartwick 2009, 130).

Against this backdrop, it is not surprising that the famous geographer Ed Soja opened his 1968 book, *Geography of Modernization in Kenya*, with an uncritical reference to the scientific achievements of colonialism: "Among the many effects of colonialization has been the spread of a world culture based on modern science and technology and specific standards of government organization and operation" (Soja 1968, 1). And even though modernization theory was increasingly criticized from the 1970s onward (Leys 1996), modernist ideas persisted, in a way slumbering below the surface for a while, before seeing a revival in the early 1990s when various authors proclaimed the "end of history" (Fukuyama 1989), a "borderless world" (Ohmae 1989), a "supply chain revolution" (Cox 1999), and the "end of poverty" (Sachs 2005). The connective project of Western capitalist globalization, spiced with some Asian values (e.g., embodied by the just-in-time principle), was the model many key actors started to accept as a blueprint for development. In this climate, the specter of modernization manifested itself in two connected developments.

First, it reappeared with the rise of the discourse on the knowledge economy and ICTs for development. In this discourse, development was reconceived as catching up to an interconnected world from which many countries in the Global South, particularly in Africa, were said to be cut off by a digital divide. The World Bank report *Knowledge for Development* (World Bank 1998) can be considered a crucial marker in this debate. The report, as one observer points out, reads much "like the communication theories of the Modernization School in the 1960s" (Schech 2002, 17), which assumed that modernization in the developing world was stalled because it lacked "access to the kind of knowledge that Western countries possess[ed]" (17; see also Kunst 2014). Hence, the report's conclusion ties well into earlier problematizations, as it situates technology and knowledge transfer as a key for overcoming this barrier: "When done right, ICT infrastructure investment and policy reform can be a key enabler of poverty reduction and shared prosperity. A 10 percent increase in high-speed internet connections is associated on average with a 1.4 percent increase in economic growth in developing countries" (World Bank 2014, n.p.).

Second, modernization theory saw a return in the rise of connectivity-oriented thinking in the realm of infrastructure and logistics. While the development of transport networks or issues of physical and geographic mobility were integral to spatial versions of 1960s modernization theory (Peet and Hartwick 2009), the increasing importance of transport infrastructure and logistics has a slightly different background. Building on the works of critical logistics studies, we see this increase of importance in light of the global logistics revolution and its geopolitical and geoeconomic underpinnings (Cowen and Smith 2009; Cowen 2014). In the realm of development, this revolution articulates itself with the New Economic Geography agenda that has had a profound effect on development economics over the past decade (see, e.g., Adam et al. 2017).[5] Among other goals, this agenda is about overcoming distance to (global) markets through efficient communication and logistics connections while increasing the permeability of borders (World Bank 2008).

Despite some clear continuities, the revival of modernization theory in the cloth of the so-called information and communication technologies for development (ICT4D) and logistics revolution also departs in some ways from earlier versions. During colonial and early postcolonial times, the state was the locus of technological development, but today private capital is increasingly central to rolling out both communication and logistics infrastructure. This discourse no longer assumes that technology will just diffuse smoothly. Rather, it is widely acknowledged that technology requires certain social orders to flourish, and that it is the self-ascribed task of private and public forces to cocreate these orders. Finally, many development economists have accepted that development will not just trickle down. Instead, they acknowledge economic development to be "naturally" spatially uneven because of agglomeration economies and other geographic barriers that impede the mobility of people, knowledge, goods, and capital. Nevertheless, the specter of modernization casts a vast shadow over the contemporary connectivist discourse in the realm of ICT4D and logistics, and similar to earlier modernist gazes, the shadow blinds the discourse to several issues that a more careful analysis of the relationship between history, social forces, technology, and adverse positionality in the global economy would highlight.

Connecting through Logistics

Logistics nowadays plays a major role in "delivering development" (Matsaert 2015; see also Stenmanns and Ouma 2015). You are unlikely to come by an African newspaper or development report these days that does not feature a piece on the importance of logistics (figures 14.1–14.2). In contrast to the 1950s and 1960s modernization transport paradigm, logistics describes the intellectually much more complex task of holistically organizing transport systems on a global scale. Whereas in the prelogistical world, firms mostly assembled and distributed commodities to national markets, the logistics revolution spatially and organizationally dissected integrated production into global supply chains (Cowen 2014). This changed sourcing and procurement tremendously and introduced new managerial imperatives. As chains are often transnational and thus involve a plethora of places, stakeholders, and national (or intrafirm) regulation across space, they in turn need to be thoroughly managed (see Busch 2007). Supply chain management, an approach taken from business management, has taken up this role of coordinating the spatially fragmented sites of sourcing, production, storage, distribution, and consumption. To achieve these logistical goals, common standards of connectivity and digital coordination, along with tracking and communication tools, have become the nuts and bolts that assemble and integrate supply chains.

At various points in history, development discourse has diagnosed African countries to be suffering from a range of "lacks": technology, capital,

Figure 14.1
Popular development discourse concerning the link between development and logistics. *Source*: *East African*, February 21, 2015, 42.

free markets, property rights, political stability, good governance, and, not least, territorial integrity. As of more recently, disconnection from, or insufficient or risky connections to, global supply chains has been added to the list, becoming a powerful meta-narrative in its own right. Knowledge initiatives by various international organizations are now reenvisioning the economic futures of African economies through the prisms of logistics, border management, and "modern supply chains" (Raballand et al. 2012; McLinden et al. 2011). In this context, African ports and transport infrastructures usually come off badly: "They are generally poorly equipped and operated at low levels of productivity. Few are capable of handling the largest of the current generation of ships, and they are generally unprepared for the dramatic changes in trade and shipping patterns that are now occurring" (Ocean Shipping Consultants 2008, n.p.). Accessibility problems, insufficient adaption of standards, and lack of infrastructural connectivity have sparked a whole range of connectivity-oriented benchmarking and development interventions.[6] Over and above that, many African ports have seen large investments by private logistics operators to overcome these barriers

Figure 14.2
The institutional development discourse concerning the link between development and logistics. *Source*: Authors' compilation.

(Debrie 2012). Geostrategic projects, such as the Chinese One Belt, One Road Initiative, or the activities of French logistics conglomerate Bolloré, aim at not only refurbishing such port infrastructures but also building up connected rail and road corridors, inland logistics centers, and manufacturing enclaves. At the core of these serial spaces lie infrastructures, such as roads, port basins, and bridges (see Easterling 2014), which are stitched together by supply chain management and its corporate and transnational standards. The art of logistics, however, is not just about material and infrastructural management of the circulation of goods. The mobility of "related information" (Danyluk 2017, 1) intrinsically creates the background for the movement of goods.

Logistics space is increasingly connected and controlled through a new geospatial information infrastructure (Kanngieser 2013). This involves tracking cargo flows (with the help of GPS), registering ingoing and outgoing goods with RFID chips, and scanning containers to combat fraud, smuggling, and trafficking (Stenmanns 2016). This surveillance logic, however, is not just embedded in cargo. From biometric identification gates to remote tracking of truck drivers' routes and pauses, surveillance logics encroach on laboring bodies in similar terms. On a much larger scale, trade facilitation, as promoted by actors such as the World Trade Organization, encompasses the techniques of supply chain management and aims to establish deeply integrated cross-border networks of so-called electronic single windows. While trade usually involves multiple national agencies, the creation of a single window is the managed and collaborative effort to include all information from the various stakeholders (such as customs declarations or trading invoices) in a common standardized framework. Connectivity, in this case, is therefore based on informational as well as infrastructural harmonization beyond borders.

Another emerging trend is the development of so-called smart ports. In 2016, a delegation of the Southern African–German Chamber of Commerce and Industry (Iyer 2016) visited the Port of Hamburg, a flagship port in terms of integrating digital governance technologies. Shortly afterward, the South African port authority contracted a multimillion-US-dollar deal with German T-Systems to digitalize the Port of Durban, Africa's largest container port. T-Systems heralds the smart port template, which includes drones, tracking, and sensor technology, as a strategic tool to boost a port's efficiency (T-Systems 2017). In such a port, the distinction between hardware and software, and between data and infrastructure, dissolves. The "sensory environment" (Halpern et al. 2013, 279) of a smart port integrates its generated data in real time with a control center dashboard, which closely monitors port operations like cargo movement, vessel berthing, and workforce performance. While such initiatives speak to the growing demands of producers, traders, and consumers in the Global South to speed up the cumbersome process of importing and exporting goods, the types of "logistical governance" (Kanngieser 2013) engendered by these cases rarely figure as objects of scrutiny or critique.

The contemporary map of logistics space in Africa indicates the massively growing importance of supply chains for national economies, but this map also reveals a severe political shift from national governments to a mix of private and parastatal modes of transport governance. This is a political space of logistics (Cowen 2014, 64), likewise built on material flow architectures and processed information, where transnational corporate actors, national governments, and international organizations shape and thus determine the conditions of connectivity, as well as broader questions of social inclusion, labor control, and political participation. Approached from this perspective, one cannot deny that the infrastructural grid of the contemporary "logistical fix" (Danyluk 2017) resembles the extractive spatiality of imperial rule. Once essential technologies of the "mission civilisatrice," seaports and connected infrastructures concentrate around the former colonial spaces of "l'Afrique utile" (see Schouten 2011). As James Ferguson has put it, "Usable Africa gets secure enclaves—noncontiguous 'useful' bits that are secured, policed, and, in a minimal sense, governed through private or semiprivate means. These enclaves are increasingly linked up, not in a national grid, but in transnational networks that connect economically valued spaces dispersed around the world in a point-to-point fashion" (Ferguson 2006, 39–40).

Logistical infrastructure thus has a severe impact on the places it traverses. Far from being merely a technological fix for decaying infrastructure, logistics space is increasingly a political space of asymmetrical connectivities and "webs of living power" (Farmer 2004, 309), delinked from the sovereign government of the nation-state and technologically shielded against political contestation.

Mobile Phone Connectivity

While ICTs clearly play a key role in extending the logistics revolution to new frontiers, they have also become an important tool for development in their own right. The rapid uptake of mobile phones, in particular, has spurred hopes for ICT-enabled development, as it finally seems to provide the connectivity needed for marginalized groups to participate in (global) social and economic relations based on knowledge, information, and communication (Castells et al. 2007; Graham, Andersen, and Mann 2015). As Ling and Horst (2011, 365) have pointed out,

> Like computers and the internet, mobile phones have captured the attention of a range of stakeholders in international development agencies, government entities, corporations and most importantly among common people. Mobile phones' relative accessibility, affordability and ease of use (compared to the PC) hold the promise of bridging the so-called digital divide. New programs such as M-Pesa in Kenya, the m-health initiatives of different agencies and a wide range of mobile information diffusion projects have all illustrated the potential of mobile phones to be utilized as instruments of development.

The history of mobile telephony in Africa began in 1987 with the first conversation by mobile phone in Zaïre, today the Democratic Republic of Congo. Market liberalization in several African countries in the late 1990s finally led to a rapid increase of mobile phone usage that still supersedes all expectations. With the availability of prepaid SIM cards and rather inexpensive secondhand mobile phones, even rural areas, as well as less affluent parts of society, have been able to participate—a development hardly any investor dared to predict, but that is now driving network operators, development agencies, and private businesses to invest in the area (figure 14.3). Economist Jeffrey Sachs even went so far as to proclaim mobile telephony as "the single most transformative technology for development" (Sachs cited in Shiner 2009, n.p.). This shows how even the latest development discourse still relies heavily on the idea that technologies are a central driver of social change (cf. Bimber 1994, 80). Moreover, as we argue above, the discourse clearly reinvigorates modernization theory's longstanding technocratic approach to development as well as its conceptualization of the Global South as "backward" (Díaz Andrade and Urquhart 2012; Graham 2008; Kunst 2014). This becomes most visible in debates about rural

Figure 14.3
Vodacom advertisement: "Nobody reaches Tanzania as we do" (left). Early Zantel advertisement in Dar es Salaam (right). *Source*: Julia Verne, 2007.

Africa, where current development programs attempt to establish "last mile" connectivity (Davis, Tall, and Guntunku 2014).

The focus on the power of new communication technologies to foster development also entails an emphasis on their transformative potential: ICT4Ds are viewed as Africa's future, catapulting it forward into the digitized mainstream and even beyond (Macata 2017), with African digital innovations envisioned as future trendsetters. Here, the so-called Silicon Savannah in Nairobi attracts specific attention, promising to turn Kenya into a major African technology hub and one of the most dynamic centers of digital development in Africa (Graham and Mann 2013; Ndemo and Weiss 2017).

The most famous mobile phone service by far is still M-Pesa, which was a crucial milestone in Kenya's dynamic ICT sector.[7] Launched by Safaricom, the largest network operator in Kenya in early 2007, as "an innovative payment service for the unbanked" (Hughes and Lonie 2007, 63), M-Pesa has since inspired multiple operators in different African countries to develop similar mobile money services:

> The product concept is very simple: an M-PESA customer can use his or her mobile phone to move money quickly, securely, and across great distances, directly to another mobile phone user. The customer does not need to have a bank account, but registers with Safaricom for an M-PESA account. Customers turn cash into e-money at Safaricom dealers, and then follow simple instructions on their phones to make payments through their M-PESA accounts; the system provides money transfers as banks do in the developed world. The account is very secure, PIN-protected, and supported with a 24/7 service provided by Safaricom and Vodafone Group. (63)

Observers soon celebrated M-Pesa as an African success story, and a prime example of the African appropriation of a foreign technology, with six million people registering for the service within only two years of its implementation, about 50 percent of all Safaricom customers at the time (Mas and Morawczynski 2009). Subsequently, penetration rates have increased further, and M-Pesa has become an everyday part of life for most Kenyan mobile phone users.

In this success story, two major aspects often remain neglected. First, these so-called African innovations are often closely entangled with powerful actors from the Global North and thus hardly represent alternative pathways of development. In this respect, only a more thorough investigation of the backstage dynamics reveals the continuation, or even perpetuation, of the uneven power relations that have often lain at the heart of the rapid expansion of digital mobile phones (Unwin 2017). Until May 2017, Safaricom, the Kenyan firm operating M-Pesa, was majority owned by the UK multinational Vodafone.[8] The clouded ownership structure of Vodafone Kenya/Safaricom, its tax practices and entanglement with global tax havens (Nguyen 2007; Turner, Mathiason, and Doward 2017), as well as its appropriation of the free labor of largely informal vendors for its operations, has received little to no attention in the celebratory accounts of M-Pesa. Critical observers also note, "Safaricom has become an integral part of Kenyan capitalism's aggressive neoliberal state–private enterprise nexus, which is marked by high concentrations of wealth and notorious corruption" (Dyer-Withford 2015, 118). On top of that, Safaricom has acquired a quasi-monopolistic market position, which only in 2017 became a matter of political debate in Kenya (Vidija 2017). Like other leading firms of the tech and platform economy, it managed to firmly entrench its extractive model in the social relations of millions of consumers in Kenya and elsewhere (Albania, Ghana, and Romania, among others).

Second, consultants, development experts, and international organizations consider technologies to be neutral apolitical tools that supposedly act on their own (i.e., as intended by their developers). This becomes evident when those promoting the use of ICTs for development assume that just by being connected to mobile banking, people will generally start managing their finances in a better way, keeping track of their accounts, paying on time, and starting to save money. Such developments indicate how the expected impact of mobile phone services is closely linked to an understanding of users as calculating subjects, always acting in their own best interests (or rather in the interest of capitalist, neoliberal logics). Neubert (2017, 306), for example, argues that mastery of the technology itself leads to a specific linear, causal understanding among its users—an idea that resonates with much earlier understandings of the power of technologies. Already in 1870, a British colonial officer in India wrote, "[Technologies] are opening the eyes of the people who are within reach of them in a variety of ways. They teach them that time is worth money, and induce them to economise that which they had been in the habit of slighting and wasting. ... Above all, they induce in them habits of self-dependence, causing them to act for themselves promptly and not to lean on others" (Rogers 1870, 112–113).

But even though numerous examples show that mobile phones have hardly proved to be straightforward technical tools for development, a more thorough investigation of the particular notions of development they entail, as well as the services and information they offer, is still missing. Celebrated as African digital innovations, most mobile phone applications developed to support e-participation, health services, education, or agricultural practices are still built around the idea that Africans need to be connected to (mainly) Western modes of knowledge. This knowledge is usually translated into concrete advice and instructions, readily applicable for the end users (e.g., Baumüller 2016; Krone et al. 2016; Mtega and Ronald 2013), sometimes accompanied by a "nudge" to further encourage the realization of intended actions (Thaler and Sunstein 2013). When apps notify farmers about price developments, they act as market devices (Muniesa, Callon, and Millo 2007) and thus as carriers of the global capitalist ethos (Thompson 2004, 21).

What is also rather neglected in discourses on the developmental benefits of mobile phone connectivity is that the implications of technologies hardly restrict themselves to those intended by their developers. Reception and use might differ considerably; the information sent or the advice given might be contested, resisted, or put to a different use. In respect to M-Pesa, for example, the increased connectivity as well as the ease with which an M-Pesa user can now transfer money from one account to another has been accompanied by a significant increase in requests for money. Thus, while connectivity might certainly come with benefits, empirical insights indicate that it can also be a social burden (Verne 2017). Overall, mobile phones are not passive containers of information. Rather, they play a much more active part depending on their role and position in complex and highly dynamic networks.

Problematizing Connectivity in the 21st Century

Building on our interpretations of the connective moment in African logistics and communication, in what follows we extend the scope of our argument by problematizing the technoliberal boosterism that not only seems to dominate these fields but also appears symptomatic of the much broader development paradigm currently at stake. In particular, we want to point out three aspects that epitomize the problematic nature of connectivity as a blueprint for transforming economies at the margins.

First, dominant actors, in their attempt to construct particular flow architectures, continue to envision development as an evolutionary, teleological process in which countries of the Global South should follow the examples of the industrial and postindustrial economies, located largely in the Global North (including the economic cores of Asia; see also Graham 2015). For instance, management systems and logistical practices from Western cities (e.g., Hamburg, Rotterdam), and increasingly from non-Western growth hubs such as Dubai, Singapore, or Shanghai, are used as templates on which to model the logistical future of African ports.[9] In the most optimistic versions of this discourse, "the local" in Africa may be given some space for autonomy while being gently elevated beyond a state of simple mimicry. This view is especially prevalent in discussions of communication technologies, where it is said that lagging countries are expected to "bypass or leapfrog institutional or infrastructural obstacles" (Wade 2002, 460), which, in some cases, may undoubtedly involve genuine innovations from the South.

Second, technology is still largely seen as a neutral solution to identified problems. Connectivity becomes a sui generis technology that acts on its own, rather than something grounded in historically grown "relationships between people, places and processes" (Graham, Andersen, and Mann 2015, 345). Like earlier versions of modernization theory, the celebratory rhetoric accompanying the connectivity discourse is often silent on the multifold global power structures and the "exploitative forms of international interconnection" (Carmody 2012, 2) in which both communication and logistics technologies are embedded. Instead, a range of domestic deficiencies explain a country's failure to catch up. While we cannot deny that many small and medium enterprises and consumers in Africa have an interest in more efficient logistics or cheaper or more comprehensive ICT technologies, the connectivity discourse usually imagines the quest for global interconnectedness in apolitical terms. This conceals the material interests of large corporate players, who see logistics or ICT in the Global South as frontiers that can enhance the turnover of capital. The apolitical view also masks the new ways in which large firms are striving to become centers of calculation and accumulation in extended networks of rent production and value extraction (Wade 2002; Murphy and Carmody 2015; Ouma 2015).[10]

If we assume, however, that technology cannot be separated from social relations, because the former stabilizes the latter, then technology ceases to be an innocent exogenous variable (Latour 1991). Rather, the social forces of power and domination materialize in technology (Headrick 1981). They are inseparable from even the most technical "techno-economic networks" (Callon 1991), especially those linking places in the North and South. This is the case for even the most promising and progressive technologies, such as the celebrated M-Pesa, and becomes even more visible in logistics, where the use of certain technologies usually enacts the geoeconomic and geopolitical interests of actors in the Global North. While actors such as the US government aim to securitize transnational supply chains against acts of illegal infiltration and terrorism, large logistics firms mobilize those technologies to increase supply chain efficiencies. The latter use technologies to extend the commodity form across all situations of the supply chain by controlling both labor and cargo in novel, panoptic ways (see also Kanngieser 2013).

But we should go beyond the tangible, material dimension of techno-economic networks when thinking critically about the quest for connectivity. Therefore, the final point we would like to make here is that the terms under which these new techno-economic networks emerge are often the product of a coloniality of knowledge, power, and being (Mignolo 2007), through which Western standards of connectivity are made universal reference points. In the realm of both logistics and ICT, adapting to the respective revolutions promised by these technologies means subscribing to a world largely shaped by the standard-setting power of large logistics firms and notions of "the economy" of particular epistemic provenance. Even the most promising local initiatives, trying to create better futures for people in many African countries, such as M-Pesa, usually do not escape this coloniality of "global value relations" (Araghi 2003).[11] These are characterized by "an embedded logic that enforces control, domination, and exploitation disguised in the language of salvation, progress, modernization, and being good for everyone" (Maldonado-Torres 2007, 245). Against this backdrop, universal standards of connectivity are not technical inevitabilities. Instead, we should always bear in mind which actors crafted them, whose interests they serve, which forms of knowledge they incorporate and which ones they exclude, and what processes they engender and render legitimate on the ground.

Conclusion

In this contribution, we have problematized the current articulation of connectivity in development discourse from three different angles. This critical engagement has emphasized that public and corporate projects envisioning and working on (global) connectivity are by no means unproblematic. While the discursive and material politics of connectivity rely heavily on the modernist assumption of teleological progress, many projects that intellectually build on this political framing represent themselves or their objectives from a merely technical stance. They provide technical solutions to socioeconomic problems. This view translates to damaging action when most of these projects ignore, neglect, or veil their entanglement with the power geometries that produce and reproduce specific structures of global inequality. In so doing, they often ensure that the historically grown global inequalities and power structures remain undisclosed. Above all, those projects entail a specific geography that, as we have shown, often remarkably resembles the extractive spatiality of colonial and imperial rule.

Finally, we have to engage with the potential issue of critical academics, especially those situated in the North, who may be quick to dismiss the desire for global connections as problematic, disregarding the legitimate aspirations of people in the Global South to be connected to the world. Acknowledging this desire, at a time when we observe a renewed interest in African futures (Mbembe 2013; 2016; Sarr 2016), we have to ask ourselves what a more progressive politics of connectivity might look like. We regard this chapter to be one contribution to the debate on "African futures," while recognizing that this is a collective struggle that must be first and foremost fought in the diverse economies of Africa.

Notes

1. As Grosfoguel (2007, 214) argues, the problematization of the African continent shifted over a period of five centuries from "people without writing" to "people without history" and merged into "people without development."

2. These are our respective fields of study: Ouma and Stenmanns have conducted research on African logistics, particularly in West Africa since 2013, from a critical geographic perspective, while Verne has done so for the case of mobile phones in East Africa since 2007 (see Stenmanns and Ouma 2015; Pfaff 2010; Verne 2017).

3. See, for example, the violent history of engineering connectivity through the construction of the Uganda Railway, which resulted in approximately 3,600 deaths of indentured laborers (Methu 2014).

4. Even the most critical accounts of empire and capitalism were not exempt from such a modernist gaze: "The political unity of India, more consolidated, and extending further than it ever did under the Great Moguls was the first condition of its regeneration. That unity, imposed by the British sword, will now be strengthened and perpetuated by the electric telegraph" (Marx 1853, cited in Seth 2009, 373).

5. New Economic Geography is an updated version of 1950s and 1960s spatial analysis thinking that was popular among geographers and regional scientists.

6. The World Bank's Logistics Performance Index (http://lpi.worldbank.org/) and the aforementioned McKinsey Global Connectedness Index are indicative of this trend.

7. The name is derived from "m" for mobile and "pesa," the Kiswahili term for money.

8. South African firm Vodacom acquired a 35 percent share in Safaricom from Vodafone in exchange for ordinary Vodacom shares (Vodafone 2017).

9. South Africa's Transnet National Ports Authority recently installed the new web-based Integrated Port Management System (IPMS) at the port of Durban and benchmarked it against the ports of Singapore and Kerlang ("SA Ports in Major smartPORT Transformation," *Port Technology*, July 28, 2015, https://www.porttechnology.org/news/sa_ports_in_major_smartport_transformation).

10. This problem is usually discussed in relation to technology companies (e.g., Apple), or, more recently, in relation to companies of the so-called platform economy (e.g., Uber), which "insert themselves between those who offer services and others who are looking for them, thereby embedding extractive processes into social interaction" (Mezzadra and Neilson 2017, 12).

11. Most observers usually treat "local initiatives" much more benevolently than the more overtly exploitative practices of large technology or gig economy firms. These initiatives are also often presented as innovations from the South, a move that departs from earlier diffusionist takes on innovation.

For us, global value relations are as much about value (i.e., a structure of accumulation and a historically grown global economic order) as they are about values (a set of historically grown norms, conventions, and discourses that shape positionality and agency) in a global matrix of power.

References

Adam, Christopher, Paul Collier, and Benno Ndulu. 2017. Introduction: Productivity, Organizations, and Connectivity. In *Tanzania: The Path to Prosperity*, edited by Christopher Adam, Paul Collier, and Benno Ndulu, 1–8. Oxford: Oxford University Press.

Araghi, Farshad. 2003. Food Regimes and the Production of Value: Some Methodological Issues. *Journal of Peasant Studies* 30:41–70.

Baumüller, Heike. 2016. Agricultural Service Delivery through Mobile Phones: Local Innovation and Technological Opportunities in Kenya. In *Technological and Institutional Innovations for Marginalized Smallholders in Agricultural Development*, edited by F. W. Gatzweiler and J. von Braun, 143–162. Cham, Switzerland: Springer.

Bimber, Bruce. 1994. Three Faces of Technological Determinism. In *Does Technology Drive History? The Dilemma of Technological Determinism*, edited by M. R. Smith and L. Marx, 79–100. Cambridge, MA: MIT Press.

Blaser, Thomas. 2013. Africa and the Future: An Interview with Achille Mbembe. *Africa Is a Country*, November 20, 2013. http://africasacountry.com/2013/11/africa-and-the-future-an-interview-with-achille-mbembe/.

Busch, Lawrence. 2007. Performing the Economy, Performing Science: From Neoclassical to Supply Chain Models in the Agrifood Sector. *Economy and Society* 36:437–466.

Callon, Michel. 1991. Techno-Economic Networks and Irreversibility. In *A Sociology of Monsters: Essays on Power, Technology and Domination*, edited by John Law, 132–161. New York: Routledge.

Carmody, Pádraig. 2012. The Informationalization of Poverty in Africa? Mobile Phones and Economic Structure. *Information Technologies and International Development* 8:1–17.

Castells, Manuel, Mireia Fernández-Ardèvol, Jack L. Qiu, and Araba Sey. 2007. *Mobile Communication and Society: A Global Perspective*. Cambridge, MA: MIT Press.

Cowen, Deborah. 2014. *The Deadly Life of Logistics: Mapping Violence in Global Trade*. Minneapolis: University of Minnesota Press.

Cowen, Deborah, and Neil Smith. 2009. After Geopolitics? From the Geopolitical Social to Geoeconomics. *Antipode* 41:22–48.

Cox, Andrew. 1999. Power, Value and Supply Chain Management. *Supply Chain Management: An International Journal* 4:167–175.

Danyluk, Martin. 2017. Capital's Logistical Fix: Accumulation, Globalization, and the Survival of Capitalism. *Environment and Planning D: Society and Space*. First published online, April 9, 2017. https://doi.org/10.1177/0263775817703663.

Davis, Alicia, Arame Tall, and Dileepkumar Guntunku. 2014. Reaching the Last Mile: Best Practices in Leveraging ICTs to Communicate Climate Information at Scale to Farmers. CCAFS Working Paper no. 70. CGIAR Research Program on Climate Change, Agriculture and Food Security (CCAFS), Copenhagen.

Debrie, Jean. 2012. The West African Port System: Global Insertion and Regional Particularities. *EchoGéo* 20:2–10.

Díaz Andrade, Antonio, and Cathy Urquhart. 2012. Unveiling the Modernity Bias: A Critical Examination of the Politics of ICT4D. *Information Technology for Development* 18:281–292.

Dyer-Withford, Nick. 2015. *Cyber-Proletariat: Global Labour in the Digital Vortext.* London: Pluto Press.

Easterling, Keller. 2014. *Extrastatecraft.* New York: Verso.

Escobar, Arturo. 1995. *Encountering Development: The Making and Unmaking of the Third World.* Princeton, NJ: Princeton University Press.

Farmer, Paul. 2004. An Anthropology of Structural Violence. *Current Anthropology* 45:303–325.

Ferguson, James. 2006. *Global Shadows: Africa in the Neoliberal World Order.* Durham, NC: Duke University Press.

Friedman, Thomas L. 2005. *The World Is Flat: A Brief History of the Twenty-First Century.* Audiobook on compact disc. Princeton, NJ: Recording for the Blind & Dyslexic.

Fukuyama, Francis. 1989. The End of History. *National Interest* 16:3–18.

Graham, Mark. 2008. Warped Geographies of Development: The Internet and Theories of Economic Development. *Geography Compass* 2:771–789.

Graham, M. 2015. Contradictory Connectivity: Spatial Imaginaries and Techno-Mediated Positionalities in Kenya's Outsourcing Sector. *Environment & Planning A* 47:867–883.

Graham, Mark, Casper Andersen, and Laura Mann. 2015. Geographical Imagination and Technological Connectivity in East Africa. *Transactions of the Institute of British Geographers* 40:334–349.

Graham, Mark, and Laura Mann. 2013. Imagining a Silicon Savannah? Technological and Conceptual Connectivity in Kenya's PBO and Software Development Sector. *Electronic Journal of Information Systems in Developing Countries* 56:1–19.

Grosfoguel, Ramon. 2007. The Epistemic Decolonial Turn: Beyond Political-Economy Paradigms. *Cultural Studies* 21:211–223.

Halpern, Orit, Jesse LeCavalier, Nerea Calvillo, and Wolfgang Pietsch. 2013. Test-Bed Urbanism. *Public Culture* 25:272–306.

Headrick, Daniel R. 1981. *The Tools of Empire: Technology and European Imperialism in the Nineteenth Century*. New York: Oxford University Press.

Hughes, Nick, and Susie Lonie. 2007. M-Pesa: Mobile Money for the 'Unbanked'. Turning Cellphones into 24-Hour Tellers in Kenya. *Innovations* 2:63–81.

Iyer, Yvonne. 2016. SA-Smart Port Delegation to Hamburg. SA German Chamber of Commerce and Industry, Durban, South Africa. http://suedafrika.ahk.de/fileadmin/ahk_suedafrika/Durban/REPORT_-_SMARTPORT_DELEGATION_2016__Yvonne_Iyer.pdf.

Kanngieser, Anja. 2013. Tracking and Tracing: Geographies of Logistical Governance and Labouring Bodies. *Environment and Planning D: Society & Space* 31:594–610.

Krone, Madlen, Peter Dannenberg, and Gilbert Nduru. 2016. The Use of Modern Information and Communication Technologies in Smallholder Agriculture: Examples from Kenya and Tanzania. *Information Development* 32:1503–1512.

Kunst, Marlene. 2014. The Link between ICT4D and Modernization Theory. *Global Media Journal: German Edition* 4 (2): 1–22.

Latour, Bruno. 1991. Technology Is Society Made Durable. In *A Sociology of Monsters: Essays on Power, Technology and Domination*, edited by. J. Law, 103–132. New York: Routledge.

Lerner, Daniel. 1958. *The Passing of Traditional Society: Modernizing the Middle East*. New York: Free Press.

Leys, Colin. 1996. *The Rise and Fall of Development Theory*. Oxford: James Currey.

Ling, Richard, and Heather Horst. 2011. Mobile Communication in the Global South. *New Media & Society* 13:363–374.

Macata, Muharram. 2017. Africa Must Embrace Digital Entrepreneurship to Transform Its Economy. *Guardian*, June 1, 2017, 12.

Maldonado-Torres, Nelson. 2007. On the Coloniality of Being. *Cultural Studies* 21:240–270.

Manyika, James, Jacques Bughin, Susan Lund, Olivia Nottebohm, David Poulter, Sebastian Jauch, and Sree Ramaswamy. 2014. *Global Flows in a Digital Age: How Trade, Finance, People and Data Connect in the World Economy*. McKinsey Global Institute, April 2014. http://www.mckinsey.com/business-functions/strategy-and-corporate-finance/our-insights/global-flows-in-a-digital-age.

Marx, K. 1853. The Future Results of British Rule in India. *New-York Daily Tribune*, August 8.

Mas, Ignacio, and Olga Morawczynski. 2009. Designing Mobile Money Services: Lessons from M-Pesa. *Innovations* 4:77–91.

Matsaert, Frank. 2015. Delivering Development: Better Logistics Critical for Africa's Growth. *East African*, February 21, 2015, 42.

Mbembe, Achille. 2016. Africa in the New Century. *Africa Is a Country*, June 29, 2016. http://africasacountry.com/2016/06/africa-in-the-new-century/.

McLinden, Gerard, Enrique Fanta, David Widdowson, and Tom Doyle. 2011. *Border Management Modernization*. Washington, DC: World Bank Group.

Methu, Soni. 2014. The Indian Migrants Who Built Kenya's "Lunatic Line." Inside Africa. *CNN*, December 11, 2014. http://www.cnn.com/2014/12/11/world/africa/kenya-railways-india/index.html.

Mezzadra, Sandro, and Brett Neilson. 2017. On the Multiple Frontiers of Extraction: Excavating Contemporary Capitalism. *Cultural Studies* 31:185–204.

Mignolo, Walter D. 2007. Introduction: Coloniality of Power and De-Colonial Thinking. *Cultural Studies* 21:155–167.

Mtega, Wulystan P., and Benard Ronald. 2013. The State of Rural Information and Communication Services in Tanzania: A Meta-Analysis. *International Journal of Information and Communication Technology Research* 3:64–73.

Muniesa, Fabian, Michel Callon, and Yuval Millo. 2007. An Introduction to Market Devices. In *Market Devices*, edited by Fabian Muniesa, Michel Callon, and Yuval Millo, 1–12. Malden, MA: Blackwell.

Murphy, J. T., and Pádraig Carmody. 2015. *Africa's Information Revolution: Technical Regimes and Production Networks*. Hoboken, NJ: Wiley.

Ndemo, Bitange, and Tim Weiss. 2017. *Digital Kenya: An Entrepreneurial Revolution in the Making. Palgrave Studies of Entrepreneurship in Africa*. London: Palgrave Macmillan.

Neubert, Dieter. 2017. Neue Technologien in Afrika. Zwischen Aneignung und der Herausforderung durch ein 'Trojanisches Pferd. In *Körper Technik Wissen. Kreativität und Aneignungsprozesse in Afrika*, edited by Markus Verne, Paola Ivanov, and Magnus Treiber, 293–314. Berlin: LIT.

Nguyen, Katie. 2007. Kenya Watchdog Probes Safaricom Ownership. *Reuters*, March 19, 2007. http://mobile.reuters.com/article/technology-media-telco-SP/idUSL1943856320070319.

Consultants, Ocean Shipping. 2008. Beyond the Bottlenecks: Ports in Sub-Saharan Africa. Background Paper 8. Prepared for the World Bank and the SSATP, June 2008. http://www.eu-africa-infrastructure-tf.net/attachments/library/aicd-background-paper-8-ports-sect-summary-en.pdf.

Ohmae, Kenichi. 1989. Managing in a Borderless World. *Harvard Business Review*, May–June 1989. https://hbr.org/1989/05/managing-in-a-borderless-world.

Ouma, Stefan. 2015. *Assembling Export Markets: The Making and Unmaking of Global Food Connections in West Africa*. Chichester, UK: Wiley-Blackwell.

Peet, Richard, and Elaine Hartwick. 2009. *Theories of Development*. 2nd ed. New York: Guilford.

Pfaff, Julia. 2010. A Mobile Phone: Mobility, Materiality and Everyday Swahili Trading Practices. *Cultural Geographies* 17 (3): 341–357.

Raballand, Gaël, Gözde Isik, Monica Beuran, and Salim Refas. 2012. *Why Does Cargo Spend Weeks in Sub-Saharan African Ports? Lessons from Six Countries*. Washington, DC: World Bank Group.

Riddell, Barry. 1970. *The Spatial Dynamics of Modernization in Sierra Leone*. Evanston, IL: Northwestern University Press.

Rogers, Everett M. 1971. *Diffusion of Innovations*. 3rd ed. New York: Free Press.

Rogers, William. 1870. The Domestic Prospects of India. In *Proceedings of the Royal Colonial Institute*, 111–135. London: Clowes.

Rostow, Walt W. 1960. *The Stages of Economic Growth: A Non-Communist Manifesto*. Cambridge: Cambridge University Press.

Sachs, Jeffrey. 2005. *The End of Poverty: Economic Possibilities for Our Time*. New York: Penguin Books.

Sarr, Felwine. 2016. *"AFROTOPIA"—Rethinking Africa: An Interview with Felwine Sarr*. Dakar: Rosa Luxemburg Stiftung West Africa.

Schech, Susanne. 2002. Wired for Change: The Links between ICTs and Development Discourses. *Journal of International Development* 14:13–23.

Schouten, Peer. 2011. Political Topographies of Private Security in Sub-Saharan Africa. In *African Engagements: Africa Negotiating an Emerging Multipolar World*, edited by Ton Dietz, Kjell J. Havnevik, Mayke Kaag, and Terje Oestigaard, 56–83. Boston, MA: Brill.

Seth, Suman. 2009. Putting Knowledge in Its Place: Science, Colonialism, and the Postcolonial. *Postcolonial Studies* 12:373–388.

Shiner, Cindy. 2009. Africa: Cell Phones Could Transform North-South Cooperation. *allAfrica*, February 16, 2009. http://allafrica.com/stories/200902161504.html.

Soja, Edward W. 1968. *The Geography of Modernization in Kenya: A Spatial Analysis of Social, Economic, and Political Change*. Syracuse, NY: Syracuse University Press.

Stenmanns, Julian. 2016. Container Scanning Unit. In *Making Things International 2: Catalysts and Reactions*, edited by Mark B. Salter, 153–165. Minneapolis: University of Minnesota Press.

Stenmanns, Julian, and Stefan Ouma. 2015. The New Zones of Circulation on the Production and Securitisation of Maritime Frontiers in West Africa. In *Cargomobilities: Moving Materials in a Global Age*, edited by Thomas Birtchnell, Satya Savitzky, and John Urry, 87–105. London: Routledge.

Thaler, Richard H., and Cass R. Sunstein. 2013. *Nudge. Wie man kluge Entscheidungen anstößt*. Berlin: Ullstein.

Thompson, Mark P. 2004. ICT, Power, and Developmental Discourse: A Critical Analysis. *Electronic Journal of Information Systems in Developing Countries* 20:1–26.

T-Systems. 2017. Drones and LTE: Port of Durban Goes "Smart." T-Systems, March 9, 2017. https://www.t-systems.com/de/en/about-t-systems/company/newsroom/news/news/smart-port-city-590512.

Turner, George, Nick Mathiason, and Jamie Doward. 2017. Revealed: How Vodafone Allowed Elites to Reap Profits of Africa's Mobile Boom. *Guardian*, November 11, 2017. https://www.theguardian.com/world/2017/nov/12/vodafone-wealthy-elites-mobile-phones-africa.

Unwin, Tim. 2017. *Reclaiming Information and Communication Technologies for Development*. Oxford: Oxford University Press.

Verne, Julia. 2017. The Mobile Phone—a Global Good? Modern Material Culture and Communication Technology in Africa. In *Routledge Handbook of Archaeology and Globalization*, edited by T. Hodos, 157–170. London: Routledge.

Vidija, Patrick. 2017. MPs Split on Proposed Bill to Break Safaricom's "Market Dominance." *Kenya Star*, March 1, 2017. http://www.the-star.co.ke/news/2017/03/01/mps-split-on-proposed-bill-to-break-safaricoms-market-dominance_c1516719.

Vodafone. 2017. Vodafone Transfers a 35% Interest in Safaricom to Vodacom in Exchange for Ordinary Shares in Vodacom. Vodafone, May 15, 2017. http://www.vodafone.com/content/index/media/vodafone-group-releases/2017/safaricom-share-transfer.html.

Wade, Robert. 2002. Bridging the Digital Divide: New Route to Development or New Form of Dependency? *Global Governance* 8:365–388.

World Bank. 1998. *World Development Report 1998/1999: Knowledge for Development*. Washington, DC: World Bank.

World Bank. 2008. *Reshaping Economic Geography. World Development Report*. Washington, DC: World Bank.

World Bank. 2014. Information and Communication Technologies: Overview. World Bank. Last updated October 2, 2014. http://www.worldbank.org/en/topic/ict/overview.

Author Affiliations

Niels Beerepoot, Amsterdam Institute for Social Science Research, University of Amsterdam

Ryan Burns, Department of Geography, University of Calgary

Jenna Burrell, School of Information, University of California, Berkeley

Julie Yujie Chen, Department of Media and Communication, Leicester University

Peter Dannenberg, University of Cologne, Institute of Geography

Uwe Deichmann, World Bank

Jonathan Donner, Caribou Digital

Christopher Foster, Information School, University of Sheffield

Nicolas Friederici, Oxford Internet Institute, University of Oxford

Hernan Galperin, Annenberg School for Communication, University of Southern California

Mark Graham, Oxford Internet Institute, University of Oxford, and Alan Turing Institute

Catrihel Greppi, Universidad Nacional de La Plata, Argentina

Anita Gurumurthy, IT for Change

Isis Hjorth, Oxford Internet Institute, University of Oxford

Lilly Irani, Communication and Science Studies, University of California, San Diego

Molly Jackman, Public Policy Research Manager, Facebook

Calestous Juma, Belfer Center for Science and International Affairs, Harvard University

Dorothea Kleine, Digital, Data and Innovation Group, SIID, University of Sheffield

Madlen Krone, University of Cologne, Institute of Geography

Vili Lehdonvirta, Oxford Internet Institute, University of Oxford

Chris Locke, Caribou Digital

Silvia Masiero, School of Business and Economics, Loughborough University, Loughborough

Hannah McCarrick, Digital, Data and Innovation Group, SIID, University of Sheffield

Deepak Mishra, World Bank

Bitange Ndemo, University of Nairobi

Jorien Oprins, Amsterdam Institute for Social Science Research, University of Amsterdam

Elisa Oreglia, Department of Digital Humanities, King's College London

Stefan Ouma, Department of Human Geography, Goethe University Frankfurt

Robert Pepper, Head of Global Connectivity Policy and Planning, Facebook

Jack Linchuan Qiu, School of Journalism and Communication, Chinese University of Hong Kong

Julian Stenmanns, Department of Human Geography, Goethe University Frankfurt

Tim Unwin, UNESCO Chair in ICT4D and Emeritus Professor of Geography, Royal Holloway, University of London

Julia Verne, Department of Geography, University of Bonn

Timothy Mwolo Waema, University of Nairobi

Index

Page numbers followed by *f* and *t* refer to figures and tables, respectively.

Aadhaar biometric system
  as a carrier of policy, 162–164
  the Indian food security system and, 154, 157–163
  purpose of, 40, 154
  as a shaper of policy, 154–157, 164–168
  trade-offs in the, 40
Abuse, digital, 45
African Development Bank, 6
AfriLabs, 203, 207
Age, Internet access and, 89
Agrawal, Ajay, 299
Agriculture
  farmers in northern rural China, 175, 178, 183–185, 187
  market price information, finding, 178–179, 187
  mobile phone use in, 175, 178, 180, 183–186
  risk-taking in, 187
  weather prediction applications, 26
  yield increases, 26–27
Agriculture, ICTs in commercial small-scale farming in East Africa
  bargaining positions of farmers, 95*t*, 96–97
  distribution channels, 92–95, 93*f*
  empirical results and discussion, 86–97
  ICT4D, bearing on, 82–83
  knowledge access, dimensions of, 90–92
  methodology, 83–86
  negative outcomes, 82–83
  outlook, 97–98
  positive aspects, 80–81
  usage types among farmers, 86–90, 87*t*, 88*t*
Agruppa, 30
Aker, J. C., 81
Amazon, 48, 140*f*
Amazon Mechanical Turk, 235, 286
Apple, 39, 48, 327–328
Artificial intelligence, 49
Arun, Shoba, 254
Asimov, Isaac, 23
Auctions, online. *See* Tea auction, Mombasa
Automation, 23
Automobile ownership statistics, 22
Autor, David H., 298

Bbun, T. M., 82
Bear, Laura, 243
BGI, 325
Bhutan, 8

Biometrics, 22, 40, 154, 161. *See also* Aadhaar biometric system
BongoHive, 203, 208
Bottom billion, 21, 44
Bright, Jake, 209
Brown, William, 286
Burawoy, Michael, 286
Buskens, I., 9, 105, 107, 108, 112

Capital fundamentalism, 173
Capitalism
  aggressive, 325
  Chinese, 326
  digital, 3, 47, 334
  enlightened, 139
  Kenyan, 352
  philanthro-, 129, 243
  post-, 330
Carmel, Erran, 249, 254, 299
Carmody, P., 82, 107, 108
Castells, Manuel, 323
CcHUB, 203, 209
Chan, Anita, 240, 243
Chew, Han Ei, 116
China
  ancient, science and technology in, 319
  electronic manufacturing growth, 325
  farmers in northern rural, 175, 178, 183–185, 187
  First Opium War, 321–322
  patent growth, 325
  post–Cultural Revolution, 324–325
  ride-sharing apps, 320–321, 330–333, 334
  shanzhai (copying) in, 3
  software industry, 325
  special economic zones (SEZs), 320, 324–325, 335–336
  worker suicides, 325
China, alternative digital in economies in Shenzen

boundaries and border crossings, 326–330
copyleft practices, 326–327
Didi and the platformization of ride services, 320–321, 330–333, 335
marginality, conceptualizing, 320–324
migrant workers, 332–333, 335
regional and national contexts, 320, 324–325
shanzhai (copying) model, 320, 326–330, 334–335
as special economic zones (SEZs), 320, 324–325
urban villages, 332–333
CloudFactory, 235–236, 238
Cloward, Richard, 284
Colonialism, 342–345, 349
Community, tech, 199–203, 205–207
Connect Africa initiative (African Development Bank), 6
Connectivity. *See also* Internet connectivity; Mobile phone connectivity
  economic development and, 24
  fiber optic in Africa, 25–28, 34
  gender gap, 31
  historically, 6, 342–345
  improvements necessary, 23–24
  for the margins, 6–7
Connectivity, in African development discourse
  historically, 342–345
  lacks in, 341, 345–346
  logistics and, 345–349, 346*f*, 347*f*
  mobile phones, 349–353, 350*f*
  problematizing, 353–356
Connectivity, universal
  challenges, 31
  data-driven approach to closing the Internet gap, 29–32
  poverty and, 22
  transformative power of, 25–28
Cordella, Antonio, 61

Corporate responsibility, 129
Crime, digital, 45
Cross, Jamie, 240
CrowdConf, 235–237
CrowdFlower, 235, 237
CrowdHack, 228, 235–237, 238, 242
Crowdsourcing, 136
Crowdsourcing infrastructures, legitimizing, 228, 235–237

Dannenberg, P., 81, 89
Delfanti, Alessandro, 244
Deng Xiaoping, 324–325
DevDesign hackathon, 227–232, 239, 241–242
Development discourse in Africa
  historically, 342–345
  lacks in, 341, 345–346
  logistics and, 345–349, 346f, 347f
  mobile phones, 349–353, 350f
  problematizing, 353–356
Development for ICTs (D4ICT), 9
Development-industrial complex, 133
Didi ride service, 334–335
Digital divide
  access and connectivity, 4
  barriers associated with the, 57
  closing the, 110, 344
  gendered, 109–110
  ICTs and the, 82
  increasing, 82
  socio-economic status in the, 44–45
Digital-financial apparatus, 48
Digital Humanitarian Network, 131
Digital Jobs Africa Initiative, 249
Disintermediation
  in agriculture, 27, 93
  in digital labor, 279–281
  in East African tea, 55
  new channels, 68
  promises of, 56, 82
  reintermediation, 72–74, 279

Disintermediation, digital
  defined, 56
  and digital labor, 278–281
  information costs in, 58–60
  in low-income countries, effects of, 56–57
  Mombasa tea auction case study, 63–73
  transaction cost models, 57–58, 59f
  transaction cost perspectives, 62
  transaction costs and, 56–63
Donner, J., 82
Doty, Andrew, 249
Duffield, Mark, 132
Duma Works, 30

East African Marine Systems, The (TEAMS), 25
East Africa Tea Trade Association (EATTA), 66–67
East India Company (EIC), 322
E-auctions. *See* Tea auction, Mombasa
Ebola, 45
Economic alternatives
  boundaries and border crossings, 326–330
  copyleft practices, 326–327
  Didi and the platformization of ride services, 320–321, 330–333, 335
  jugaad, 3
  marginality, conceptualizing, 320–324
  migrant workers, 332–333, 335
  regional and national contexts, 320, 324–325
  shanzhai (copying) model, 320, 326–330, 334–335
  special economic zone (SEZ), 320, 324–325
  urban villages, 332–333
Economic growth and development
  foundations of, 24
  GDP value in understanding, 8
  ICTs and, 79, 81

Economic growth and development (cont.)
  Internet connectivity and, 29–31, 344
  spatial unevenness, 345
Economics
  development, 344–345
  economies' relation to, 174
  of Internet access, 89
  of mobile phone connectivity, 22
  neoclassical, 58
  new institutional, 180
Economy
  global, 29–30
  moral, 134
  organization, new architecture of, 48
Economy, digital. *See also* Innovation hubs, rise of in Africa
  Eurocentric assumptions of, 319
  at the margins, 9–15, 43–45, 47–50
  platform players in the, 48–49
  twin pillars of newly emerging, 3–4
  unqualified optimism, questioning, 49
  women in, 104
Edges, geographic, 323
Education
  apps, 226
  Internet access and, 89
  telecoms universities, 35–36
Einstein, Albert, 27
Electricity, poverty and access to, 22
Elwood, Sarah, 134
Elyachar, Julia, 240
Emerson, Jed, 252
Employment, 23, 30. *See also* Labor, digital
Empowerment
  female, 103, 108, 121–122
  of the poor through tech, 24, 44–45
Entrepreneurs, female
  desires of, 115, 117, 121
  motivation of, 116–117
  neoliberal, 119–123
  risk-taking by, 118–119
  self-identification, 114
Entrepreneurship, female
  barriers to, 116–117
  benefits of, 111
  constraints on, 113–114, 118
  ICTs and, 105, 110–111
  men's role in, 117
  widening the discourse of, 111–112
  within ICT4D projects, 109
Entrepreneurship, female, case studies
  Chile, 113, 118–121
  Zanzibar, Tanzania, 112–118
Entrepreneurship, youth, 35
Equality
  future of, 49
  ICTs and, 9–10, 23
Escobari, M. X., 82
Essien, Mark, 209
Ethics, 41
Ethnic cleansing, 45
Evans, Peter, 173

Facebook
  connectivity plans, 5f
  economic benefits of, 30
  financial services, 48
  hackathons, 225–226
  platform, 39
  services, 40
Feminist visions of the network society, 8–9
Ferguson, James, 9, 349
Fiber optics, 25, 34
Fieldhouse, D. K., 321
52RD.com, 329–330
Financial crisis (2007–2008), 47
Fintech-philanthropy-development complex, 48
First Opium War, 321–322
Fishing/fish trade, Uganda
  information requirements, 178–179

mobile phone use, 175, 180, 183–185, 187
risk-taking, 187
trade relations, 180–181
Flat world, 295, 342–345
Foxconn, 325
Friederici, Nicolas, 223
Friedman, Thomas, 295
From Silicon Valley to Shenzhen (Lüthje et al.), 319

Gallup, John, 7
Game of Phones: Deloitte's Mobile Consumer Survey, 27
Gates Foundation, 226
Gefen, David, 299
Gender
 connectivity gap, 31
 in development discourse, 106–107
 equality, ICTs support of, 108
 ICT4D and, 106–110
 Internet access and, 89
Gender and development (GAD) discourse, 107–108
Geography of Modernization in Kenya (Soja), 343
Ghani, Ejaz, 299
Gino, Francesca, 252
GlobalGAP standard, 83
Globality, digital, 48–49
Globalization, democratizing, 29
Goldin, Claudia, 300
Google, 5f, 39, 48
 Android, 39
 Crisis Response Team, 139
 Google-bombing, 286
 Google.org, 146n9
 Project Loon, 4
Governance, digital, 153
Granovetter, Mark, 180
Greenspan, Anna, 326, 329
Gregg, Melissa, 226
Gro Intelligence, 26

Gross domestic product (GDP), value in understanding welfare or development, 8
Gross national happiness (GNH), 8
Gunboats, steam-powered, 321–322
Gurumurthy, Anita, 8

Hackathons
 background, 225
 benefits of, 225–227
 cultural knowledge, accessing, 226
 inclusion and, 237–239
 labor, accessing free, 226–227
 leveraging the local, 239–241
 organizational form, 226–227
 participants, 241–242, 244
 platform dependence, cultivating, 242
 popularizing infrastructure with, 224
 purposes of, 225–226
Hackathons, case studies
 CrowdHack, 242
 Delhi design studio (DevDesign), 227–232, 239, 241–242
 Silicon Valley crowdsourcing startup, 228, 235–237
 World Bank, 242
 World Bank water hackathons, 224, 228, 233–234
HackEd, 226
Harvey, David, 269, 282
#heforshe campaign (UN), 107
Headrick, Daniel R., 322
Heeks, Richard, 153, 155, 254
Hersman, Erik, 198–199, 202–203
Ho, Josephine, 327
Ho, Karen, 243
Horst, Heather, 349–353
Howe, Jeff, 235
Hruby, Aubrey, 209
Huawei, 325
Hub London, 198
Humanitarianism
 neoliberal reforms, 133

Humanitarianism (cont.)
  philanthro-capitalism and, 133–135
  primary motive, 143–144
Humanitarianism, digital
  austerity and innovation in, 136–139
  extended case method, 135
  philanthro-capitalism and, 139, 142–145
  private sector involvement, 138–139, 140f
  social origins of, 131–133
Humphrey, J., 60, 81, 82
Hyman, Richard, 286

Iacono, Suzanne C., 155
Identification systems, digital, 22. *See also* Biometrics
Identity, digital, 39–41
iHub Nairobi, 193, 197–202, 205–206, 210–212
iHub Research, 203, 208
Ilavarasan, Vigneswara, 116
Impact sourcing
  defined, 249
  ICT4D objectives, fulfilling, 252
  literature review, 251–253
  premise, 249
  scholarship on, 250
  service providers, 253
  service workers, benefits to, 253–255, 260–261
Impact sourcing in the Philippines, case study
  balancing commercial and social welfare logics, 257–259
  research methodology, 255–256
  service workers, profiling, 259–261
  Visaya KPO, 250–252, 256–261
Imperialism, 321–322, 342–345, 349
Inclusion
  crowdsourcing and, 236
  digital, platforms and, 39–41
  hackathons and, 237–239
  industrial, digital services and, 33–36
  mobile, 34–35
Inclusion gap, data-driven approach to closing the, 29–32
Income gap, 29
Incubators, 194, 196, 208–210
India
  Aadhaar, 153
  hackathons in, 226
  jugaad (innovative hacks) practices, 3
India, Public Distribution System (PDS)
  biometric controls, 154, 157–163
  corruption and leakage in, 160, 162–165
  digitalization of in Kerala, 160–161
  entitlement to, 158
  Keralite system, 158–160
  purpose of, 157–158
  Ration Card Management system, 161
  supply chain, 158
India Stack, 39–41
Indigo Trust, 203
Industrial ecosystems, 35–36
Industrial Technology Research Institute, 34–35
Information
  as boundary object in MIS, 185–188
  costs, 58–60
  digital, questions surrounding, 3
  as extractable good, 183
  impersonal nature of, 179
  mobile phones and, 173
  poverty reduction and, 174, 187
  reputational, 183
Information and communications technologies (ICTs)
  access to, 43–45, 109–110
  in agricultural practices, 26–27
  of ancient China, 319
  benefits of, 79, 81, 82
  dangers of, 9
  equality and, 9–10, 23

Index

female empowerment and, 103, 108, 121–122
female entrepreneurship and, 105, 110–111
financial inclusivity and, 26
gender equality and, 108
inclusiveness, 27
negative outcomes, 82–83
optimistic perspective, drivers of, 79
poverty reduction and, 21–22, 344
private sector focus for the, 44–45
public services, transforming, 26
transformative power of, 25–28
universal access, benefits of, 108
Information and communications technologies (ICTs) in commercial small-scale farming in East Africa
bargaining positions of farmers, 95t, 96–97
distribution channels, 92–95, 93f
empirical results and discussion, 86–97
ICT4D, bearing on, 82–83
knowledge access, dimensions of, 90–92
methodology, 83–86
outlook, 97–98
positive aspects, 80–81
usage types among farmers, 86–90, 87t, 88t
Information and communication technologies for development (ICT4D)
criticisms of, 108
digital platforms, addressing, 39–41
failures, 9
gender and, 103–104, 106–112
general objective of, 252
modernization theory and, 344–345
potentials of, debate over, 79
primary focus, 9
purposes, economic, 8–9
small-scale farming, relevance for, 80–83
technology as central agent of change, 9
Information scarcity, 173
Information technology (IT), poverty reduction and, 21–22
Information technology (IT) hubs, emergence of, 35
Infosys, 226
Infrastructure revolution, Africa, 34
Infrastructures
crowdsourcing, 235–237
geospatial information, 347–348
logistical, 345–349
popularizing, 223–224
promises of, 223
transport, 344–349
Innovation from the periphery, 321–322
Innovation hubs, rise of in Africa
aspirations for, 194, 196
diffusion, 202–203, 205
discussion, 210–212
examples, 195–196
historically, 198–202
incubator expectation, 208–210
map of, 204f
methodology, 197–198
network infrastructure expectation, 205–208
outlook, 212–214
statistics, 193
Intelligence, digital, 48
Interest paradigm, 106–107
Internalization, 27
International Association of Outsourcing Professionals (IAOP), 285
Internet
hope and hype of the, 223
the Victorian, 6
Internet access
costs, 89

Internet access (cont.)
  for the poor, 233
  statistics, 22, 31
  use of, 223–224, 233
Internet connectivity. *See also*
    Connectivity
  benefits of, 29–31, 43, 344
  black spots, 5f
  dangers of, 45
  disintermediation and, 56–57
  economic growth and, 344
  goal of, 6
  inequality in, 44–45
  investment in, 5–6
  in low-income countries, 269
  penetration percentages, 3–4, 5f
  for the poor, 43–45
  poverty reduction and, 29, 43
  promises of/benefits promised, 21, 223–224
  statistics, 269
  transformative power of, 29–31
Internet users
  FFV farmers, 86–90, 87t, 88t
  location by income, 4–5, 4f
  penetration growth, 5f
  socio-economic status and, 44
  statistics, 1, 3, 4f, 31

Jugaad (innovative hacks) practices, India, 3

Kagame, Paul, 6–7
Kagwe, Mutahi, 25
Kallinikos, Jannis, 157
Kaufman, David, 142
Kenya Tea Development Agency (KTDA), 70–71
Kerr, William, 299
Kibaki, Mwai, 25
Kibra (Kenya) digital map, 26
Kirkpatrick, Robert, 138
Kirschenmann, Joleen, 300

Kiva, 134
KLab, 195–196
Kleine, Dorothea, 8, 109
Knowledge, economic, 173–188
Knowledge for Development (World Bank), 344
Ko, Tin-yau, 328–329
Kramer, Mark R., 252, 257
Kuek, Siou Chew, 273

Labor, digital
  grand narrative of, 49
  potentials of, 295–296
  rise of, 269–270, 297–301
  scholarship on, 297–298
  unionizing, 286
  vulnerabilities associated, 314
  workers rights, protecting, 314
Labor, digital, development and
  bargaining power concerns, 271–276
  concerns, 282–284
  demand versus supply, 272–273f, 273–274
  discrimination concerns, 277–278
  dollar inflow, 274f
  economic exclusion concerns, 276–278
  empirical foundation, 270–271
  hourly pay, median requested, by country, 273–275, 274f
  intermediation concerns, 278–281
  labor rights strategies, 285–287
  market-based strategies, 284–285
  political economy strategies, 287–288
  regulatory strategies, 287
  skill and capability development concerns, 281–282
Labor, digital, geographic discrimination in
  data descriptive results, 301–306
  data set, 296, 301
  digital work, emergence of, 297–301
  empirical strategy, 296

hiring penalty, findings, 296, 307–309, 308t
information-related frictions, findings on, 296
methods and results, 306–307
platform design implications, 314
policy implications, 314
regulatory strategies, 314
wage premium, findings, 309–312, 310–311t
Labor activism, 331–333
Labor supply, SMEs and, 30
Lacetera, Nicola, 299
Lacity, Mary C., 249, 254
Lagarde, Christine, 205
Levy, Mark R., 116
Li, David, 326, 329
Lindtner, Silvia, 326, 329
Ling, Richard, 349–353
Literacy, 22, 23
Logistics, development and, 345–349, 346f, 347f
Logistics revolution, 344, 345
Lüthje, Boy, 319
Lyons, Elizabeth, 299

Madon, Shirin, 254, 259
Malik, Fareesa, 254, 259
Mao Tse Tung, 324
Mapping, digital, 26
Marginality, conceptualizing, 320–324
Margins
  the center and, 321–324
  connecting the, 6–7
  digital economy at the, 9–15, 43–45, 47–50
  platforms at the, 39–41
Market information systems (MIS), 174, 185–188
Market price information
  as boundary object in the ICTD community, 176–177
  delivery via mobile phones, 187

Market price information, myths of
  as boundary object in MIS, 185–188
  counter-narrative, 175, 177
  criticality in trade-related decision making, 179–182
  market efficiency improvements from mobile phones, 183–184
  obtaining is most valued application of mobile phone in trade, 184–185
  scarcity of price information, 177–179
Massey, Doreen, 7, 322
Maurer, Bill, 134
Mbiti, I. M., 81
McGoey, Linsey, 134
Meena, R., 111
Mellinger, Andrew, 7
Meltwater School of Technology (MEST), 208
Mezzadra, Sandro, 323
Mill, Roy, 299
Mission 4636, 131
Mitchell, Katharyne, 133
Mobile broadband network coverage, 3
Mobile-cellular network coverage, 3
Mobile phone connectivity. *See also* Connectivity
  benefits of, 22
  development discourse, Africa, 349–353, 350f
  promises of/benefits promised, 21
Mobile phone use
  in agriculture, 175, 185
  costs of, 89
  fishing/fish trade, Uganda, 175, 180, 183–185, 187
  growth, 223, 233
  for mobile money, 21–22, 26, 27, 33, 351
  statistics, 3, 22
Mobile revolution, 33–34, 36
Modernization theory, 342–345

Molyneux, Maxine, 105, 106–107
Money, mobile, 21–22, 26, 27, 33, 351
Moody, Kim, 286
Moon, Ban Ki, 194
Morgan, Sharon, 254, 259
M-Pesa, 33, 351–352
Mukhebi, A., 81
Munro, Rob, 138
Murphy, J. T., 82, 107, 108

National Civic Day of Hacking, 226
Neckerman, Kathryn M., 300
Neoliberalism, 108–111, 133
Network society, feminist visions of the, 8–9
New Economic Geography, 344
Newman, Katherine S., 300
Ngũgĩ wa Thiong'o, 336
Nicholson, Brian, 253, 254, 259
Norris, John, 133
Nubelo, 296–297, 301–306, 302t, 303–306f

Offshoring, 251, 256, 299
Okihiro, Gary, 324
Oluwagbemi, Michael, 208–210
Open governance hackathons, 227–232
Orlikowski, Wanda, 155
Outsourcing, 235–236, 251

Pallais, Amanda, 300
Peck, Jamie, 133, 284
Philanthro-capitalism, 130–131, 133–135, 139, 142–145
Piven, Frances Fox, 284
Platforms
  global policies, role in, 48–49
  at the margins, 39–41
Poor, the
  empowering through technology, 24, 44–45
  food security (see India, Public Distribution System [PDS])
  Internet connectivity for, 43–45
Pornography, online, 45
Porter, Michael E., 252, 257
Positionalities
  altering, 6–8
  defined, 6
  economic, 6–7
Poverty
  estimating levels of, satellite photos for, 26
  statistics, 21
  universal access and, 22
Poverty alleviation
  automation, impact on, 23
  ICTs and, 21–23, 43, 344
  information and, 174, 187
  Internet connectivity and, 29, 43
  technology in, 154–169
  tool-and-effect logic in, 153, 155
Power-geometries, 7
Project Loon (Google), 5f
Property rights approach to transaction costs, 57–58, 59f, 62

Qiang, C., 81
Qiu, Jack L., 335

Rajesh, Kalpana, 30
Random Hacks of Kindness (RHoK), 239–241
Rasmussen, Eric, 138
Rationality, myth of, 49
Ravishankar, M. N., 253, 257
Ride-sharing apps, China, 320–321, 330–333, 334
Risk taking, farmers and fishermen, 181–182, 187
Rockefeller Foundation, 249, 252
Rottman, Joseph W., 254
Rouse, Cecilia, 300
Roy, Anaya, 134

Rusimbi, R., 111
Rwanda, 7

Sachs, Jeffrey, 7, 350
Safaricom, 351–352
Samasource, 236, 253
Sandeep, M. S., 253, 257
Scarry, Elaine, 143
Schoen, Donald, 240
Scott, James, 324
Sen, Amartya, 8, 254
Service workers
  benefits of impact sourcing to, 253–255, 260–261
  in the Philippines, profiling, 259–261
Sexual harassment, online, 45
Shanzhai (copying) in China, 3
Sharanappa, Sandesh, 254, 259
Sheppard, Eric, 6
Silicon Valley model, emulation attempts, 3
Small and medium enterprises (SMEs), labor supply, 30
Smart ports, 348
Smith, Adam, 7
Social policy design, 156
Social safety, 156
Socioeconomic modeling, 49
Söderberg, Johan, 244
Software production events. *See* Hackathons
Soja, Ed, 343
Sparke, Matthew, 133
Staats, Bradley R., 252
Standage, Tom, 6
Standby Task Force, 131, 140*f*
Stanton, Christopher T., 299
Starbucks' Ethos Water, 130
Stark, David, 239
Surborg, B., 82
Sustainable Development Goals (SDGs), monitoring progress toward, 27, 28

Tea auction, Mombasa
  background, 55
  brokers role, 64–67
  collusion allegations, 70–71
  constraints in transactions, 71–72
  digitization, demand for, 55
  disintermediation, new channels of, 68–70, 69*f*
  e-auction, 66–68, 71
  evolution of the, 63–64
  methodology, 63–71
  pressures to reform, 65–66
  strategic actions and transaction costs, 72–73
  tensions, 64–66
  unpredictability in the, 70
Technology
  community, 199–203, 205–207
  empowering the poor, 24, 44–45
  internalization, 27
  modernization theory and, 342–343
Technology, digital
  as a carrier of policy, 153–154, 156, 162–164
  inequality, increasing, 23
  positionalities, altering, 6–8
  as a shaper of policy, 154, 155–157, 164–168
  unfulfilled potential of, 21–24
Tecno, 328–329, 334
Telecenters, 110
Tencent, 325
Text users, FFV farmers, 86–90, 87*t*, 88*t*
Thomas, Catherine, 299
Thornton, A., 82
Tickell, Adam, 133
TOMS, 130
Trade facilitation, 347–349
Trade relations, mobile phone use in, 180–181
Transaction costs
  information costs in, 58–60
  models, 57–58, 59*f*, 62

Transactions
  conditions influencing internal versus external trade, 60
  digitizing, externalities of, 61
  institutional frameworks of, 60–61
Transportation infrastructures, 344–349
Tsing, Anna, 243, 323

Uber, 39, 320–321, 330, 334
Uganda, 26
UNICEF venture fund, 139, 141f
Unionizing digital workers, 286
United Nations, #heforshe campaign, 107
United Nations' Global Pulse, 26
Universities, telecom, 35–36
Unwin, Tim, 9
Ushahidi, 131

Victorian era, 6
Visaya Knowledge Process Outsourcing (Visaya KPO), 250–252, 256–261
Vodafone, 352
Voice and text users, FFV farmers, 86–90, 87t
Voice-only users, FFV farmers, 86–90, 87t, 88t

Wallerstein, Immanuel, 319, 324
Walsham, Geoff, 154, 169
Wang, H., 332
Water Hackathon: Lessons Learned (World Bank), 234, 238
Weather prediction, 26
Webb, A., 9
WeChat, 325
Wennovation Hub, 208–210
WhatsApp, 27
Williams, Raymond, 323, 329
Women, online sexual harassment of, 45

Women in development (WID) discourse, 106–109. *See also* Entrepreneurship, female
Women's economic empowerment (WEE), 104. *See also* Empowerment, female
Women's entrepreneurship development (WED), 110–111. *See also* Entrepreneurship, female
World Bank
  connectivity infrastructure investment, 6
  hackathons, 224, 226, 228, 233–234, 238, 242
  Random Hacks of Kindness (RHoK), 239
World Is Flat, The (Friedman), 295
Wright, Chris F., 286

Yanagisako, Sylvia, 243

Žižek, Slavoj, 130
ZTE, 325